I0888204

THE 2017 A201 DESKBOOK

Editors
Peter W. Hahn
Joseph C. Kovars
Amanda S. MacVey

FORUM ON CONSTRUCTION LAW
Building the Best Construction Lawyers

Cover design by Tamara Kowalski/ABA Design

Printed in the United States of America.

21 20 19 18 17 5 4 3 2 1

Library of Congress Cataloging-in-Publication Data

Names: Hahn, Peter W., editor. | Kovars, Joseph C., editor. | MacVey, Amanda Schermer, editor. | Sink, Charles M. A201 deskbook.
Title: The 2017 A201 deskbook / edited by Peter W. Hahn, Joseph C. Kovars, and Amanda Schermer MacVey.
Description: First edition. | Chicago : American Bar Association, 2017.
Identifiers: LCCN 2017022229 | ISBN 9781634259439 (print)
Subjects: LCSH: Construction contracts—United States. | Architectural contracts—United States. | Standardized terms of contract—United States.
Classification: LCC KF902 .A15 2017 | DDC 343.7307/8624—dc23
LC record available at https://lccn.loc.gov/2017022229

Table of Contents

About the Editors

Peter W. Hahn is a partner in the construction law practice group of Dinsmore & Shohl's Columbus, Ohio, office. He advises owners and contractors on issues and liability avoidance at every phase of a construction project, from the drafting and negotiation of contracts through post-project issues and construction defect claims. He also regularly advises public authorities throughout Ohio on such topics as construction procurement, project delivery methods, contract terms, regulatory compliance, and claims. He is active in the ABA Forum on Construction Law, where he is a member of the Steering Committee for Division 13 (Government Construction). Mr. Hahn has a bachelor's degree from the College of Wooster and a law degree from Washington University in St. Louis.

Joseph C. Kovars is a shareholder in the Baltimore office of Baker Donelson, where he cochairs the firm's practice group on construction law. An experienced civil litigator, Mr. Kovars concentrates his practice in construction and public contracts law. He represents contractors, subcontractors, sureties, and owners in contract formation and construction disputes involving many types of construction projects. Mr. Kovars graduated magna cum laude from Boston University with a BA and received his law degree, with honors, from George Washington University Law School. Active in the ABA Forum on the Construction Industry, he is currently the Chief Finance Officer and is a former member of its Governing Committee. Mr. Kovars writes and lectures regularly on construction topics. He is a coeditor of *Construction Subcontracting* (ABA 2012). He is the editor in chief of the *Maryland Construction Desk Book* (MSBA 2d ed. 2017). He has been listed since 2003 in The Best Lawyers in America, Construction Litigation category, and was named Best Lawyers' Baltimore Litigation—Construction "Lawyer of the Year" (2015). In 2013, Mr. Kovars was elected to be a fellow of the American College of Construction Lawyers. For more information, please visit http://www.bakerdonelson.com/joseph-c-kovars.

Mandy Schermer MacVey is a partner in the Washington, D.C., office of Schiff Hardin LLP. Her practice focuses on full-time boots-on-the-ground legal services on megaprojects. She has expertise in developing contracting strategy, executing complex contract negotiations, and implementing construction industry best practices to effectively prevent, mitigate, and resolve disputes that arise over the life cycle of construction projects. She received her BS and her JD from the University of Illinois Champaign-Urbana. She is a frequent lecturer and author, as well as adjunct professor of construction law at John Marshall Law School. Ms. MacVey volunteers her time as outside general counsel for Street Law, Inc., and was recognized in the 2015 and 2016 Emerging Lawyer Network.

About the Authors

Cathy Altman is a partner with the Dallas law firm Carrington Coleman Sloman & Blumenthal, LLP, and chair of the construction group. One of the inaugural board-certified Texas attorneys in construction law, Ms. Altman has represented owners, contractors, subcontractors, construction professionals, and suppliers in transactions and disputes involving both public and private projects. Ms. Altman is active with the ABA Forum on Construction Law, serving as past chair for the Owners & Project Finance Division, the Trial Academy, and the 2015 annual meeting "The Construction Lawyer's Bucket List." She has been recognized by Super Lawyers, Best Lawyers in America, and D Magazine Best Lawyers in Dallas for her proficiency as a construction lawyer. She obtained her undergraduate degree from Vanderbilt University and her JD from the University of Kentucky.

Jeremy S. Baker, a partner at Schiff Hardin LLP and a resident of its Chicago office, is a member of its national construction law practice. Mr. Baker is an adjunct professor of law at John Marshall Law School, where he teaches an elective construction law course to JD and LLM students through the Center for Real Estate Law. Mr. Baker helps his clients complete challenging projects by emphasizing dispute avoidance and early, cost-efficient resolution of claims. He represents owners, developers, design professionals, and contractors in all aspects of design and construction. Mr. Baker's practice involves project structuring across many project delivery methods, plus contract review and negotiation. He has prosecuted and defended lawsuits in well over 30 state and federal court venues—but he prefers to resolve disputes using cost-efficient alternative dispute resolution. Mr. Baker is a graduate of the Tulane University Law School and the University of California, Santa Barbara.

Levi W. Barrett is a senior associate in the construction transactions group at Peckar & Abramson, PC, in New York, New York. His practice focuses on all aspects of public and private construction contracts, including contract negotiation, dispute avoidance, and contract administration. Mr. Barrett frequently advises owners and general contractors in connection with large-scale domestic and international construction projects. He also has extensive experience representing contractors and subcontractors in complex disputes and construction claims. He has served the ABA Forum on Construction Law as a member of its Division 13 (Government Contracting) Steering Committee and as a speaker and editor. Mr. Barrett is also involved with the Associated General Contractors of America (AGC). He is a member of the AGC's Contract Documents and Alternative Documents committees and actively participates in drafting and revising ConsensusDocs construction contracts. He graduated with a degree in economics from San Francisco State University and obtained his law degree from Syracuse University College of Law.

Daniel S. Brennan is one of the founding partners of Laurie & Brennan, LLP, in Chicago, IL. Before starting his own firm, Mr. Brennan practiced with DLA Piper and Schiff Hardin in Chicago. He concentrates his practice on advising owners, designers, and contractors on risk management in the construction process; prosecuting and defending construction claims including professional negligence, cost overrun, and delay claims; negotiating and drafting construction and design contracts for architects, engineers, and owners; and advising clients with respect to design and construction of accessible facilities under Title III of the Americans with

Disabilities Act and the Fair Housing Act. Mr. Brennan is a fellow of the American College of Construction Lawyers. Mr. Brennan is also an active member of the ABA Forum on Construction Law and served on the Governing Committee of the forum. He is also a member of the ABA Public Contract Section and the ABA Tort Trial and Insurance Practice Section. Mr. Brennan received his BA from the University of Notre Dame and his JD from Georgetown University Law Center.

Anatoly Darov is the founding member of Darov & Associates US PLLC, providing legal counsel and consulting services to real estate and construction clients. He acts on behalf of clients on the delivery of capital infrastructure and building projects with a particular focus on project development, government contracts and procurement, and dispute resolution. Mr. Darov advises clients on insurance products and coverage matters involving commercial liability, property insurance, professional liability, and wrap-up programs, as well as emerging risk issues relevant to construction and real estate. He is a registered professional engineer in the Commonwealth of Massachusetts and has worked in an engineering capacity on a variety of major infrastructure projects. Mr. Darov received his JD from New England College of Law and his MS and BS from the University of Massachusetts Amherst.

Paul E. Davis is a principal in the law firm of Conner Gwyn Schenck PLLC in Raleigh and Greensboro, North Carolina. Mr. Davis counsels public owners, construction managers, contractors, subcontractors, and sureties in relation to all of their construction industry needs, including project delivery systems, procurement, contract negotiation and drafting, occupational licensing, risk management, and the mediation, arbitration, and litigation of complex claims in state and federal courts. Mr. Davis also regularly drafts EPC agreements and energy service agreements, particularly for the solar energy industry. Mr. Davis is a former chair of the North Carolina Bar Association Construction Law Section. He has authored and coauthored numerous books and articles on a wide variety of construction topics. Mr. Davis received his BA and JD from the University of North Carolina at Chapel Hill.

Kimberly A. Davison is the managing shareholder and vice president of Griffith Davison & Shurtleff, P.C., in Dallas, Texas. Ms. Davison has developed expertise in the drafting and preparation of contracts, resolution of construction disputes, and preparation and analysis of lien and bond claims. Ms. Davison earned a BS from Vanderbilt University and a JD from Baylor Law School, where she was a member of the *Baylor Law Review*. Ms. Davison has served as a council member of the Construction Law Section of the State Bar of Texas and is ranked as a "Leader in Their Field" by Chambers & Partners USA, is rated "AV® Preeminent™," and has been selected for the Bar Register of Preeminent Women Lawyers by Martindale-Hubbell®. Ms. Davison founded the Monthly Construction Ladies Luncheon, an organization for women in the construction industry, representing over 12 years of networking, leadership, and education. Ms. Davison is the first attorney in Texas to be board certified in both Construction Law and Commercial Real Estate by the Texas Board of Legal Specialization.

Robert J. Dietz, of counsel with BrigliaMcLaughlin, PLLC, in Northern Virginia, has over a decade of experience with construction law, government contracts, and surety law. He earned his undergraduate degree from the University of Florida and his law degree from Washington & Lee University School of Law in 2004, where he served on the law review. Throughout his career, Mr. Dietz has focused his practice on representing contractors, subcontractors, sureties, and public and private owners on a variety of complex matters and has successfully litigated cases in favor of these clients in federal and state courts throughout Virginia, Maryland, and the District of Columbia.

Neal M. Eiseman, a partner in the New York City office of Goetz Fitzpatrick LLP, counsels corporate clients on commercial, construction, and real estate matters and represents them in litigations, arbitrations, and mediations. Best Lawyers selected Mr. Eiseman as the 2017 "Lawyer of the Year" in New York City for Litigation-Construction Law, and Super Lawyers named him as one of the "Top 100 Lawyers" in 2016 in the New York metropolitan area. He is an arbitrator on the American Arbitration Association's Construction Arbitration Master Panel and serves as a mediator on the AAA's Master Mediation Panel. Mr. Eiseman trains mediators for the AAA and teaches master courses in construction and real estate law and negotiation and dispute resolution at New York University. He is a member of NYU's Schack Institute of Real Estate/Construction Management Advisory Board, a member of the AAA's Arbitration and National Advisory Committee, and a fellow and board member of the College of Commercial Arbitrators.

David L. Fine is a partner and chair of the construction industry practice group at Mirick, O'Connell, DeMallie & Lougee, LLP, a full-service law firm with offices in Boston, Westborough, and Worcester, Massachusetts. Mr. Fine represents project owners and developers, general contractors, construction managers, specialty subcontractors, suppliers, and manufacturers in all phases of construction, including adverse proceedings such as litigation and arbitration, consulting regarding risk management, contract preparation and negotiation, and providing general business advice. He works closely with several construction trade and bar organizations on both a local and a national level, and he often lectures and writes about topics pertinent to the construction industry, such as project delivery, contracts, teaming, indemnity and warranty concerns, insurance coverage, green building and solar issues, bond claims, mechanics' liens, and dispute resolution methods and procedures. Mr. Fine is an active member of the ABA Forum on Construction Law, serving in multiple leadership positions. He has been recognized as a Massachusetts "Rising Star" in the category of Construction Litigation by *Boston* magazine and *Law & Politics*.

Kate Glaze is an attorney at Carrington Coleman Sloman & Blumenthal, LLP, in Dallas, TX. Her practice includes construction, real estate, and lending. She represents contractors and owners in drafting construction contracts, resolving disputes, and placing or removing liens. Her extensive experience in negotiating construction contracts and navigating roadblocks makes her an asset to her clients. Ms. Glaze received her BA from the University of Texas and her law degree from Southern Methodist University Dedman School of Law.

Kelli E. Goss is a principal in the law firm of Conner Gwyn Schenck PLLC in Raleigh and Greensboro, North Carolina. Ms. Goss began her career as an architect and is currently a North Carolina licensed architect. Prior to attending law school, she was a partner in a prominent architectural firm in Raleigh, where she served as director of commercial and corporate architecture. Ms. Goss's construction law practice includes drafting, review, and negotiation of design and construction contracts and subcontracts. Her practice also includes surety defense and litigation and arbitration of delay and acceleration claims, scope of work disputes, and differing site conditions. She has experience in state and federal court and before the Civilian Board of Contract Appeals and the U.S. Court of Federal Claims. She received her BA in visual design from Dartmouth College, her MArch from North Carolina State University School of Design, and her JD from the University of North Carolina at Chapel Hill.

Richard F. Greenleaf is special counsel to Berg Hill Greenleaf Ruscitti LLP. He has more than 30 years of experience specializing in construction law matters. He has worked on projects in 41 states and internationally. Mr. Greenleaf arbitrates and mediates large complex construction disputes with the American Arbitration Association, International Centre for Dispute Resolution, the CPR International Institute for Conflict Prevention and Resolution, and the International Chamber of Commerce. He is a fellow of the College of Commercial Arbitrators, the Chartered Institute of Arbitrators, the National Academy of Distinguished Neutrals, the American College of E-Neutrals, and the Dispute Board Foundation. He devotes most of his time to serving as a neutral in complex and high-dollar disputes and has served as an arbitrator through final award on over 200 cases with claims amounts in excess of $300 million and projects with values exceeding $6 billion. Mr. Greenleaf was the founding cochair of the College of Commercial Arbitrators Construction Arbitration Section. He obtained his BA from the University of South Florida and his JD from South Texas College of Law–Houston. He regularly speaks and writes on topics of interest to the construction industry.

Nancy Greenwald has more than 35 years of experience in the construction industry as in-house counsel and CFO of a design-build company. Since 2010, she has used this expertise to help parties resolve their disputes as an arbitrator and mediator. She serves on the American Arbitration Association's Construction Industry Panel of Arbitrators and Mediators and has been appointed as a neutral by numerous private clients. Ms. Greenwald is also the executive director of the University of Hartford's Construction Institute, a nonprofit association whose mission is to provide leadership in cross-industry collaboration. Its member companies are regional, national, and international and include public and private owners, facility managers, architects, engineers, contractors, subcontractors, manufacturers, governmental agencies, lawyers, and accountants, among others. Ms. Greenwald publishes and speaks frequently on best practices in construction project communication, arbitration, mediation, and dispute systems design and is active in many construction industry groups, including the ABA Forum on Construction. She received her JD from Harvard Law School (cum laude) in 1981. You can learn more about Ms. Greenwald at construction-dispute-solutions.org.

Claramargaret H. Groover, of counsel to Becker & Poliakoff in its Orlando office, has over 36 years of experience in working with construction law issues and the full range of construction clients. She is a Florida Bar board certified construction lawyer and a Florida certified circuit civil and appellate mediator. She also has experience as a Florida qualified arbitrator and is a panel arbitrator with the American Arbitration Association. She served as director of legal affairs for University City Development Partnership in its $2.5 billion addition to Universal Studios in Orlando. She received her BA from Stetson University and her JD from the University of Florida, College of Law. She serves on the Governing Committee of the ABA Forum on Construction Law and has served as chair of its Division 12—Owners and Project Finance. She is a frequent lecturer and author, as well as adjunct professor of construction law and ethics in the legal studies program for paralegals at Seminole State College.

Keith Heath is an attorney at Griffith Davison & Shurtleff, P.C., located in Dallas, Texas. Over the years, Mr. Heath has held numerous positions in the construction industry ranging from general laborer and heavy equipment operator to vice president and corporate counsel for a site contracting company. He has also served as claims counsel for a national surety company. Mr. Heath brings this diverse industry experience to bear in his representation of owners, general contractors, design-builders, construction managers, and subcontractors on the varied legal issues impacting them, including the drafting, review, and negotiation of various construction contracts and other business documents, lien claims, and project defaults across multiple states and jurisdictions. Mr. Heath received both his BA and his JD from the University of Louisville and is licensed to practice law in Texas and Kentucky.

J. Wallace Irvin is special counsel in Lewis Thomason's Nashville office, where he focuses his practice on construction law with an emphasis on the representation of architects and engineers in all phases of construction contracting and against allegations of professional negligence. Mr. Irvin also represents contractors in a wide variety of business matters, including contract negotiations, project administration, closeout, construction defect defense, and payment disputes, including lien and bond claims. Additionally, Mr. Irvin advises contractors in administrative matters before the Tennessee Board for Licensing Contractors, Tennessee Occupational Safety and Health Administration (TOSHA), and Tennessee Department of Transportation (TDOT). Mr. Irvin is a Tennessee Supreme Court Rule 31 listed Civil Mediator. Mr. Irvin is a member of the National Society of Professional Engineers and the American, Tennessee, and Nashville Bar Associations. Mr. Irvin also serves on the boards for the Tennessee Association of Construction Counsel and the Associated General Contractors of Middle Tennessee Construction Leadership Council. Mr. Irvin obtained a BS in civil and environmental engineering from Tennessee Technological University and a JD from the University of Memphis.

William E. Kelley, Jr., LEED AP, is a partner with Drewry Simmons Vornehm, LLP, in the Indianapolis and Carmel, Indiana, offices, and he focuses his legal practice in the areas of construction, real estate, and professional liability. On the transaction side, Mr. Kelley assists a wide range of clients with drafting and negotiating construction and design contracts, mechanic's liens, real estate purchase and development agreements, and lease agreements. On the litigation side, Mr. Kelley represents design professionals, contractors, owners, developers, and insurers with construction, architect/engineer professional liability, and real estate matters, including pre-litigation claims, loss prevention assistance, insurance coverage issues, mediation, arbitration, and litigated matters in state and federal court. He graduated from Hanover College and the Robert H. McKinney School of Law, Indiana University. Mr. Kelley is a liaison to the American Institute of Architects (AIA) Documents Committee and is a former member of the Associated Builders and Contractors (ABC) National Green Building Committee. He has been named by Super Lawyers an "Indiana Rising Star" from 2009 to 2017 in the area of Construction Litigation.

Mike Lapicola is a partner in the construction practice of Faegre Baker Daniels LLP, located in Denver, CO. He has a national construction law practice, representing clients in dozens of states across the United States. He has significant expertise drafting and negotiating all types of construction contracts, including design agreements, design-build contracts, EPC and BOP agreements, purchase orders, bid documents, procurement contracts, indemnity agreements, joint venture agreements, and IPD agreements. Mr. Lapicola provides daily consultation to clients regarding project structure, procurement, licensing, liens, bonds, insurance, indemnity, delays, defects, changes, payment, termination, and other construction project matters. He also represents clients in settlement discussions, mediations, arbitrations, litigation, and other dispute resolution activities involving construction issues. He is on the Steering Committee for Division 2 (Contract Documents) of the ABA's Forum on Construction Law.

Sam Laurin is a partner and the chair of the construction and litigation practice groups at Bose McKinney & Evans LLP in Indianapolis, Indiana. He litigates construction disputes and has been lead counsel in trials in state and federal courts and in arbitration proceedings. He also negotiates construction contracts for projects throughout the United States. Mr. Laurin is a frequent author on construction law issues, and many of these articles focus on insurance issues that arise in the construction industry. He has been recognized in the areas of construction litigation, real estate litigation, and commercial litigation by Best Lawyers®, Chambers USA, and Indiana Super Lawyers®. Best Lawyers® also recognized him as Indianapolis Litigation—Construction Lawyer of the Year in 2012 and 2016 as well as Indianapolis Litigation—Real Estate Lawyer of the Year in 2014 and 2016. Mr. Laurin has been a longtime active member of the Forum on Construction Law and is currently a member of Division 7 Steering Committee on Insurance Surety and Liens.

Aileen M. Leipprandt is a shareholder at Hilger Hammond in Grand Rapids, Michigan. For the past 25 years, her practice has focused on construction law, including complex commercial and real estate litigation, arbitration, and mediation. Ms. Leipprandt's experience spans all manner of construction law, including extensive construction contract drafting and negotiation, contract administration, claim negotiation and litigation (delay, default, defect, change orders, indemnity), payment and performance bond claims, construction lien claims, bid protests, Builders Trust Fund Act claims, regulatory licensing, and OSHA compliance. Her clients include contractors, construction managers, subcontractors, and suppliers. She maintains strong professional affiliations with many construction industry trade organizations. She is a member of the Associated General Contractors (Legal Advisory Committee), Associated Builders and Contractors (past board president), American Subcontractors Association of West Michigan, Construction Financial Management Association (board member), Home Builders Association of Greater Grand Rapids, and National Association of Women in Construction (past board member).

Steven B. Lesser is a shareholder and chairs the construction law group at Becker & Poliakoff PA in Fort Lauderdale, Florida. He is board certified in construction law by the Florida Bar. Mr. Lesser is a Florida certified circuit civil mediator and certified arbitrator on construction issues for the American Arbitration Association. He is AV Rated Preeminent by Martindale-Hubbell and was named "Construction Lawyer of the Year" for 2016 for South Florida by Best Lawyers in America. Mr. Lesser has been ranked in Tier 1 and recognized as a top construction lawyer in Florida by Chambers USA America's Leading Lawyers for Business. He is past chair of the ABA Forum on Construction Law and is a fellow of the American College of Construction Lawyers.

Tamara J. Lindsay is the U.S. commercial & contracts director for MWH Constructors in Broomfield, Colorado, and was previously an attorney in the Denver and Las Vegas offices of Holland & Hart, where she consulted and advised commercial and industrial clients on a full spectrum of construction-related issues, from contract negotiation through dispute resolution. Her broad range of experience includes casinos, water treatment facilities, chemical and manufacturing facilities, hydropower facilities, renewable energy projects, airports, pipelines, rail systems, hospitals, schools, and multifamily development. Ms. Lindsay graduated magna cum laude from Spring Hill College in Mobile, Alabama, where she was a four-year starter on the women's basketball team, including two years as team captain. She received her law degree from the University of Tennessee in 2008. Prior to practicing law full-time, Ms. Lindsay worked for the Department of Defense training Marines for mental and physical combat readiness. She was selected as one of ENR's Top 20 under 40 in 2016 and is a frequent author and speaker for numerous construction industry organizations.

Nora E. Loftus is a partner in the construction group at Frantz Ward LLP. Ms. Loftus represents general and specialty contractors and subcontractors, owners, design professionals, sureties, and others in the construction industry in all facets of construction law, including contract drafting and negotiation, bid protests, project documentation, risk assessment and management, insurance coverage, claims processing, and dispute resolution. Ms. Loftus practices in state and federal courts nationwide and has resolved multimillion-dollar disputes throughout the country. She takes pride in working with clients to understand and grow their businesses and find the best resolution to any conflict. She earned her BA from Miami University and her JD from the Cleveland-Marshall College of Law.

Peter J. Marino is a partner with Smith, Anderson, Blount, Dorsett, Mitchell & Jernigan, LLP, in Raleigh, NC, and chairs the firm's construction and real estate development group. Mr. Marino's practice includes all facets of construction law, including disputes arising out of public and private projects. He is the incoming chair of the Contract Documents Division (Division 2) of the ABA Forum on Construction Law and is a former chair of the NCBA Construction Law Council. Mr. Marino has served as editor and a coauthor of the *North Carolina Construction Law Deskbook*. He has published and presented on many topics, including construction contracting, lien law, and insurance issues in construction. He is listed for construction law in Woodward/White Inc.'s The Best Lawyers in America, Business North Carolina's "Legal Elite," Hall of Fame (Construction Law), and North Carolina Super Lawyers (Top 100 Lawyers Honoree). Mr. Marino received his BA from LeMoyne College and his JD with honors from the University of North Carolina at Chapel Hill.

Stephen D. Marso is a member of the law firm of Whitfield & Eddy, PLC, in its Des Moines, Iowa, office. Mr. Marso's primary areas of practice are construction law and litigation, and he is licensed to practice in the states of Iowa and Nebraska as well as several federal courts. Mr. Marso graduated magna cum laude from Hope College in 1996 with a degree in Spanish, and he graduated with high distinction from the University of Iowa College of Law in 1999. Prior to joining Whitfield & Eddy in 2001, Mr. Marso served as judicial law clerk to Judge Harold Vietor of the U.S. District Court for the Southern District of Iowa from 1999 to 2001. Mr. Marso is an arbitrator on the American Arbitration Association's roster of construction arbitrators, he has given many construction-related presentations to clients and construction industry groups, and he is the author of various publications, such as *Public*

Construction Liens in Iowa: A History and Analysis of Iowa Code Chapter 573, 60 DRAKE L. REV. 101 (fall 2011). He is a member of the Polk County (Iowa), Iowa State, Nebraska State, and American Bar Associations and is a member of several construction-industry organizations and associations.

Matthew J. Ninneman is special counsel at Hall & Evans, LLC, in Denver, Colorado, and is a member of its construction and design practice group representing construction, design, insurance, and surety professionals on commercial, residential, and governmental projects including professional licensure, bid protests, contract drafting and negotiation, lien prosecution, and litigation in state and federal courts, arbitration, and appeals. He is a frequent author and lecturer on issues concerning construction claims, contracts, liability mitigation, and various issues affecting the construction industry. Mr. Ninneman is the coauthor of the quarterly "Hard Hat Case Notes" article in the ABA Forum's *Construction Lawyer* and a contributing editor. He has also coauthored several ABA Forum publications on construction fundamentals, subcontracting, professional licensure, and local/municipal projects. He is also a member of the steering committee for the ABA Forum's Division 3—Design, a member of its Division 2—Contract Documents, and a member of several other ABA Forum committees. Outside of the law, he spends his time enjoying the Rocky Mountains with his wife, Cindy, and three children, Abigail, William, and Charlotte.

Elliotte Quinn is an associate in Parker Poe Adams & Bernstein LLP's Charleston, South Carolina, office whose practice focuses on construction, energy and regulatory, and commercial litigation. He received his bachelor's degree from the College of Charleston and his law degree from Emory University.

Allison Taller Reich is a member of the construction group at Frantz Ward LLP. She focuses on all areas of construction law, including drafting and negotiating contracts; assisting in claim management; and resolving disputes through mediation, arbitration, dispute review board hearings, and litigation. Ms. Reich primarily represents contractors and subcontractors with issues related to sewer and road projects; educational and residential projects; healthcare and hospital projects; and retail, mixed-use, and development projects. Her interest in and passion for construction law dates back to her days at The Ohio State University, where she obtained a BS in architecture before attending Cleveland-Marshall College of Law in Cleveland, Ohio.

Richard S. Robinson is a senior counsel in the New York, NY, office of Peckar & Abramson, P.C. After 15 years of litigation practice focused primarily on construction cases, Mr. Robinson joined the firm's contract review group in 2009 and has since reviewed well over a thousand construction agreements (and drafted or negotiated a substantial number of those agreements) for international, national, and regional general contractors, owners, and P3/infrastructure consortiums. Mr. Robinson draws on his substantial construction litigation experience to assist transactional clients in avoiding or addressing issues during the drafting and negotiation phases of their projects and also provides clients with project-specific risk management training. Mr. Robinson received his BA with honors from the University of Pennsylvania and his law degree from Seton Hall University School of Law.

Sean Ryan is a partner at Peckar and Abramson P.C., where he chairs the firm's construction transactional practice. He leads a team that has been involved in over 6,000 contracts in the past decade. His practice focuses on drafting, negotiating, and advising clients on managing risks in contracts and other transactional documents. Mr. Ryan conducts training programs for clients and has written on topics such as lean construction, integrated project delivery and procurement strategies, and delivery methods on supertall buildings. He is experienced in all types and forms of contracts, including AIA, ConsensusDocs, DBIA, CMAA, CCDC, and FIDIC. His practice spans the globe, including projects in the United States, Mexico, Canada, South America, Europe, and Asia, where he often teams with local counsel to represent clients. He represents U.S. clients on international projects and international clients doing business in the United States and in foreign jurisdictions. Projects have included office and mixed-use buildings, schools, hotels, hospitals, condominiums, stadiums, dams, power plants, process plants, and townships.

Emily Wurtenberger Schott is an associate attorney in the Cincinnati office of Vorys, Sater, Seymour and Pease LLP, where she practices in the real estate, finance, and energy group. Ms. Schott has provided legal counsel to owners and contractors in contract negotiation. She is an active member of Division 12 of the ABA Forum on Construction Law. Ms. Schott received her BBA from the University of Notre Dame and her JD from Vanderbilt University Law School.

Michael Shull is a member of Frost Brown Todd LLC, concentrating his practice on the construction industry and public procurement. His clients include both public and private owners as well as general contractors, subcontractors, and suppliers. He has previously contributed to publications by the ABA, Aspen Publishers, Thomson/West, and the Associated General Contractors. Mr. Shull holds a BS in civil engineering as well as his JD, both from the University of Kentucky. Prior to practicing law, Mr. Shull was a research engineer with the Kentucky Transportation Center. He is admitted to practice in Colorado, Georgia, and Kentucky.

Introduction

The American Institute of Architects (AIA) first published standardized general conditions—the precursor to the A201—in 1888. Since then, the AIA Document A201, *General Conditions of the Contract for Construction*, has become uniformly regarded as the industry standard for balancing risks and responsibilities and managing the relationship between the owner and contractor and between the owner and architect. It is perhaps the most widely-used single form of contract for commercial construction projects in the United States. By virtue of the A201's longevity and ubiquity, the expectations of the contracting parties, and the perceived fair allocation of risk, the parties' rights and obligations as embodied in the A201 reflect current industry standards. Often, courts and practitioners will impute these standards into contractual relationships even in the absence of the parties' adoption of the A201 for their project. Indeed, even other publishers of competing contract families have borrowed the spirit if not the letter of many of the A201's provisions. Love it or hate it, the A201 is a foundational document for the construction industry.

As with any major document, the creation and modification of the A201 has not been immune to pressures from both inside and outside the industry. In order to adapt to changes in industry practices, the AIA reviews and modifies its project delivery "families" of documents on a ten-year cycle. In 2017, the AIA released updated versions of the A201 family of documents, which apply to the design-bid-build delivery model.[1] The 2017 revisions to the A201 family of documents were the result of a multi-year review process by the AIA Documents Committee, aided by AIA staff attorneys and outside legal counsel. While the document has always been drafted and published by the AIA, the evolution of the A201 terms are a result of the cooperative input and experience of the AIA Documents Committee with input from other industry experts and stakeholders, including a variety of construction industry associations and attorney groups representing the interests of all construction project participants. Obviously, not all voices and positions can prevail in the final draft of the document, but the AIA Documents Committee carefully considers all comments submitted. Accordingly, despite its longstanding use and influence in the industry, it remains an often-modified document by project participants savvy enough to hire competent construction counsel. There are many reasons why standard construction contracts may be revised, including project- or party-specific considerations, changes in the economy, jurisdiction-specific court interpretations, specific technology use on projects, improved efficiencies or preferences in project management, and so on.

While the rest of this Deskbook contains a comprehensive discussion of all of the changes in the A201, we highlight the following major 2017 changes to the A201:

- *Insurance.* Most of Article 11's insurance provisions have been moved into a separate exhibit, allowing the parties greater flexibility over provisions that are frequently tailored for specific projects. The new exhibit is more complex with multiple check boxes, which require more choices to be made than before. The change in location of these terms to an exhibit will facilitate communication with insurance advisors or brokers. Additionally, it

1. The AIA publishes coordinated sets of agreements for the following delivery methods: design-bid-build; design-build; construction manager as constructor (also referred to as CM at risk); construction manager as advisor; and integrated project delivery.

will make it easier to update insurance requirements, as necessary, based on industry or market changes.

- *Terminations for Convenience*. Historically, owners were free to terminate the contract for convenience but in doing so owed the contractor for costs incurred through the date of termination plus reasonable overhead and profit on unperformed work. The 2017 version of the A201 eliminates the contractor's automatic entitlement to profit for unperformed work and instead requires the parties to negotiate a reasonable termination fee.
- *Dates of Commencement and Substantial Completion*. The date of commencement of the work is now a check-box format with options of (1) the date of the Agreement, (2) the issuance of a notice to proceed by the owner, or (3) a different date as agreed by the parties. The default date is the date of the Agreement. A check box also has been added so the parties can select whether substantial completion occurs as of a specified calendar date or within a certain number of days from the date of commencement.
- *Owner's right to carry out the Work*. Prior versions of the A201 provided that the owner, upon the contractor's default, could correct the contractor's default and issue an appropriate change order deducting the owner's reasonable costs from payments then or thereafter due the contractor. A change order, however, requires an "agreement" by both the owner and contractor with respect to its terms. Because complete consensus might not be reached when an owner undertakes part of the contractor's work, the A201–2017 now provides that if the owner cures, the Architect may, pursuant to section 9.5.1 (Decision to Withhold Certification), withhold or nullify a certificate for payment in whole or in part, to the extent reasonably necessary to reimburse the owner. If the contractor disagrees with the actions of the owner or the architect, or the amounts claimed as costs to the owner, the contractor can pursue a claim.
- *Differing Site Conditions*. Prior editions of the A201 provided that upon encountering differing site conditions, the contractor was to promptly provide notice to the owner and architect, before the conditions are disturbed, and in no event later than 21 days after the conditions were first observed. A201–2017 shortens the time for notice from 21 to 14 days.
- *Digital Data*. Digital practice and the use of building information modeling (BIM) is an expanding and evolving area of the construction industry. While discussion of the AIA standard forms governing the roles and responsibilities of the project participants when exchanging information in digital format is beyond the scope of this book, the A201 has been modified to clarify that the contractor may maintain project documentation at the site in electronic format.
- *Liquidated Damages*. Both the 1997 and 2007 versions of the AIA contract documents included a mutual waiver of consequential damages. Prior versions of the A201 contemplated that the parties might include liquidated damages, early completion bonuses, cost savings, or other incentives by including a parenthetical fill point within the Contract Time article. The 2017 revision adds a specific fill point to prompt discussion and negotiation of these terms.
- *Notice Provisions*. The A201–2017 requires that all notice be written, but it also permits electronic transmission of the notice, with the exception of notice of claims, which may be served only by certified or registered mail, or by courier providing proof of delivery. The default terms for the

electronic format of notice is the use of the AIA Document E203™–2013, Building Information Modeling and Digital Data Exhibit, which is now a Contract Document. The parties are free to specify alternate terms regarding electronic notice.
- *Contractor's Means and Methods*. The A201–2017 changes the approach from the 2007 documents. The A201–2007 provided that if the owner or architect ultimately directed the contractor to proceed with work in spite of the contractor's safety concerns, the owner was responsible for any loss or damage. The 2017 version provides that the contractor remains solely responsible for, and in control over, construction means and methods. If a contractor has a safety concern with any instructions or procedure in the Contract Documents, the contractor is required to give notice and propose an alternative. The architect's review is limited to conformance with the design intent. Absent an objection, the contractor is to perform the work using the alternative it proposed.
- *Direct Communication Between the Owner and Contractor.* Prior editions of the A201 provided that the owner and contractor should communicate with each other through the architect. The 2017 version recognizes that sometimes it is appropriate for the owner and contractor to communicate without the involvement of the architect. The A201–2017 now provides that the owner and contractor are required to include the architect only in communications that relate to or affect the architect's responsibilities and services. For all other communications between the owner and the contractor, it is the owner's responsibility to promptly notify the architect of the substance of such communications.

This Deskbook identifies and analyzes all of these changes and more. What are some key terms whose substance has not changed in the 2017 version? The use of arbitration as optional to litigation for dispute resolution remains in a check box. So does the opportunity to designate an Initial Decision Maker (IDM), who is different than the architect, although many report that most still choose the architect as the IDM. Also the A201 once again does not require that a CPM be used for scheduling the project. Finally, the disclosure of the owner's financial arrangements during the project continues to be required only when there is a material change in the contract sum, a payment failure, or reasonable evidence of a material concern over the ability to pay. The clause has been tinkered with to provide that, when a duty to disclose arises, the data must be provided within 10 days.

Previous editions of this Deskbook have focused almost exclusively on the changes to the newly published A201. This time, however, the editors decided to go a step beyond just a description of the changes to the new A201, which we cover in detail. For this edition, our goal was to create a single comprehensive point of reference for all sections of the A201. This book delivers a section-by-section analysis of how courts and commentators around the country have interpreted the A201 over the years and attempts to give a current state of the law on the most significant sections. Obviously, a lot of the commentary and court decisions will relate to previous editions of the A201, and we include in our analysis the effect, if any, the 2017 changes might have on the interpretation of modified sections. In undertaking this project, the editors and authors have been careful not to take sides favoring the owner, contractor, or architect. And while we recognize the standard the A201 has set in the industry, we have not assumed that the AIA's language is right or wrong, fair or unfair, balanced or tilted in favor of any party. Our goal was to identify the changes in the 2017 edition of the A201 and to analyze how the A201's individual sections have been interpreted over the years. We hope this is the first place you will look when negotiating or litigating a disputed term in the A201. While the editors and authors have tried to be as thorough as possible, it was not possible to include every court decision that has ever considered the A201, nor is it possible to adequately represent the nuances of every viewpoint. Practitioners are cautioned to conduct their own research within the relevant jurisdiction and not to rely solely on the analysis in this Deskbook. This book is not intended to, nor does it, provide legal advice with respect to any specific project or particular jurisdiction. The list of those individuals who provided invaluable assistance in the re-imagined form of the popular Deskbook is too long to print. The editors and authors specifically thank Sarah Forbes Orwig with ABA Book Publishing and Susan Van Bell with the AIA for their review, comments, and suggestions. The opinions expressed in this book are solely those of the respective authors and do necessarily reflect the opinions or reviews of the American Bar Association, the American Bar Association Forum on Construction Law, or the American Institute of Architects.

—Peter W. Hahn
—Joseph C. Kovars
—Amanda S. MacVey

CHAPTER 1:
Article 1 General Provisions

J. Wallace Irvin

ARTICLE 1 GENERAL PROVISIONS

Article 1 contains general provisions of the Contract that define basic terms and identify the Contract Documents. Article 1 also contains a severability provision that ensures the validity of the agreement in the event a court finds a particular provision or section void and unenforceable. The Article then addresses the ownership, transfer, and use of the Instruments of Service and building information models. Finally, it describes the acceptable means of transmitting notices required by the Contract.

The primary modifications to Article 1 are small additions addressing notice and the use of the Instruments of Service and building information models. Otherwise, the changes to this Article are merely grammatical corrections, such as the addition of serial commas. This chapter will explain all changes to the text of Article 1 and analyze the case law pertaining to these provisions.[1]

§ 1.1 BASIC DEFINITIONS

§ 1.1.1 THE CONTRACT DOCUMENTS

The Contract Documents are enumerated in the Agreement between the Owner and Contractor (hereinafter the Agreement) and consist of the Agreement, Conditions of the Contract (General, Supplementary and other Conditions), Drawings, Specifications, Addenda issued prior to execution of the Contract, other documents listed in the Agreement, and Modifications issued after execution of the Contract. A Modification is (1) a written amendment to the Contract signed by both parties, (2) a Change Order, (3) a Construction Change Directive, or (4) a written order for a minor change in the Work issued by the Architect. Unless specifically enumerated in the Agreement, the Contract Documents do not include the advertisement or invitation to bid, Instructions to Bidders, sample forms, other information furnished by the Owner in anticipation of receiving bids or proposals, the Contractor's bid or proposal, or portions of Addenda relating to bidding or proposal requirements.

The AIA made minor changes to this section from the AIA® Document A201™–2007, including punctuation changes and a clarification of those documents that are not part of the Contract Documents.

Section 1.1.1, and similar provisions in other current or past AIA® documents, defines the scope of the agreement between the parties by listing all of the documents that are included in the definition of the term "Contract Documents." These may include the contract, drawings, specifications, addenda, modifications issued after execution, as well as other documents identified in the agreement. If using the AIA suite, the parties should identify the specific version of the general conditions (for example, state that the General Conditions are the AIA® Document A201™–2017). In practice, it is imperative to specify because some courts previously found the provision ambiguous.[2] The general conditions may be amended by the supplementary conditions. The phrase "other Conditions" generally refers to federal, state, or local laws or private conditions that are usually specified by the owner.

1. The majority of the sections in Article 1 contain basic definitions and general provisions that have not been the subject of much litigation.
2. *See, e.g.*, Axthelm & Swett Constr., Inc. v. Caudill, No. 35995-9-1, 1997 Wash. App. LEXIS 735 (May 12, 1997).

Absent extraordinary conditions, courts uniformly uphold these provisions to allow for enforcement of extra-contractual terms contained within the incorporated documents. For instance, in discussing the AIA® Document A105™–1993's[3] incorporation of the AIA® Document A205™–1993, *General Conditions of the Contract for Construction of a Small Project*, the Appellate Court of Connecticut stated "[i]n the present case, the language of the contract clearly and unambiguously refers to A205 as part of the contract. . . . We cannot strain ourselves to import ambiguity into the words of this contract."[4]

However, courts have found the failure to specifically identify documents prevents those documents from becoming a part of the agreement.[5] Accordingly, if any documents included in the definition of Contract Documents are not physically attached or separately signed by the parties, practitioners should consider modifying this section to explicitly state the particular document (or version) that the parties intend to be included in the Contract Documents.

This section also lists information and documents that are explicitly excluded from the Contract Documents. The purpose of excluding these documents is to eliminate a "battle of the forms" and prevent a contractor from attempting to enforce terms or conditions contained in its proposal that materially differ from the Contract Documents. In practice, some public owners require the bidding documents to be included within the Contract Documents. To avoid the ambiguities and conflicts referenced above, modifications to this section are suggested to provide that any bid documents are superseded when the public entity awards the contract for construction. In the event the bid documents contain statutorily required language, that language may be included in the supplementary conditions.

§ 1.1.2 THE CONTRACT

The Contract Documents form the Contract for Construction. The Contract represents the entire and integrated agreement between the parties hereto and supersedes prior negotiations, representations, or agreements, either written or oral. The Contract may be amended or modified only by a Modification. The Contract Documents shall not be construed to create a contractual relationship of any kind (1) between the Contractor and the Architect or the Architect's consultants, (2) between the Owner and a Subcontractor or a Sub-subcontractor, (3) between the Owner and the Architect or the Architect's consultants, or (4) between any persons or entities other than the Owner and the Contractor. The Architect shall, however, be entitled to performance and enforcement of obligations under the Contract intended to facilitate performance of the Architect's duties.

Section 1.1.2 contains an integration clause, restricts modifications of the contract, and clarifies the contractual relationships between the parties on the construction project. An integration clause serves to eliminate claims that terms in prior negotiations or agreements are incorporated into the final agreement. This section

3. AIA® Document A105™–1993, Owner/Contractor Agreement—Small Project, Stipulated Sum § 1.2.

4. 566 New Park Assocs. v. Blardo, 906 A.2d 720 (Conn. App. Ct. 2006) (reversing trial court's finding that general conditions were not incorporated into contract, but finding plaintiff-owner failed to engage in consultation with architect and did not give defendant-contractor requisite notice prior to termination which were contained in general conditions, preventing recovery).

5. *See, e.g.*, *Axthelm & Swett Constr.*, 1997 Wash. App. LEXIS 735, at *9–10 ("The AIA® materials, which were part of the bid information, suggest that the AIA has more than one general conditions form. Thus, even if it was clear that the general conditions document referred to an AIA® form, further specification would be necessary to determine which AIA® form was a part of this particular contract.").

supplements the listing of information and documents in Section 1.1.1 that form the Contract Documents by clarifying the contract supersedes all other negotiations, representations, or agreements.

Once the Contract Documents are established, they may only be altered by a modification. The definition of a modification is clearly intended to encourage the parties to memorialize all changes to the Contract Documents in writing. On most projects, changes are routinely made in the field by project superintendents, architects, and other project representatives. Due to the number of typical changes and the time constraints under which most projects are constructed, it is common for the parties to orally agree to changes without any further documentation. These changes are frequently the subject of litigation. Whether a particular court will enforce an oral modification depends entirely on the facts of each case. Parties are, therefore, encouraged to reduce all changes, however small, to writing.

Section 1.1.2 clarifies that the Contract Documents do not create any third-party beneficiaries or create any contractual relationship between parties that are not signatory to the agreement. The AIA A201® document family is based on the idea that construction contractual relationships should be two-party based. Accordingly, the AIA maintains owner/architect, owner/contractor, contractor/subcontractor, and architect/consultant agreements. Each of the parties to these agreements is deemed to be in privity of contract only with the other party to the agreement. Most jurisdictions find provisions such as Section 1.1.2 prevent an owner from maintaining an action against a subcontractor on a third-party beneficiary theory as well as preventing a subcontractor from maintaining a breach of contract action against an owner on such a theory.[6] For example, under a similar non-AIA® provision, the Michigan Court of Appeals found where an owner contracted separately with a construction manager and a general contractor, despite the fact the construction manager's duties to supervise and manage the project would affect and possibly benefit the general contractor, these effects were incidental only and would not overcome the contractual limitation on third-party beneficiary status.[7] The California Court of Appeals came to the same conclusion under a similar scenario and found an architect could not rely on an indemnification provision contained in a construction management contract where that agreement provided no rights or benefits would be conferred on parties that were not part of the agreement.[8]

The only exception to the exclusion of third-party beneficiaries is contained in the final sentence of this section, which notes the architect is entitled to the performance and enforcement of the contractor's obligations. The architect may, therefore, seek enforcement of the contractor's obligations, independent of the owner. These obligations include the contractor's duty to provide access to the work as well as the contractor's obligations to purchase insurance, waive subrogation rights, and indemnify the architect against certain claims. Additionally, courts uphold the documents' incorporation despite the fact they are not separately signed.[9] For instance, in reviewing a provision similar to Section 1.1.2, a Pennsylvania court found a contract that referenced and incorporated prior agreements subjected

---

6. Pierce Associates, Inc. v. Nemours Found., 865 F.2d 530 (3d Cir. 1988); E.C. Ernst, Inc. v. Manhattan Constr. Co. of Texas, 551 F.2d 1026 (5th Cir. 1977), *cert. denied*, 434 U.S. 1067 (1978) (barring subcontractor third-party beneficiary claim under general contract); Honey v. George Hyman Constr. Co., 63 F.R.D. 443 (D.D.C. 1974) (barring subrogation action by insurer against subcontractor based on third-party beneficiary theory); Tucker Materials (Georgia), Inc. v. Devito Contracting & Supply, Inc., 535 S.E.2d 858 (Ga. Ct. App. 2000).

7. Dynamic Constr. Co. v. Barton Malow Co., 543 N.W.2d 31 (Mich. Ct. App. 1995).

8. Ratcliff Architects v. Vanir Constr. Mgmt., Inc., 106 Cal. Rptr. 2d 1 (Ct. App. 2001).

9. *In re* Premont Indep. Sch. Dist., 225 S.W.3d 329, 335 (Tex. App. 2007), *reh'g denied*, No. 04-06-00810-CV, 2007 Tex. App. LEXIS 4943 (Feb. 23, 2007) ("[N]othing in the contract required the Supplementary Conditions to be signed. . . .").

claims arising out of those prior agreements to an arbitration provision contained only in the referencing Agreement.[10]

> **§ 1.1.3 THE WORK**
>
> The term "Work" means the construction and services required by the Contract Documents, whether completed or partially completed, and includes all other labor, materials, equipment, and services provided or to be provided by the Contractor to fulfill the Contractor's obligations. The Work may constitute the whole or a part of the Project.

The term "Work" is used throughout the AIA Documents and describes the contractor's duty to provide all necessary labor, materials, equipment, and services necessary to fulfill its obligations under the Contract.[11] The work of subcontractors is included within the definition of the contractor's work. It is important to recognize the term "Work" is not coextensive with the definition of "Project" in Section 1.1.4. Because the owner may utilize multiple contractors or self-perform other work that is part of the project, the work of the contractor may be, and frequently is, less than the work required to construct the project.

Aside from distinguishing between the work required to complete the project and the contractor's work, a detailed scope of work is necessary to determine what property is covered by insurance required under the contract or whether damages to property are covered by the standard subrogation waiver contained in the AIA Documents. For example, the Missouri Court of Appeals looked to the definition of work in finding that a waiver of claims set forth in Section 11.4 did not bar claims to existing structures that were clearly outside the scope of the contractor's work.[12] In practice, clearly defining the scope of the contractor's work provides clear expectations of what work the contract requires of the contractor.

> **§ 1.1.4 THE PROJECT**
>
> The Project is the total construction of which the Work performed under the Contract Documents may be the whole or a part and which may include construction by the Owner and by ~~separate contractors~~Separate Contractors.

The AIA did not make any substantive changes to this section from the AIA® Document A201™–2007, except to make the term "Separate Contractors" a defined term. Section 6.1.1 states "the term 'Separate Contractor(s)' shall mean other contractors retained by the Owner under separate agreements." The addition clarifies the project may include not only the work performed by the contractor, but also the work of separate contractors on the site, further distinguishing the work from the project.

10. Huegel v. Mifflin Constr. Co., 796 A.2d 350 (Pa. Super. Ct. 2002). *See also* Nova Corp. v. Joseph Stadelmann Elec. Contractors, Inc., No. 07-cv-1104, 2008 U.S. Dist. LEXIS 21040 (D.N.J. March 18, 2008) (arbitration agreement within AIA A201, which was incorporated by reference into project manual, which was itself incorporated by reference into purchase order, found binding on parties).

11. *See, e.g.*, Eagle Fire Protection Corp. v. First Indem. of Am. Ins. Co., 678 A.2d 699 (N.J. 1996).

12. Knob Noster R-VII Sch. Dist. v. Dankenbring, 220 S.W.3d 809 (Mo. Ct. App. 2007). *See also* Fidelity & Guar. Ins. Co. v. Craig-Wilkinson, Inc., 948 F. Supp. 608 (S.D. Miss. 1995), *aff'd without opinion*, 101 F.3d 699 (5th Cir. 1996).

§ 1.1.5 THE DRAWINGS

The Drawings are the graphic and pictorial portions of the Contract Documents showing the design, location and dimensions of the Work, generally including plans, elevations, sections, details, schedules, and diagrams.

Although a layperson may generally consider drawings to be the large paper drawings produced by a design professional, drawings may be in both physical and electronic form. In addition to a physical construction set of drawings, drawings may also be found in addenda, change orders, construction change directives, minor changes in the work, or responses to a contractor's request for information.

As technology has advanced, parties on a construction project frequently transmit drawings electronically. And as building information modeling continues to increase in usage, the transmission of information on construction projects is performed almost exclusively through the Internet. Accordingly, parties should review the AIA's 2013 Digital Practice Documents to establish procedures to protect the proprietary information contained in the drawings.

§ 1.1.6 THE SPECIFICATIONS

The Specifications are that portion of the Contract Documents consisting of the written requirements for materials, equipment, systems, standards and workmanship for the Work, and performance of related services.

The specifications are written descriptions that qualitatively define the contractor's work. Most specifications are written according to the 16 divisions of MASTERFORMAT, a publication of the Construction Specifications Institute. Each of the 16 divisions is subdivided to address particular items of work. While the scope of work contained in the agreement provides a broad overview of the work, the specifications set forth the detailed requirements for individual portions of the work, such as the products to be used.

§ 1.1.7 INSTRUMENTS OF SERVICE

Instruments of Service are representations, in any medium of expression now known or later developed, of the tangible and intangible creative work performed by the Architect and the Architect's consultants under their respective professional services agreements. Instruments of Service may include, without limitation, studies, surveys, models, sketches, drawings, specifications, and other similar materials.

The Instruments of Service represent the professional services provided by the architect and its consultants, regardless of form. The purpose of this definition is to make clear that the physical or electronic documents cannot be separated from the services performed by the architect.

§ 1.1.8 INITIAL DECISION MAKER

The Initial Decision Maker is the person identified in the Agreement to render initial decisions on Claims in accordance with Section 15.2 ~~and certify termination of the~~

~~Agreement under Section 14.2.2.~~ . The Initial Decision Maker shall not show partiality to the Owner or Contractor and shall not be liable for results of interpretations or decisions rendered in good faith.

The dispute resolution process set forth in Article 15 requires the parties to submit most claims to an initial decision maker as a first step. The architect traditionally performed this service, but in 2007, the AIA modified the general conditions to allow the parties to name another person or company to serve in this capacity. In 2017, the AIA added a new second sentence to Section 1.1.8, to clarify the initial decision maker shall not show partiality to either the owner or the contractor and will not be liable for the results of interpretations or decisions that are rendered in good faith. The addition tracks the language found in Section 4.2.12, which addresses the architect's interpretations and decisions, generally.

The reason for this modification is the fact the initial decision maker acts in a capacity similar to an arbitrator. An arbitration award may generally only be challenged on one of a few select grounds, including partiality of the arbitrator. Because many architects are not intimately familiar with the legal requirements surrounding the arbitration process, the change provides additional guidance to remind the architect that although the architect's fee is borne by the owner, the architect may not show partiality to either party. Instead, the initial decision maker must remain unbiased, rendering a decision on the facts presented by the parties.

§ 1.2 CORRELATION AND INTENT OF THE CONTRACT DOCUMENTS

§ 1.2.1 The intent of the Contract Documents is to include all items necessary for the proper execution and completion of the Work by the Contractor. The Contract Documents are complementary, and what is required by one shall be as binding as if required by all; performance by the Contractor shall be required only to the extent consistent with the Contract Documents and reasonably inferable from them as being necessary to produce the indicated results.

As discussed previously, the AIA A201 document family is based on the presumption that all relationships on a construction project are only between two parties. However, because a construction project requires the participation of numerous parties performing under various agreements, this section provides a means for tying all of the agreements together in the event of any ambiguity. Oftentimes, information may be included on one of the contract documents but excluded from another. To address the practical realities of construction, the AIA documents require the contractor to make reasonable inferences from the contract documents necessary to complete the project as intended.

§ 1.2.1.1 The invalidity of any provision of the Contract Documents shall not invalidate the Contract or its remaining provisions. If it is determined that any provision of the Contract Documents violates any law, or is otherwise invalid or unenforceable, then that provision shall be revised to the extent necessary to make that provision legal and enforceable. In such case the Contract Documents shall be construed, to the fullest extent permitted by law, to give effect to the parties' intentions and purposes in executing the Contract.

Section 1.2.1.1 is new to the AIA® Document A201™–2017. The AIA added a severability clause to avoid situations where the entire contract, or portions thereof, could be rendered invalid because a single section of the contract is determined to be invalid or unenforceable. Instead, in cases where a court finds portions of the contract to be void or otherwise unenforceable, this section allows the court to read the offending section out of the contract, maintaining the validity and enforceability of the remainder of the agreement.

> **§ 1.2.2** Organization of the Specifications into divisions, sections and articles, and arrangement of Drawings shall not control the Contractor in dividing the Work among Subcontractors or in establishing the extent of Work to be performed by any trade.

The AIA A201 document family repeatedly makes clear that the contractor is solely responsible for the means and methods of construction, including the assignment or subcontracting of any portion of the work. This section makes clear that the contractor may not rely on the subdivision of the work into divisions or sections within the specifications or drawings for the purpose of assigning a portion of the work to any particular subcontractor based upon where that work is depicted within the specifications or drawings.

> **§ 1.2.3** Unless otherwise stated in the Contract Documents, words that have well-known technical or construction industry meanings are used in the Contract Documents in accordance with such recognized meanings.

To limit the length of the contract documents, terms that have well-defined technical, industry, or trade meanings are not separately defined within the contract documents. Instead, this section makes clear those terms shall be used in accordance with their recognized meanings.

§ 1.3 CAPITALIZATION

> Terms capitalized in these General Conditions include those that are (1) specifically defined, (2) the titles of numbered articles, or (3) the titles of other documents published by the American Institute of Architects.

This section clarifies that terms that are capitalized within the contract documents have a specialized definition that is set forth within the contract documents. To the extent those terms may have other meanings in general usage, the definitions contained in the contract documents control.

§ 1.4 INTERPRETATION

> In the interest of brevity the Contract Documents frequently omit modifying words such as "all" and "any" and articles such as "the" and "an," but the fact that a modifier or an article is absent from one statement and appears in another is not intended to affect the interpretation of either statement.

Similar to Section 1.2.3, this section clarifies that to prevent the contract documents from becoming too lengthy, certain modifiers are omitted from the contract documents. This section attempts to limit any claim the contract documents are ambiguous simply because a modifier is not included.

§ 1.5 OWNERSHIP AND USE OF DRAWINGS, SPECIFICATIONS, AND OTHER INSTRUMENTS OF SERVICE

> **§ 1.5.1** The Architect and the Architect's consultants shall be deemed the authors and owners of their respective Instruments of Service, including the Drawings and Specifications, and ~~will~~ retain all common law, statutory, and other reserved rights in their Instruments of Service, including copyrights. The Contractor, Subcontractors, Sub-subcontractors, and ~~material or equipment~~ suppliers shall not own or claim a copyright in the Instruments of Service. Submittal or distribution to meet official regulatory requirements or for other purposes in connection with ~~this~~the Project is not to be construed as publication in derogation of the Architect's or Architect's consultants' reserved rights.

As the author and owner of the intellectual property contained in the instruments of service, this section makes clear the architect and its consultants retain all common law, statutory, and other reserved rights in the instruments of service, including copyrights. Because of the evolution of computer aided drafting and the transmission of documents electronically, parties should consider reviewing the AIA's 2013 Digital Practice Documents to establish procedures to protect the proprietary information contained in the instruments of service.

Courts have generally found this provision unambiguous, enforcing copyright infringement claims by design professionals and affirming the architect's and its consultants' ownership rights in the instruments of service.[13]

> **§ 1.5.2** The Contractor, Subcontractors, Sub-subcontractors, and ~~material or equipment~~ suppliers are authorized to use and reproduce the Instruments of Service provided to them, subject to any protocols established pursuant to Sections 1.7 and 1.8, solely and exclusively for execution of the Work. All copies made under this authorization shall bear the copyright notice, if any, shown on the Instruments of Service. The Contractor, Subcontractors, Sub-subcontractors, and ~~material or equipment~~ suppliers may not use the Instruments of Service on other projects or for additions to ~~this~~the Project outside the scope of the Work without the specific written consent of the Owner, Architect, and the Architect's consultants

Although the architect and its consultants retain all ownership rights in the instruments of service, the agreement provides the owner with a limited license to use the instruments of service in relation to the project. This license authorizes the contractor and its subcontractors to also use the instruments of service to construct the project. The restrictions on the use of the instruments of service protect the interests of the owner, architect, and the architect's consultants as well as the general public, which could be harmed by the unauthorized use and modification of the instruments of service by parties other than the architect and its consultants.

13. *See, e.g.*, Eiben v. A. Epstein & Sons, Int'l, No. 97 C 8448, 1999 U.S. Dist. LEXIS 11651 (N.D. Ill. July 26, 1999); Guillot-Vogt Assocs., Inc. v. Holly & Smith, 848 F. Supp. 682 (E.D. La. 1994).

§ 1.6 NOTICE

§ 1.6.1 Except as otherwise provided in Section 1.6.2, where the Contract Documents require one party to notify or give notice to the other party, such notice shall be provided in writing to the designated representative of the party to whom the notice is addressed and shall be deemed to have been duly served if delivered in person, by mail, by courier, or by electronic transmission if a method for electronic transmission is set forth in the Agreement.

§ 1.6.2 Notice of Claims as provided in Section 15.1.3 shall be provided in writing and shall be deemed to have been duly served only if delivered to the designated representative of the party to whom the notice is addressed by certified or registered mail, or by courier providing proof of delivery.

In the AIA® Document A201™–2017, the AIA moved the section on written notice from Article 13 and placed it in Article 1. This revised section requires the parties to provide all notices under the contract in writing. An addition to the provision also allows the parties to provide notice through electronic means, such as e-mail, attempting to keep up with common industry practices. However, it is important to distinguish between typical notices required under the contract and a notice of claim, which must be delivered to the designated representative of the party to whom the notice is addressed by certified or registered mail, or by courier providing proof of delivery.

§ 1.7 DIGITAL DATA USE AND TRANSMISSION ~~OF DATA IN DIGITAL FORM~~

~~If the~~The parties ~~intend to transmit~~shall agree upon protocols governing the transmission and use of Instruments of Service or any other information or documentation in digital form~~, they shall endeavor~~. The parties will use AIA Document E203™–2013, Building Information Modeling and Digital Data Exhibit, to establish ~~necessary protocols governing such transmissions, unless otherwise already provided in the Agreement or the Contract Documents~~the protocols for the development, use, transmission, and exchange of digital data.

Formerly Section 1.6 in AIA® Document A201™–2007, this modified section requires the parties to develop protocols for the transmission of instruments of service or other data in digital form. Specifically, unless the parties agree otherwise, the revised section requires the parties to utilize AIA® Document E203™–2013, Building Information Modeling and Digital Data Exhibit, for this purpose.

§ 1.8 BUILDING INFORMATION MODELS USE AND RELIANCE

Any use of, or reliance on, all or a portion of a building information model without agreement to protocols governing the use of, and reliance on, the information contained in the model and without having those protocols set forth in AIA Document E203™–2013, Building Information Modeling and Digital Data Exhibit, and the requisite AIA Document

G202™–2013, Project Building Information Modeling Protocol Form, shall be at the using or relying party's sole risk and without liability to the other party and its contractors or consultants, the authors of, or contributors to, the building information model, and each of their agents and employees.

A new section to AIA® Document A201™–2017, this provision requires the parties to develop protocols for the use of, and reliance on, a building information model. Unless the parties establish protocols, the use of, or reliance on, a building information model will be at the using or relying parties' sole risk. The provision requires that the parties utilize the AIA companion documents AIA® Document E203™–2013, Building Information Modeling and Digital Data Exhibit, and AIA® Document G202™–2013, Project Building Information Modeling Protocol Form.

CHAPTER 2:
Article 2 Owner

Steven B. Lesser
Claramargaret H. Groover

ARTICLE 2 OWNER

Article 2 focuses on the role and responsibilities of the owner in the construction process. Generally, this Article addresses a variety of issues such as: (1) the individual who may act for the owner; (2) the deadline for the owner to furnish essential information necessary for the contractor to protect its rights under the lien law; (3) the process in which the owner provides financial information to demonstrate capacity to pay the contractor; (4) the requirement for the owner to pay for essential permits; (5) the owner requirement to engage a licensed architect for design who is to be replaced, if terminated; (6) the owner requirement to provide reports of known conditions (i.e., surveys, soils reports, and other such documented information) timely; (7) the owner's right to stop work for the contractor's repeated failure to perform timely and properly or its repeated failure to correct nonconforming work; and (8) the owner leverage to remove portions of the original scope of work from a nonperforming contractor. Recognizing the pivotal role of the owner in the fluent relationship with the contractor and architect, this Article clarifies the duties and obligations of the owner during construction.

The AIA expanded certain financial obligations of the owner by incorporating several changes to Section 2.2—*Evidence of the Owner's Financial Arrangements.* These revisions require the contractor to request financial information in writing and further provides the contractor with greater latitude to refuse to commence and/or continue performance if the owner fails to provide "reasonable evidence" of additional funding in the event that the owner i) fails to make a payment, ii) materially changes the scope of work, or iii) merely identifies its insecurity that the owner may not pay. New for 2017, the AIA requires that financial information disclosed under this provision be maintained as confidential. As addressed below, this chapter will also discuss how the financial disclosure revisions may further complicate an already problematic process for both the owner and the contractor, especially given the fact that the contractor has valuable contract rights to protect itself from owner nonpayment in other provisions within A201–2017, as well as critical statutory rights under the lien law of the jurisdiction.

Article 2 also allows the owner to complete the project in those instances when a contractor fails to perform in accordance with the Agreement. As discussed below, the owner must perfect its rights with great care to avoid extinguishing valuable rights and remedies.

§ 2.1 GENERAL

> **§ 2.1.1** The Owner is the person or entity identified as such in the Agreement and is referred to throughout the Contract Documents as if singular in number. The Owner shall designate in writing a representative who shall have express authority to bind the Owner with respect to all matters requiring the Owner's approval or authorization. Except as otherwise provided in Section 4.2.1, the Architect does not have such authority. The term "Owner" means the Owner or the Owner's authorized representative.

Section 2.1.1 for the 2017 edition of A201 is a straightforward provision. The purpose is to clarify the authority of the owner and recognize who may act on its behalf. The AIA A101–2017 Agreement identifies the owner's legal entity and indicates that the owner shall designate its representative in writing (Agreement—Miscellaneous Provisions). Only the owner's designated representative shall have the authority to act on behalf of the owner to communicate the myriad decisions to be addressed during the course of construction. The owner's representative should exercise this authority in good faith in rejecting the contractor's performance.[1] In *Mann v. Philip Vizzini & Son, Inc.*, the contractor was installing sod, mowing, cleaning up, and performing other landscaping duties. The city inspector rejected the work and the contractor remedied the rejected work only to have the inspector reject it multiple times in a repair-and-rejection-cycle, never to accept the contractor's work. The contractor's evidence was not sufficient to show that the inspector's decisions to reject the work were arbitrary and capricious or based upon fraud.[2]

The goal of A201 is to clarify rights and obligations as opposed to promoting ambiguity and fostering disputes. Historically, an architect or engineer could be considered to act as the owner's agent. The professional's duties to administer the contract included the obligation to review the work to determine its conformance to drawings and specifications. Certainly in the dynamic governed by A201 and common law, such action by the architect or engineer may be interpreted as actions taken on behalf of the owner. If the contract is silent relative to identifying the owner's representative, such an individual may appear to act on behalf of the owner and establish liability for its acts and/or omissions. Consider, for example, the Louisiana case of *Diamond B. Construction Co. v. City of Plaquemine*.[3] A trial court entered judgment against an owner, ordering the owner to pay the contractor's retainage with interest based in part on the theory that the engineer acted as the owner's agent in accepting the contractor's work. On appeal, however, the court reasoned that the "owner's representative," identified in the contract, was the individual authorized to withhold payment, and that person was not the engineer. The engineer reviewed the work to determine whether the contractor failed to perform in accordance with drawings and specifications. Since the engineer's actions did not bind the owner, the court held that the owner had the right to withhold payment under the contract.

PRACTICE TIP

The owner's best interest is served by a specific, written identification of the individual having authority to act on the owner's behalf during construction. The owner should be counseled to designate a representative as opposed to leaving the placeholder blank.[4] If the owner's agent is not specifically identified, the contractor may reasonably rely upon another who merely appears to act as the owner's agent but in reality may not have authority. Defining the designated representative will avoid confusion and allow construction to proceed with confidence that proper approvals have been acquired from those with authority. This scenario will help avoid a future legal battle over concepts of agency and apparent authority.

To address personnel changes during construction, it is also recommended that the contract language provide that the owner's representative may be changed by written notice to the contractor and architect. This is particularly important for multi-year projects.

1. Mann v. Philip Vizzini & Son, Inc., 283 A.2d 577 (Md. Ct. App. 1971).
2. *Id.* at 582–83.
3. 673 So. 2d 636 (La. Ct. App. 1996).
4. AIA A101–2017 Agreement–(Miscellaneous Provisions).

In many instances, owners may prefer to separate the engineering and design work from other construction administration functions. For example, the owner may elect to have others certify payment applications, approve performance sufficient to release retainage, and issue Certificates of Substantial Completion and Final Payment. Various reasons may justify this approach, including assigning this responsibility to more qualified and skilled professionals. For example, an architect may be capable of reviewing Applications for Payment based upon work completed but not be properly experienced to interpret construction schedules and updates. Other reasons may exist to delegate certain traditional duties of the architect to others, such as to save money or to prevent the architect from binding the owner if the relationship has not been solidified by trust and past experience.

Some owners may elect to delineate tiered or multiple representatives with specific, separate authority to act on the owner's behalf. There are risks to having multiple individuals with different substantive areas or degrees of authority. The better approach is for the owner to designate a single point of contact responsible for official communication to and from the contractor and architect. Internal to the owner's organization, the tiered authority can function as described previously, but the owner's final decisions/directions should be transmitted to the contractor by the individual designated as the owner's representative. Keeping it simple and straightforward for project management minimizes the risk for confusion and inconsistent messaging and is a best practice.

§ 2.1.2 The Owner shall furnish to the Contractor, within fifteen days after receipt of a written request, information necessary and relevant for the Contractor to evaluate, give notice of, or enforce mechanic's lien rights. Such information shall include a correct statement of the record legal title to the property on which the Project is located, usually referred to as the site, and the Owner's interest therein.

With the exception of the added commas, the provision remains substantively unchanged for 2017. Here, the owner is obligated to provide an accurate legal description and notice of the owner's interest in the property such as fee simple, leasehold, or other such ownership interest. This provision is critical for the contractor and its subcontractor to protect their respective rights to payment and to secure their rights to a pre-judgment mechanic's or construction lien.

§ 2.2 ~~INFORMATION AND SERVICES REQUIRED~~EVIDENCE OF THE ~~OWNER~~OWNER'S FINANCIAL ARRANGEMENTS

§ 2.2.1 Prior to commencement of the Work~~,~~ and upon written request by the Contractor ~~may request in writing that~~, the Owner ~~provide~~shall furnish to the Contractor reasonable evidence that the Owner has made financial arrangements to fulfill the Owner's obligations under the Contract. ~~Thereafter, the~~The Contractor ~~may only request~~ shall have no obligation to commence the Work until the Owner provides such evidence. If commencement of the Work is delayed under this Section 2.2.1, the Contract Time shall be extended appropriately.

The AIA made subtle but precise refinements to this Section for 2017. Specifically in this provision, the contractor has the affirmative obligation to trigger the owner's duty to provide "reasonable evidence" of proper project financing. Only upon issuance of the contractor's written demand will the owner have a duty to provide it.

The owner's duty to provide adequate financing contains within it an implied covenant to act in good faith and to deal fairly.[5] Discharging that duty, however, is problematic given the enduring vagueness in the phrase "reasonable evidence." A debate over what is reasonable evidence could result in delayed commencement or even stopping the job. While it is fundamental that the owner must pay for improvements under the contract, the exact nature of "reasonable evidence that the owner has made financial arrangements" is less clear.

PRACTICE TIP

The parties can determine what will constitute "reasonable evidence" that the owner could provide that it has secured financing and that will satisfy the contractor. For example, would a statement of certified funds, the amount of the construction financing, or other statement from the lender be sufficient? If the owner is unable to get the contractor to agree to strike the provisions of Section 2.2.2, discussed below, such reasonable evidence may also apply to satisfy the contractor in case the owner fails to pay or fails to provide the contractor with adequate assurance of its ability to pay for the construction or makes a change that would materially change the contract sum for construction.

§ 2.2.2 Following commencement of the Work and upon written request by the Contractor, the Owner shall furnish to the Contractor reasonable evidence that the Owner has made financial arrangements to fulfill the Owner's obligations under the Contract only if (1) the Owner fails to make payments to the Contractor as the Contract Documents require; (2) ~~a change in the Work materially changes the Contract Sum; or (3)~~ the Contractor identifies in writing a reasonable concern regarding the Owner's ability to make payment when due~~. The Owner shall furnish such evidence as a condition precedent to commencement or continuation of the Work or the portion of the Work affected by a material change. After the Owner furnishes the evidence~~; or (3) a change in the Work materially changes the Contract Sum. If the Owner fails to provide such evidence, as required, within fourteen days of the Contractor's request, the Contractor may immediately stop the Work and, in that event, shall notify the Owner that the Work has stopped. However, if the request is made because a change in the Work materially changes the Contract Sum under (3) above, the Contractor may immediately stop only that portion of the Work affected by the change until reasonable evidence is provided. If the Work is stopped under this Section 2.2.2, the Contract Time shall be extended appropriately and the Contract Sum shall be increased by the amount of the Contractor's reasonable costs of shutdown, delay and start-up, plus interest as provided in the Contract Documents.

The most significant addition for 2017 is the contractor's right to stop work if the owner fails to provide "reasonable evidence" that it can make payments when due. This revision provides the contractor a valuable right to protect itself in the event of financial instability of the owner.

5. EKE Bldrs, Inc. v. Quail Bluff Assocs., 714 P.2d 604 (Okla. Ct. App. 1985).

The 1997 version of this provision allowed the contractor to refuse to *commence or continue* work in the absence of financial assurances.[6] The 2007 version introduced the additional circumstances when a contractor may acquire financial assurances after commencement of the work if one of three potential circumstances exist, namely, (1) owner's failure to pay as the contract documents require, (2) a change in the Work that materially changes the contract sum, or (3) contractor has a reasonable concern regarding the owner's ability to pay when payment is due.[7]

Disclosure of financial arrangements following commencement of the work would have been difficult for a contractor to demand successfully under the 2007 version, if the basis was the owner's failure to pay, unless the owner completely lacked justification for its nonpayment.[8] As for the second circumstance, the phrase "materially changes the Contract Sum" is a question of degree and case law has provided some additional guidance. For example, in what instances does a material change rise to the level of a cardinal change? To avoid ambiguity and establish certainty, the parties may wish instead to establish a dollar value threshold that must be satisfied before a change will be considered as material.

Despite obstacles inherent in two of the three opportunities, the third provides the contractor with great latitude. Utilizing this option, contractor must only demonstrate that a "reasonable concern" exists to justify this request and may even be based upon a newspaper article suggesting that the owner is having financial difficulty or a bankruptcy filing is contemplated. Furthermore, this provision provides the contractor with a basis to stop the work for self-serving reasons. It is conceivable that if the contractor were facing a material delivery delay that would expose the contractor to liquidated damages, this provision could be triggered to relieve him of liability. For instance, by simply sending the notice, the contractor could gain an advantage to extend the time for performance while delay caused by the contractor could be extinguished. This provision is likely to prompt the dispute resolution process identified in the contract because it offers no guidance as to who decides what constitutes a "reasonable basis." Most importantly, the provision fails to require the contractor to continue to perform its work pending resolution of a dispute over the sufficiency of the financial evidence produced by the owner[9] and, instead, allows the contractor to stop work.

PRACTICE TIP

This section is riddled in potential controversy, especially when a contractor feels uneasy about its new relationship with an owner who has formed a single-purpose, single-asset limited liability company for the project. In this regard, the contractor may be justified to acquire financial assurances at every opportunity provided by Article 2.2.2. As the term "reasonable concern" is not defined, the owner's legal counsel should evaluate opportunities to limit those instances when financial assurances will be required and detail the specific documentation that the owner will produce in response to serve as "reasonable evidence" of such financing. For example, financial assurances would be required from an owner when i) a specified

6. Section 2.2.1 of AIA A201–1997.

7. Section 2.2.1 of AIA A201–2007. The reasonable concern must be addressed in writing. Note that the reasonable evidence financial arrangements required to be produce relates to the owner's obligations under the entire contract as opposed to perhaps, the one payment that the owner failed to pay.

8. Section 9.5.1.1 to .7 of AIA A201–2007—bases for withholding payment by owner.

9. Section 15.1.3 of AIA A201–2007 provides that pending final resolution of a claim "Contractor shall proceed diligently with performance of the Contract and the Owner shall continue to make payments in accordance with the Contract Documents."

monetary threshold for change orders has been exceeded; ii) delays caused by design changes result in the contract time being exceeded by a set number of days beyond the original duration specified in the contract; or iii) a bankruptcy proceeding is filed, as opposed to a suggestion in a newspaper article. The contractor should also consider negotiating a modification to this provision to affirmatively require periodic production of financial statements and other related information without the necessity to send a written request to the owner.

The contractor will have little risk to its position since the ill-defined "reasonable evidence" is all that is required. As stated above, the parties are encouraged to define what they would agree is "reasonable evidence" in order to forestall potential disputes. "Reasonable evidence" may only be reasonable to the owner providing it, not the contractor receiving it. Even though the contractor could instead exercise its right to terminate upon notice for the owner's nonpayment or record a mechanic's or construction lien to protect its rights pending resolution of the evidence debate, the contractor's right to "reasonable evidence" could bring the project to a halt. Failing receipt of adequate assurance, the contractor may stop work and upon resumption be entitled to its costs and an extension of time to perform. This provision gives the contractor significant contractual authority supported by financial recourse.

Both parties need to have adequate funding for design and construction of the project. Article 2 currently and historically, however, provides only inspirational guidance regarding the quality and sufficiency of financial documents provided while also allowing such owner-deliverable to serve as a condition precedent to the contractor's duty to perform. An owner may not wish for the contractor to exercise this additional control over the project given the fact that the contractor already has a significant right to terminate upon the first sign that the owner cannot or will not pay a progress payment (Section 14.1).

> **§ 2.2.3** After the Owner furnishes evidence of financial arrangements under this Section 2.2, the Owner shall not materially vary such financial arrangements without prior notice to the Contractor.

The AIA did not substantively change this provision for 2017. The only modification was to reformat the first clause from the preceding paragraph in Section 2.2.2 of A201–2007 to clarify when the owner had a duty not to vary the financial arrangements, i.e., after furnishing evidence of such arrangements.

PRACTICE TIP

Most owners would consider striking this provision so that their refinancing arrangements and other private funding matters remain confidential and their ability to make management decisions remains unfettered by third-party involvement. The owner's duty to pay for performance is immutable through contract completion.

> ~~§ 2.2.2~~**§ 2.2.4** Where the Owner has designated information furnished under this Section 2.2 as "confidential," the Contractor shall keep the information confidential and shall not disclose it to any other person. However, the Contractor may disclose "confidential" information, after seven (7) days' notice to the Owner, where disclosure is

required by law, including a subpoena or other form of compulsory legal process issued by a court or governmental entity, or by court or arbitrator(s) order. The Contractor may also disclose "confidential" information to its employees, consultants, sureties, Subcontractors and their employees, Sub-subcontractors, and others who need to know the content of such information solely and exclusively for the Project and who agree to maintain the confidentiality of such information.

This provision is new for 2017. The contractor and its subcontractors have the express duty to not to disclose those financial documents, marked as confidential, except upon certain exceptions listed in the new Section 2.2.4. From the owner's standpoint, once the funding is in place the owner does not want to provide extensive and often confidential details of its finances. From the contractor's standpoint, it does not want to mobilize, commence, and immediately encounter cash flow problems that will obstruct its ability to prosecute the work.

PRACTICE TIP

The parties may wish to negotiate a shorter period between notice and disclosure of confidential information. Will the contractor be required to get individual confidentiality agreements? An owner may require that confidential documents be returned after the project is completed. Given more use of electronic document management, the owner may have more practical control over such sensitive project information during construction and after completion.

§ 2.3 INFORMATION AND SERVICES REQUIRED OF THE OWNER

§ 2.3.1 Except for permits and fees that are the responsibility of the Contractor under the Contract Documents, including those required under Section 3.7.1, the Owner shall secure and pay for necessary approvals, easements, assessments and charges required for construction, use or occupancy of permanent structures or for permanent changes in existing facilities.

This provision read in conjunction Section 3.7.1 requires the contractor to provide the building permit, as well as individual permits for major subcontractors such as mechanical, electrical, or plumbing. The owner is obligated to obtain its land use and development approvals from all local planning agencies, including those of city, county, state, and federal. The owner must also arrange for all easements to be recorded and pay third parties for any assessments and other charges necessary for construction. Also, the owner must arrange for evidence of other entitlements showing its right to occupy any existing structures or to make permanent changes in facilities already on the project site.

These provisions become critical when considering the early impact of delays on a project attributed to acquiring permits, securing approvals, and timely paying for all necessary and proper entitlements.[10] In a Virginia case, a public owner

10. Paulus v. City of St. Louis, 446 S.W.2d 144 (Mo. Ct. App 1969).

terminated a contractor for its failure to construct a bridge in a timely manner. The contractor defended by claiming that the owner failed to obtain a necessary access easement, but the court found that the contractor was to have obtained it under the contract. The contractor was rightfully terminated for failure to perform.[11]

PRACTICE TIP

The parties can investigate, schedule, and allocate responsibility to pay for the various permits as part of the schedule of values that the contractor is required to create and maintain for the project. The owner may seek to modify this provision to require that the contractor initiate and expedite the filing of all documents to secure the building permit and otherwise cooperate with the owner to secure the permit. The schedule of values, AIA Document G703–1992, serves as a continuation sheet to the Application for Payment, AIA Document G702–1992, and it identifies each scope of work, allocating various price components necessary to monitor balances paid for actual work performed and for work remaining to be performed.

§ 2.3.2 The Owner shall retain an architect lawfully licensed to practice architecture, or an entity lawfully practicing architecture, in the jurisdiction where the Project is located. That person or entity is identified as the Architect in the Agreement and is referred to throughout the Contract Documents as if singular in number.

§ 2.3.3 If the employment of the Architect terminates, the Owner shall employ a successor to whom the Contractor has no reasonable objection and whose status under the Contract Documents shall be that of the Architect.

Sections 2.3.2 and 2.3.3, above, were relocated from Section 4.1. and remain substantively unchanged for 2017. The contractor's best interests are served by requiring the owner to retain a professionally qualified (licensed) architect to design the project and to remain available to provide any design services required at all times.

PRACTICE TIP

At the outset of preparation of the Agreement, the first step should be to research the licensing status of the design professionals engaged to prepare and submit the construction drawings for permitting to confirm the compliance with this requirement.

§ 2.3.4 The Owner shall furnish surveys describing physical characteristics, legal limitations and utility locations for the site of the Project, and a legal description of the site. The Contractor shall be entitled to rely on the accuracy of information furnished by the Owner but shall exercise proper precautions relating to the safe performance of the Work.

11. D.C. McClain, Inc. v. Arlington County, 452 S.E.2d 659 (Va. 1995).

Just as the *Spearin*[12] doctrine stands for the proposition that the owner impliedly warrants the adequacy of the drawings and specifications to a contractor, the information provided by the owner under this section may similarly be presumed to be adequate unless adequacy and accuracy are expressly disclaimed. If the owner provides reports and test data without disclaiming accuracy, it could be held liable when the contractor relies upon such data to its detriment.[13] Where language in the contract documents advises that the contractor is presumed to have inspected the site and made detailed surveys required for construction, this disclaimer language may defeat the contractor's damage claim arising from inaccurate contract documents.[14]

A design professional has a similar right to rely upon the accuracy of the owner-provided information under a similar provision in the AIA B-series documents.[15]

Even though a disclaimer on the transmittal or, even better, a disclaimer included as a negotiated revision to the A201 may prove legally useful, an owner's reliance upon broad disclaimer language may be misplaced, prohibiting the owner from escaping liability for subsurface conditions. In one instance, a public owner became obligated to pay additional costs and extend the completion date where the contractor discovered subsurface conditions similar to those known to the owner.[16] The public owner required the contractor to examine the contract documents carefully and disclaimed guarantees of accuracy as to the borings. The case turned on the fact that the public owner knew of similar conditions encountered in another area dredged by a separate contractor on a separate contract. As such, the owner could not deny that it knew of the nature of the subsurface conditions. The owner had a duty to disclose all project-related tests and reports so as not to mislead bidders. In some instances, an owner may be obligated to share expensive tests and studies with the contractor, knowing that a reasonable contractor might not expend such sums to repeat the testing/studies. In such instance, the owner may expect that the contractor may be entitled to rely upon the expensive test results commissioned by the owner.[17]

On another public works project, an owner had misstated the existence of an easement that would allow the contractor unimpeded access to the site. The contractual no-damages-for-delay provision may be overcome if the owner is found to have misrepresented such information necessary for the contractor's full performance.[18] The appellate court in *Franklin* concluded that any disclaimer of site conditions referred to the physical attributes of the site and of actual demolition activities and not to access to the site.[19] Reference to an easement allowing site access, though provided orally at a project construction meeting, was found to have been incorrect and obtaining site access thereafter delayed the contractor's commencement.[20]

PRACTICE TIP

The owner will want to make it clear that it has transferred the risk of site access, subsurface conditions, and other site conditions to the contractor by adding provisions that disclaim the owner's liability. For example, the owner may add similar

12. United States v. Spearin, 248 U.S. 132 (1918).
13. Warner Constr. Corp. v. City of Los Angeles, 466 P.2d 996 (Cal. 1970).
14. Bilotta Construction Corp. v. Village of Mamaroneck, 604 N.Y.S.2d 966 (App. Div. 1993).
15. Taylor v. DeLosso, 725 A.2d 51 (N.J. Super. Ct. App. Div. 1999); Cadral Corp. v. Solomon, Cordwell, Buenz & Assoc., 497 N.E. 2d 1285 (Ill. App. Ct. 1986).
16. Jacksonville Port Authority v. Parkhill-Goodloe Co., 362 So. 2d 1009 (Fla. Dist. Ct. App. 1978).
17. *Id.*
18. Franklin Contracting Co. v. State, 365 A.2d 952 (N.J. Super. Ct. App. Div. 1976).
19. *Id.* at 955.
20. *Id.* at 953, 954.

language to the contract: "Contractor has examined the Project site to its full satisfaction, is familiar with local conditions under which the Work is to be performed, and has correlated its observations in order to perform the Work in accordance with the Contract Documents. Owner is not liable for inaccuracies shown on its Owner Reports generated by third-party professionals. Contractor shall give written notice to Owner of any discrepancies, errors, or omissions that it observes in the Owner Reports attached as an exhibit."

§ 2.2.43.5 The Owner shall furnish information or services required of the Owner by the Contract Documents with reasonable promptness. The Owner shall also furnish any other information or services under the Owner's control and relevant to the Contractor's performance of the Work with reasonable promptness after receiving the Contractor's written request for such information or services.

PRACTICE TIP

The owner may consider providing information relevant to the site and the work of the contract on its own initiative, i.e., whether the contractor asks for it or not. Implied in every contract is a duty to cooperate and to deal fairly. Not disclosing pertinent information that would assist the contractor to construct the project properly is not in the owner's self-interest and could be viewed as a misrepresentation as to the condition of the site.

§ 2.2.53.6 Unless otherwise provided in the Contract Documents, the Owner shall furnish to the Contractor one copy of the Contract Documents for purposes of making reproductions pursuant to Section 1.5.2.

The owner must properly communicate the detail, scope, and intent of the project as stated in the contract documents. With advances in technology over the past decade, electronic transmittal of contract documents has become more cost effective, faster, and more accurate. This provision confirms the owner's duty to inform the contractor of specifically what is required to complete and facilitate the contractor's ability to reproduce the contract documents, impliedly at its own expense, in order to inform its subcontractors.

PRACTICE TIP

Transmitting electronic contract documents triggers the need for the contractor to execute the AIA E202-Digital Data Protocol to regulate liability for transfer of information and to acknowledge continuing ownership of the instruments of service.

§ 2.34 OWNER'S RIGHT TO STOP THE WORK

If the Contractor fails to correct Work that is not in accordance with the requirements of the Contract Documents as required by Section 12.2 or repeatedly fails to carry out Work in accordance with the Contract Documents, the Owner may issue a written order to the Contractor to stop the Work, or any portion thereof, until the cause for such order has been eliminated; however, the right of the Owner to stop the Work shall not give rise to a duty on the part of the Owner to exercise this right for the benefit of the Contractor or any other person or entity, except to the extent required by Section 6.1.3.

A201–2017 and prior editions provided the owner the right, but not the duty except in a narrow circumstance, to stop work in order to remediate improperly performed work. *Bruner & O'Connor on Construction Law* notes that the right to stop work is for qualitative reasons.[21] Specifically, the architect may reject work under Section 12.2.1, A201–2017, that fails to conform to the contract documents, and the owner will be entitled to recover the cost of correcting such rejected work and other expenses made necessary by the contractor's failure to perform. A201 Section 12.2.1 also makes clear that the owner does not have to wait until project completion to enforce its rights to acquire defect-free work.

The owner may also stop the work if the contractor is not timely performing. The right to stop work becomes more critical for the participants when the owner must act for its separate contractors under Section 6.1.3, A201, and, in that instance, this right elevates to the level of a duty. In this setting, the owner must act to avoid interference and maintain coordination with the project schedule or it may face claims from both its contractor and any separate contractor impacted.

The recent federal case of *Ramirez v. Alabama Power Co.*[22] clarified that the owner has no duty for safety under the A201, particularly in light of the myriad duties undertaken by the contractor.

In earlier versions of the A201, the architect had the right to stop work, and failing to exercise such right led to many cases against architects who faced allegations that they were liable for job site safety.[23] As case law developed, the AIA limited the architect's duty relative to safety. The A201 intended to make clear that the architect's duty is to guard against nonconforming work and to ensure satisfactory progress, not to ensure safety standards were met.

Owners were frequently faced with similar exposure and sought to shift liability to the contractor. In that respect, cases interpreting the common law and statutory indemnity ultimately resulted in finding the contractor liable.[24] In the Illinois case of *Moore v. Clearing Industrial District, Inc.*,[25] an injured worker sued the owner, who in turn sought common law indemnity from the contractor. The court found for the injured worker based upon the owner's contract right to terminate the contractor for persistent failures and code violations and because the owner was found to have made frequent inspections and to have remedied deficient performance. The court also awarded the owner common law indemnity from the contractor. The rationale for liability remaining on the contractor is the fact that the contractor controls the means, methods, techniques, and sequencing of the work; is obligated to complete the work timely; is expressly obligated to protect persons and property under A201; and is not the owner's agent.[26]

The owner can become liable for the contractor's reasonable recovery if, as in one case, the contractor left the job in response to the owner's notice that the work

21. 2 Philip L. Bruner, Patrick J. O'Connor, Bruner & O'Connor on Construction Law § 5:44 (2002).
22. 898 F. Supp. 1537 (M.D. Ala. 1995), *aff'd without opinion*, 86 F.3d 1170 (11th Cir. 1996).
23. Hanna v. Huer, Johns, Neel, Rivers & Webb, 662 P.2d 243 (Kan. 1983).
24. Pugh v. Butler Tel. C., 512 So. 2d 1317 (Ala. 1987).
25. 380 N.E.2d 1063 (Ill. App. Ct. 1978).
26. Kraft Gen. Foods, Inc. v. Maxwell, 464 S.E.2d 639 (Ga. Ct. App. 1995).

required remediation. The contractor mistook the notice for a termination notice. The contract was found to have been rescinded by the parties.[27] The stop-work notice needs to be factually clear and specific including the specific provision as authority for the owner's action.[28]

PRACTICE TIP

Giving written notice and carefully drafting the written notice become critical. Care should be exercised to prevent the stop-work notice from being interpreted as a termination notice.

As part of the process leading up to the stop-work notice, the owner should require the architect's rejection of the work in writing, as it is a condition precedent to the owner's right to correct the work under A201 Section 12.2.1. Acting on impulse is not recommended since stopping the work without unimpeachable grounds for doing so can risk liability for the owner. This is especially true because this section also allows stopping the work due to contractor's repeated failure to perform the work in accordance with the contract documents. When relying on this provision, the owner should establish a written record of several instances of the contractor's failure to perform the work in accordance with the contract documents to satisfy the "*repeatedly fails*" standard described above.

§ 2.~~4~~5 OWNER'S RIGHT TO CARRY OUT THE WORK

> If the Contractor defaults or neglects to carry out the Work in accordance with the Contract Documents and fails within a ten-day period after receipt of ~~written~~ notice from the Owner to commence and continue correction of such default or neglect with diligence and promptness, the Owner may, without prejudice to other remedies the Owner may have, correct such ~~deficiencies. In such case an appropriate Change Order shall be issued deducting from payments then or thereafter due the Contractor~~default or neglect. Such action by the Owner and amounts charged to the Contractor are both subject to prior approval of the Architect and the Architect may, pursuant to Section 9.5.1, withhold or nullify a Certificate for Payment in whole or in part, to the extent reasonably necessary to reimburse the Owner for the reasonable cost of correcting such deficiencies, including Owner's expenses and compensation for the Architect's additional services made necessary by such default, neglect, or failure. ~~Such action by the Owner~~If current and ~~amounts charged to the Contractor are both subject to prior approval of the Architect. If~~ future payments ~~then or thereafter due the Contractor~~ are not sufficient to cover such amounts, the Contractor shall pay the difference to the Owner. If the Contractor disagrees with the actions of the Owner or the Architect, or the amounts claimed as costs to the Owner, the Contractor may file a Claim pursuant to Article 15.

This renumbered provision replaces "deficiencies" with "default or neglect" as the standard for determining when and whether the owner has a right to carry out the work. By changing to the more specific failure of the contractor, i.e., that of "default or neglect," the revised provision raises the standard of contractor nonperformance

27. Francavilla v. Nagar Constr. Co., 542 N.Y.S.2d 557 (App. Div. 1989).
28. *See also* Mor-Wood Contractors, Inc. v. Ottinger, 562 N.E.2d 1247 (Ill. App. Ct. 1990).

before the owner is entitled to take over and carry out the work. This provision serves as a useful interim measure to provide the owner with leverage over the contractor during construction. This right allows the owner to carry out the contractor's work to correct contractor default or neglect and withhold payment to the extent reasonably necessary to reimburse the owner for correcting the work. Arguably, it allows the owner to "de-scope" work from the defaulting or negligent contractor without formally terminating the entire contract. Even though the owner has the right to carry out the work, the owner has no duty to do so and exonerate the contractor from liability to perform the terms of the contract.[29]

The owner may carry out the work of a nonperforming contractor but must first secure the architect's approval. Although the provision does not require that the architect's approval be in writing, good practice may dictate that it should be. Only where the parties have secured the architect's approval as a requirement may the owner proceed to take over and remedy the nonconforming work.[30]

An owner who fails to follow the contract procedures for carrying out the work may find itself without recovery for its costs expended to correct and complete.[31] The current revisions clarify that the architect may withhold or nullify certification of payment in whole or in part "to reimburse the Owner for the reasonable cost of correcting" the work taken over. The owner is well advised to issue a deductive change order for the "de-scoped" work. In *State Surety Co. v. Lamb Construction Co.*, the owner failed to credit the value of the landscape work that was not performed by the contractor and was not acknowledged under the contract by deductive change order. The court only awarded damages for the original costs of the landscaping under the contract, rather than the actual costs to correct and complete.[32]

Written notice to the nonperforming contractor prior to taking action will protect the owner from the contractor's claim that it did not have a fair opportunity to mitigate the damages resulting from its nonconforming work.[33]

CONCLUSION

With the exception of Sections 2.2.1, 2.2.2, and 2.5, A201–2017 only reflects modest changes to Article 2. The owner's requirements are, for the most part, specific and directive. Parties must be aware of those provisions that require the owner to produce financials as a condition to the contractor's continued performance. The phrase *reasonable evidence* is not a clear standard and a contractor must be cautious in exercising and enforcing this provision. In this way the parties have a better chance to resolve any financial incapacity and avoid improper or pre-textual mid-course work cessations that might serve only to mask larger matters that threaten timely and competent completion.

29. Burlington v. Arnold Constr. Co., 727 S.W.2d 241 (Tenn. Ct. App. 1987).
30. Lebco, Inc. v. MacGregor Park Nat. Bank of Houston, 500 S.W.2d 698 (Tex. Civ. App. 1973).
31. State Sur. Co. v. Lamb Const. Co., 625 P.2d 184 (Wyo. 1981).
32. *Id.* at 193.
33. Denver Ventures, Inc. v. Arlington Lane Corp., 754 P.2d 785 (Colo. Ct. App. 1988).

CHAPTER 3:
Article 3 Contractor

Levi W. Barrett
Richard S. Robinson
Sean P. Ryan

ARTICLE 3 CONTRACTOR

Many of the contractor's obligations are found in Article 3, one of the larger articles in the A201. Article 3 includes sections dealing with such significant matters as warranty, site conditions, and indemnification, as well as provisions allocating obligations and responsibilities for taxes, permits, allowances, submittals, compliance with laws, and intellectual property.

§ 3.1 GENERAL

> **§ 3.1.1** The Contractor is the person or entity identified as such in the Agreement and is referred to throughout the Contract Documents as if singular in number. The Contractor shall be lawfully licensed, if required in the jurisdiction where the Project is located. The Contractor shall designate in writing a representative who shall have express authority to bind the Contractor with respect to all matters under this Contract. The term "Contractor" means the Contractor or the Contractor's authorized representative.

The legal definition of "Contractor" contained in this section does not exist in a vacuum. Rather, it exists in the context of various state statutory definitions. One state definition of "Contractor" includes "any person who undertakes to . . . construct . . . any building . . . or other structure."[1] Other state statutory definitions include far greater detail.[2]

Why must the contractor be licensed? The primary reasons are protection of the public and some assurance of competency.[3] One state's licensing requirement is based upon the rationale that "licensing schemes are enacted to promote safe and workmanlike construction for the protection of the public."[4] Another state licensing requirement has been interpreted as serving to "provide minimal assurance that all persons offering such services . . . have the requisite skill and character, understand applicable local law and codes, and know the rudiments of administering a contracting business."[5]

What are the consequences if the contractor is unlicensed? As a practical matter, the unlicensed contractor may be statutorily prohibited from obtaining permits necessary for construction.[6] In the event of a payment dispute, many state statutes bar the unlicensed contractor from enforcing an agreement against the owner.[7] However, states are less uniform in their views on whether the unlicensed contractor may recover against a subcontractor or defend itself against owner claims on the merits. While one state permits the unlicensed contractor to enforce a contract against a subcontractor where "no injury to the public is apparent from enforcement of the subcontract between the parties to it," California statute strictly prohibits the unlicensed contractor from retaining any compensation whatsoever.[8]

1. Cal. Bus. & Prof. Code § 7026 (West 2016).
2. Wash. Rev. Code § 18.27.010 (West 2016); La. Rev. Stat. Ann. § 37:2150.1 (2016). For purposes of the A201, the term "contractor" is defined in the A101.
3. La. Rev. Stat. Ann. § 37:2167 (2016); Minn. Stat. Ann. § 326B.805 (West 2016); Ala. Code § 34-8-1 (2016); Ariz. Rev. Stat. Ann. § 32-1101 (2016).
4. Midland Fire Prot. Inc. v. Clancy & Theys Constr. Co., 623 S.E.2d 369 (N.C. Ct. App. 2006).
5. Hydrotech Sys., Ltd. v. Oasis Waterpark, 803 P.2d 370 (Cal. 1991).
6. Fla. Stat. Ann. § 553.79 (West 2016).
7. Epps v. 4 Quarters Restoration LLC, 872 N.W.2d 412 (Mich. 2015); Mortise v. 55 Liberty Owners Corp., 469 N.E.2d 529 (N.Y. 1984).
8. Vogel v. Supply Co., 177 S.E.2d 273, 282 (N.C. 1970); Judicial Council of Cal. v. Jacobs Facilities, Inc., 191 Cal. Rptr. 3d 714 (Ct. App. 2015).

Also, as at least one state Supreme Court has noted, since licensing statutes are intended to be "shields, not swords," the unlicensed contractor may not be barred from defending itself against owner claims on the merits.[9]

§ 3.1.2 The Contractor shall perform the Work in accordance with the Contract Documents.

In the context of both public and private construction projects, it has long been held that if the contractor relies upon the plans and specifications provided by the owner and performs the work in accordance with such contract documents, the contractor will not be liable for latent defects contained in those documents.[10] The 2017 change in the language of Section 3.12.10.1 acknowledges the long-standing legal underpinning for the contractor's entitlement to reliance upon the architect's plans and specifications for the work. However, if the contractor "should reasonably have known" or even "had reason to believe" that plans and specifications contained defects and failed to raise its concerns prior to performing work, it may be held at least partially liable for resulting construction defects.[11] One federal court defined a "patent" defect (one that should have been discovered by the contractor) to mean an "obvious" defect.[12] In deciding whether a defect should be "obvious" to the contractor, courts have considered a number of factors, such as 1) time available to the contractor for price estimation and preparation;[13] 2) the size of the product and quantity of bid documents;[14] 3) the contractor's expertise, experience, and special knowledge of technical field;[15] 4) the dollar value of the defect or discrepancy;[16] 5) whether or not the contractor's bid was negotiated;[17] and 6) whether the owner or designer knew about the defect discrepancy and ease of clarification.[18]

§ 3.1.3 The Contractor shall not be relieved of its obligations to perform the Work in accordance with the Contract Documents either by activities or duties of the Architect in the Architect's administration of the Contract, or by tests, inspections or approvals required or performed by persons or entities other than the Contractor.

This provision in part protects the owner from the risk of contractor arguments that the architect's or other third party's actions revise the requirements of the contract documents. While this section makes clear that a contractor may not rely upon the

9. *Epps*, 872 N.W.2d at 419.

10. United States v. Spearin, 248 U.S. 132 (1918); Alpert v. Commonwealth., 258 N.E.2d 755 (Mass. 1970); Newberg Constr. Co. & Del Constr. Co. v. State, 36 Ill. Ct. Cl. 93 (1984); Sherman R. Smoot Co. v. Ohio Dep't. of Admin. Serv., 736 N.E.2d 69 (Ohio Ct. App. 2000); Los Angeles Unified Sch. Dist. v. Great Am. Ins. Co., 234 P.3d 490 (Cal. 2010); Gordon v. Heritage Fellowship, 784 S.E.2d 265 (Va. 2016).

11. Craig Johnson Constr., L.L.C. v. Floyd Town Architect, P.A., 134 P.3d 648 (Idaho 2006); Morgan v. Lafourche Recreation Dist. No. 5, 822 So. 2d 716 (La. Ct. App. 2002).

12. 3 Philip L. Bruner, Patrick J. O'Connor, Bruner & O'Connor on Construction Law § 9.6.4 (2002 & Dec. 2016 Update) (citing Carl Garris & Son, Inc. v. United States, 918 F.2d 187 (Fed. Cir. 1990) (unpublished)).

13. Appeal of Arnell Constr. Corp., No. 86-1 BCA P 18550, 1985 WL 17121 (Gen. Servs. Admin. B.C.A. Nov. 8, 1985).

14. D & L Constr. Co. & Assocs. v. United States, 402 F.2d 990 (Ct. Cl. 1968).

15. ETS-Hokin Corp. v. United States, 420 F.2d 716 (Ct. Cl. 1970).

16. Mountain Home Contrs. v. United States, 425 F.2d 1260 (Ct. Cl. 1970); Brezina Constr. Co. v. United States, 449 F.2d 372 (Ct. Cl. 1971).

17. Consumers Ice. Co. v. United States, 475 F.2d 1161 (Ct. Cl. 1973).

18. Gorn Corp. v. United States, 424 F.2d 588 (Ct. Cl. 1970).

architect's actions or failures to act regarding the architect's contract administration duties,[19] one court has noted that, conversely, the architect may not rely upon this provision to escape liability if the architect arguably fails to properly carry out its contract administration duties.[20]

§ 3.2 REVIEW OF CONTRACT DOCUMENTS AND FIELD CONDITIONS BY CONTRACTOR

§ 3.2.1 Execution of the Contract by the Contractor is a representation that the Contractor has visited the site, become generally familiar with local conditions under which the Work is to be performed, and correlated personal observations with requirements of the Contract Documents.

While this provision generally requires the contractor to acknowledge that it has made itself familiar with the conditions under which the work will be performed, it may also give rise to responsibility for differing conditions and diminish the contractor's ability to rely solely on owner provided information.[21] It has been held that a site inspection clause places a duty on a contractor to exercise "professional skill" in inspecting the site and estimating the cost of the work.[22] Under some circumstances, it has even been held that reliance on the owner's specifications may be unreasonable where a contractor has an independent duty to inspect the premises.[23] Nevertheless, while an inspection clause may bind the contractor to obtain knowledge of certain existing conditions, the contractor's inspection need only be "reasonable" under the circumstances.[24]

§ 3.2.2 Because the Contract Documents are complementary, the Contractor shall, before starting each portion of the Work, carefully study and compare the various Contract Documents relative to that portion of the Work, as well as the information furnished by the Owner pursuant to Section 2.2.3.4, shall take field measurements of any existing conditions related to that portion of the Work, and shall observe any conditions at the site affecting it. These obligations are for the purpose of facilitating coordination and construction by the Contractor and are not for the purpose of discovering errors, omissions, or inconsistencies in the Contract Documents; however, the Contractor shall promptly report to the Architect any errors, inconsistencies or omissions discovered by or made known to the Contractor as a request for information in such form as the Architect may require. It is recognized that the Contractor's review is made in the Contractor's capacity as a contractor and not as a licensed design professional, unless otherwise specifically provided in the Contract Documents.

In addition to having the contractor actually perform the construction of the work, the owner is interested in having the contractor properly plan and lay out the

19. Norair Eng'g Corp. v. Saint Joseph's Hosp., Inc., 249 S.E.2d 642 (Ga. Ct. App. 1978).
20. Rusnak v. Brent Wagner Architects, No. 64A03-1510-PL, 2016 WL 3057429 (Ind. Ct. App. May 31, 2016).
21. Compare to AIA A201–2017 § 2.3.4.
22. Pinkerton & Laws Co. v. Roadway Express, Inc., 650 F. Supp. 1138, 1146 (N.D. Ga. 1986); Green Constr. Co. v. Kan. Power & Light Co., 1 F.3d 1005, 1009 (10th Cir. 1993).
23. Brant Constr. Co. v. Metro. Water Reclam. Dist., 967 F.2d 244, 248 (7th Cir. 1992) (citing W.H. Lyman Constr. Co. v. Gurnee, 403 N.E.2d 1325 (Ill. App. Ct. 1980)).
24. Triest & Earle, Inc. v. United States, 84 Ct. Cl. 84 (1936); John Collins v. United States, 26 C.C.F. ¶ 83, 110 (Ct. Cl. 1979); Turnagain Paint & Constr. Co., ASBCA No. 17884, 74-2 BCA ¶ 10, 872 (1974).

construction work. This is accomplished by imposing an obligation on the contractor to familiarize itself with the project site and the contract documents and requiring that the contractor take all field measurements. The stated goal of this paragraph is to ensure that the contractor, prior to each portion of the work, reviews the relevant contract documents and cross-references the contract documents with then-existing conditions and field observations for purposes of facilitating and coordinating construction. To balance this obligation, the contractor's liability for errors it does not discover is limited insofar as it is not expected to review the documents as a design professional. That said, the contractor is responsible for promptly reporting to the architect any errors, inconsistencies, or omissions that it discovers when performing the analysis required by this section. An owner who furnishes plans and specifications impliedly warrants their sufficiency.[25] Such implied warranty is not overcome by the general clauses requiring a contractor to examine the site, to check the plans, and to assume responsibility for the work until completion and acceptance.[26] Moreover, where a "positive" representation is made by an owner, that party impliedly warrants that it has made "full disclosure."[27] This is not to say that a contractor could not be held liable for defects in design documents; however, to impose liability on a contractor, defects would have to be "so glaring and out of the ordinary as to bring home to the contractor that it was [constructing] something which would be likely to cause injury."[28]

> **§ 3.2.3** The Contractor is not required to ascertain that the Contract Documents are in accordance with applicable laws, statutes, ordinances, codes, rules and regulations, or lawful orders of public authorities, but the Contractor shall promptly report to the Architect any nonconformity discovered by or made known to the Contractor as a request for information in such form as the Architect may require.

This provision makes it obvious that the contractor is not required to ensure that the contract documents are legally compliant, but must report any noncompliance that it observes. This is a concise recapitulation of the generally understood notion that the architect, and not the contractor, bears responsibility for designing a legally compliant project.

> **§ 3.2.4** If the Contractor believes that additional cost or time is involved because of clarifications or instructions the Architect issues in response to the Contractor's notices or requests for information pursuant to Sections 3.2.2 or 3.2.3, the Contractor shall ~~make~~submit Claims as provided in Article 15. If the Contractor fails to perform the obligations of Sections 3.2.2 or 3.2.3, the Contractor shall pay such costs and damages to the Owner, subject to Section 15.1.7, as would have been avoided if the Contractor had performed such obligations. If the Contractor performs those obligations, the Contractor shall not be liable to the Owner or Architect for damages resulting from errors, inconsistencies or omissions in the Contract Documents, for differences between field

25. United States v. Spearin, 248 U.S. 132 (1918).
26. *Id.* at 137.
27. Alpert v. Commonwealth, 258 N.E.2d 755, 763–64 (Mass. 1970); Christie v. United States, 237 U.S. 234, 239–42 (1915); Hersey Gravel Co. v. State Highway Dep't, 9 N.W.2d 567 (Mich. 1943).
28. Soave v. Nat'l Velour Corp., 863 A.2d 186 (R.I. 2004) (citing Ryan v. Feeney & Sheehan Bldg. Co., 239 N.Y. 43 (1924)). *See also* Romano v. Rossano Constr. Co., 171 N.E.2d 853 (Mass. 1961); Gee v. City of New York, 758 N.Y.S.2d 157 (App. Div. 2003); Loconti v. Creede, 564 N.Y.S.2d 823 (App. Div. 1991); Pioli v. Town of Kirkwood, 499 N.Y.S.2d 266 (App. Div. 1986).

measurements or conditions and the Contract Documents, or for nonconformities of the Contract Documents to applicable laws, statutes, ordinances, codes, rules and regulations, and lawful orders of public authorities.

This provision ties a contractor's obligation to seek additional time and/or costs associated with clarifications or instructions to the claims provisions contained in Article 15. Moreover, this provision provides that the contractor will be liable to the owner for costs and damages where it fails to, among other things, verify field conditions (Section 3.2.2) and/or report known nonconformities (Section 3.2.3). Although the contractor will be liable for failing to abide by its obligations under Sections 3.2.2 and 3.2.3, Section 3.2.4 acknowledges the *Spearin* Doctrine and provides the contractor with a remedy as long as the contractor fulfills its obligations to review the contract documents and report errors, omissions, or legal nonconformities it discovers.[29]

The addition of the "subject to Section 15.1.7" language applies the mutual waiver of consequential damages to limit the damages payable to the owner for the contractor's breach of Sections 3.2.2 or 3.2.3.

§ 3.3 SUPERVISION AND CONSTRUCTION PROCEDURES

§ 3.3.1 The Contractor shall supervise and direct the Work, using the Contractor's best skill and attention. The Contractor shall be solely responsible for, and have control over, construction means, methods, techniques, sequences, and procedures, and for coordinating all portions of the Work under the Contract~~, unless the Contract Documents give other specific instructions concerning these matters~~. If the Contract Documents give specific instructions concerning construction means, methods, techniques, sequences, or procedures, the Contractor shall evaluate the jobsite safety thereof and~~, except as stated below,~~ shall be ~~fully and~~ solely responsible for the jobsite safety of such means, methods, techniques, sequences, or procedures. If the Contractor determines that such means, methods, techniques, sequences, or procedures may not be safe, the Contractor shall give timely ~~written~~ notice to the Owner and Architect, and shall ~~not proceed with that portion of the Work without further written instructions from the Architect. If the Contractor is then instructed to proceed with the required~~propose alternative means, methods, techniques, sequences, or procedures ~~without acceptance of changes~~. The Architect shall evaluate the proposed ~~by the Contractor, the Owner shall be~~alternative solely ~~responsible~~ for ~~any loss or damage arising solely from those Owner-required~~conformance with the design intent for the completed construction. Unless the Architect objects to the Contractor's proposed alternative, the Contractor shall perform the Work using its alternative means, methods, techniques, sequences, or procedures.

The contractor is responsible for supervising and directing the work, which includes means and methods of construction, as well as jobsite safety. In many instances this means that the contractor will be liable for workplace injuries to the exclusion of the owner.[30] The owner, in essence, has relinquished its duty of care, and the

29. *Spearin*, 248 U.S. 132.

30. Hymas v. UAP Distrib., Inc., 272 P.3d 889 (Wash. Ct. App. 2012) ("A general contractor's supervisory authority places the general in the best position to ensure compliance with safety regulations.") (citing Stute v. P.B.M.C., Inc., 788 P.2d 545 (Wash. 1990)); Dalton v. 933 Peachtree, L.P., 661 S.E.2d 156 (Ga. Ct. App. 2008); Saunders v. Indus. Metals & Surplus, Inc., 646 S.E.2d 294 (Ga. Ct. App. 2007) ("owner may be liable only if he retains the right to direct or control the time and manner of executing the independent contractor's work or interferers with the work to a sufficient degree") (citing Grey v. Milliken & Co., 539 S.E.2d 186 (Ga. Ct. App. 2000)).

associated liability for a breach of that duty to the contractor.[31] It has been held that the contractor's duty is to conduct periodic inspections, as needed to ensure that "all work [is] performed properly, resulting in a building free from defects."[32]

PRACTICE TIP

As the contractor's responsibility relates only to the work, it may be possible to narrow the definition of "Work," or exclude certain responsibilities using the contractor's clarifications and assumptions (in the case of a guaranteed maximum price contract). It may also be prudent for a contractor to ensure that it has the ability to suspend work where it encounters safety hazards.

Under the 2007 version of the A201, an owner who elected to direct certain aspects of the work, over the contractor's timely objection as to their safety, would assume additional responsibility in the form of sole liability for harm arising solely from those instructions.[33] As revised, the contractor is responsible for identifying any "specific instruction" and evaluating the safety thereof. The 2007 version of the A201 document provided that the contractor would be excused from responsibility for means and methods that it believed unsafe, where the owner required the contractor to proceed in light of its objection to the safety of the owner's instructions. An owner who directed a contractor to proceed with means and methods that the contractor deemed objectionable would be subject to additional liability. The revisions to Section 3.3.1 have eliminated any reference to an assumption of liability for means and methods by the owner. Further, a contractor faced with specific instructions that it finds unsafe must identify and evaluate the instruction and propose alternative procedures, which are to be evaluated by the architect. Unless the architect objects, the contractor may proceed with its proposed alternative means and methods. Moreover, the architect's evaluation of the contractor's proposed alternative is done only for purposes of conformance with the design intent. If the architect objects to the alternative, it is unclear how the contractor is to proceed to resolve an impasse—as the contractor is no longer obligated to proceed with the owner's direction, with liability shifted to the owner. Although an owner may bear common law liability for directing the contractor to proceed with means and methods over the contractor's objection regarding safety, this is no longer an express condition of the A201 form.[34]

PRACTICE TIP

Contractors may seek to modify this provision to add a clear entitlement to an increase in the contract sum or contract time resulting from the use of the alternative means and methods. Even in the absence of such a modification, the contractor may assert an entitlement to an adjustment of the contract sum or contract time based on an argument that the owner's original means and methods were an error in the contract documents, which should result in a change in the scope of work.

31. McCorkle v. North Point Chrysler Jeep, Inc., 703 S.E.2d 750 (N.C. Ct. App. 2010).
32. Rivnor Prop. v. Herbert O'Donnell, Inc., 533 So. 2d 735 (5th Cir. 1994).
33. AIA A201–2007 § 3.3.1.
34. Corsetti v. Stone Co., 483 N.E.2d 793 (Mass. 1985) ("One who entrusts work to an independent contractor, but who retains the control of any part of the work, is subject to liability for physical harm to others for whose safety the employer owes a duty to exercise reasonable care, which is caused by his failure to exercise his control with reasonable care."); Ferrero v. Best Modular Homes, Inc., 823 N.Y.S.2d 477 (App. Div. 2006), *leave to appeal dismissed*, 862 N.E.2d 784 (N.Y. 2007).

> **§ 3.3.2** The Contractor shall be responsible to the Owner for acts and omissions of the Contractor's employees, Subcontractors and their agents and employees, and other persons or entities performing portions of the Work for, or on behalf of, the Contractor or any of its Subcontractors.

The contractor, as the owner's chief point of contact for the performance of the work, will be responsible for the acts and omissions of its employees and those parties with whom it subcontracts portions of the work.[35] The contractor, however, is generally not responsible for acts or omissions of the owner, the owner's personnel, separate contractors, or the architect, unless it assumes a separate and distinct contractual obligation to review their work.

PRACTICE TIP

The contractor should "flow-down" this responsibility to its subcontractors to ensure that the contractor has the ability to recoup losses from the parties actually responsible for the harm.

> **§ 3.3.3** The Contractor shall be responsible for inspection of portions of Work already performed to determine that such portions are in proper condition to receive subsequent Work.

The contractor's obligation is to review the work-in-place to ensure that it was performed correctly and is in appropriate condition to receive additional work. For example, the contractor will be required to inspect a building's framing, rough plumbing, and electrical work before moving forward with the installation of sheetrock. It should be noted that the architect also bears responsibility for inspecting the work and is required to visit the site to "determine in general if the Work is being performed in a manner indicating that the Work, when fully completed, will be in accordance with the Contract Documents."[36]

§ 3.4 LABOR AND MATERIALS

> **§ 3.4.1** Unless otherwise provided in the Contract Documents, the Contractor shall provide and pay for labor, materials, equipment, tools, construction equipment and machinery, water, heat, utilities, transportation, and other facilities and services necessary for proper execution and completion of the Work, whether temporary or permanent and whether or not incorporated or to be incorporated in the Work.

The contractor's obligation to provide and pay for labor extends to temporary facilities necessary for the execution of the work. Implicit in its obligation to furnish

35. Calderon v. Residential Homes of Am., Inc., 885 N.E.2d 1138 (Ill. App. Ct. 2008) ("[O]ne who entrusts work to an independent contractor will not be liable for the acts or omissions of that independent contractor. . . ." A "general contractor, by retaining control over the operative details of its subcontractor's work may become vicariously for the subcontractor's negligence; alternatively, even in the absence of such control, the general contractor may be directly liable for not exercising his supervisory control with reasonable care.") (citations omitted).

36. AIA A201–2017 § 4.2.2.

materials and equipment is an obligation to ensure the safety of same.[37] A California court has used this provision (in conjunction with Section 3.3.1) to extinguish the owner's liability to a laborer who fell as a result of a contractor's failure to provide appropriate safety harnesses. This case is distinguishable from a similar set of factual circumstances where a worker was injured using an owner-furnished forklift that was not properly equipped, in which case the owner was subject to liability.[38]

Moreover, the contractor's obligation to provide temporary facilities may implicitly give rise to design obligations, which require the contractor to observe the licensing requirements of Section 3.12.10. As part of its obligation to provide temporary facilities, the contractor may, for example, be required to design and construct scaffolding systems. As set forth in Section 3.12.10, where the contractor is obligated to provide such services, it will be required to use an appropriately licensed design professional.

PRACTICE TIP

As a practical matter, where water, heat, and utilities are active at the project site, it may be prudent to require the owner to provide these services. It is common for this provision to be negotiated between the owner and the contractor and modified based on project need and other practical considerations. The time of year and the specific nature of the construction can also significantly impact the quantity and quality of tools, equipment, and services required to perform the work. Where the construction of a building involves the use of the building utilities during the course of construction, it may also be advisable to require the contractor to clean and/or restore any utilities impacted by construction (e.g., HVAC ducts).

§ 3.4.2 Except in the case of minor changes in the Work ~~authorized~~approved by the Architect in accordance with ~~Sections~~Section 3.12.8 or ordered by the Architect in accordance with Section 7.4, the Contractor may make substitutions only with the consent of the Owner, after evaluation by the Architect and in accordance with a Change Order or Construction Change Directive.

This provision protects the owner and contractor by establishing a procedure for authorization of substitutions. Owners want the contractor to use the products, materials, equipment, and other requirements specified in the contract documents. Contractors may have commercial, technical, or other reasons for preferring to substitute alternate products, materials, and equipment. Generally, the contractor may not depart from the requirements of the contract documents unless a change order or construction change directive is issued or, with the architect's authorization, in the case of a "minor change." A "minor change" is a change that is consistent with the intent of the contract documents and does not involve an adjustment to the contract sum or an extension of the contract time—and must be made in writing.[39]

37. Lopez v. C.G.M. Dev., Inc., 124 Cal. Rptr. 2d 227 (Cal. Ct. App. 2002) ("[A] [property owner] owes no duty of care to an employee of a subcontractor to prevent or correct unsafe procedures or practices to which the [owner] did not contribute by direction, induced reliance, or other affirmative conduct.") (citing Hooker v. Dep't of Transp., 38 P.3d 1081 (Cal. 2002)).
38. McKown v. Wal-Mart Stores, Inc., 38 P.3d 1094 (Cal. 2002).
39. AIA A201–2017 § 7.4.

PRACTICE TIP

Where the architect makes a minor change that it fails to issue in a written directive, it is advisable to memorialize the change in meeting minutes or through a confirming letter or e-mail.

§ 3.4.3 The Contractor shall enforce strict discipline and good order among the Contractor's employees and other persons carrying out the Work. The Contractor shall not permit employment of unfit persons or persons not properly skilled in tasks assigned to them.

During construction, there can be a high volume of people on the project site. Both owners and contractors have an interest in seeing the work progress in a safe and orderly fashion. This paragraph imposes a duty of care and an obligation on the contractor to hire only skilled personnel and to replace employees who are "unfit." This obligation is part and parcel of the contractor's duty to control means, methods, and safety. Where an employee causes damage, this provision may give rise to a claim against the contractor for negligently hiring, training, or supervising its employees.

§ 3.5 WARRANTY

§ 3.5.1 The Contractor warrants to the Owner and Architect that materials and equipment furnished under the Contract will be of good quality and new unless the Contract Documents require or permit otherwise. The Contractor further warrants that the Work will conform to the requirements of the Contract Documents and will be free from defects, except for those inherent in the quality of the Work the Contract Documents require or permit. Work, materials, or equipment not conforming to these requirements may be considered defective. The Contractor's warranty excludes remedy for damage or defect caused by abuse, alterations to the Work not executed by the Contractor, improper or insufficient maintenance, improper operation, or normal wear and tear and normal usage. If required by the Architect, the Contractor shall furnish satisfactory evidence as to the kind and quality of materials and equipment.

The longstanding debate regarding the rationale for extension of the contractor's construction warranty to the architect—a non-party to the agreement between the owner and contractor—remains relevant due to the lack of change in the 2017 document.[40]

Although Article 12 references a one-year construction warranty period (commonly referred to as the "call back period"), this section does not reference a one-year construction warranty period or any other time limitation. The contractor may want to consider negotiating addition of reference to a one-year construction warranty period here to limit warranty liability. Conversely, the owner may want to avoid such a limitation. However, adding a one-year (or otherwise limited) warranty duration here will not necessarily insulate the contractor from claims, sounding

40. 3 Philip & O'Connor, *supra* note 12, § 5:58, ¶ 3.5.1 (2002 & Dec. 2016 update); Jonathan J. Sweet, Sweet on Construction Industry Contracts: Major AIA Documents § 12.6 (6th ed. 2017 Supplement).

in breach of contract rather than breach of warranty, that are brought after the one-year construction warranty period but within the statute of limitations for contract claims.[41]

Of equal importance is the fact that there is no language, either in this section or Article 12, limiting the contractor's warranty obligations to express contractual warranties. Therefore, the contractor may be subject to "cumulative remedy" provisions elsewhere in the agreement and exposed to liability for breach of common law or statutory implied warranties such as "fitness for particular purpose,"[42] "warranty of merchantability,"[43] "implied builder's warranty of good quality and workmanship,"[44] "an implied duty to perform the work or services, skillfully, carefully, diligently, and in a workmanlike manner,"[45] or simply "that he [the contractor] will do a workmanlike job."[46]

However, if the contractor seeks to limit its warranty exposure by disclaiming implied warranties, the exact wording of an express warranty is important. Even statutory immunity enjoyed by the contractor against defects contained in plans and specifications can be overcome if the contractor provides an extended or overly broad express warranty or guaranty.[47] The contractor should also consider that it may be held liable for defective construction work performed by subcontractors "inherited" from the owner if it does not object to use of those subcontractors prior to conclusion of bidding.[48]

Owner and contractor alike should note that, with regard to the "free from defects" language, courts have held that the contractor's warranty obligations do not arise if the owner-provided contract documents specify a product that contains a latent defect.[49] The parties should also be aware that the "conform to the requirements of the Contract Documents" language may obligate the contractor to extended warranties for various materials or equipment that are contained in the plans, specifications, or other documents drafted by the architect.

> ~~§ 3.6~~ **§ 3.5.2** All material, equipment, or other special warranties required by the Contract Documents shall be issued in the name of the Owner, or shall be transferable to the Owner, and shall commence in accordance with Section 9.8.4.

41. Norair Eng'g Corp. v. Saint Joseph's Hosp., Inc., 249 S.E.2d 642 (Ga. Ct. App. 1978); Burton-Dixie Corp. v. Timothy McCarthy Constr. Co., 436 F.2d 405 (5th Cir. 1971); 1000 Virginia Ltd. P'ship v. Vertecs Corp., 112 P.3d 1276 (Wash. Ct. App. 2005).

42. N.Y. U.C.C. Law § 2-315 (2016) ("Where the seller at the time of contracting has reason to know any particular purpose for which the goods are required and that the buyer is relying on the seller's skill or judgment to select or furnish suitable goods, there is unless excluded or modified under the next section an implied warranty that the goods shall be fit for such purpose.").

43. N.Y. U.C.C. Law § 2-314 (2016)

"(1) Unless excluded or modified (Section 2-316), a warranty that the goods shall be merchantable is implied in a contract for their sale if the seller is a merchant with respect to goods of that kind. Under this section the serving for value of food or drink to be consumed either on the premises or elsewhere is a sale.

(2) Goods to be merchantable must be at least such as
 (a) pass without objection in the trade under the contract description; and
 (b) in the case of fungible goods, are of fair average quality within the description; and
 (c) are fit for the ordinary purposes for which such goods are used; and
 (d) run, within the variations permitted by the agreement, of even kind, quality and quantity within each unit and among all units involved; and
 (e) are adequately contained, packaged, and labeled as the agreement may require; and
 (f) conform to the promises or affirmations of fact made on the container or label if any.

(3) Unless excluded or modified (Section 2-316) other implied warranties may arise from course of dealing or usage of trade."

44. Healy v. Silverhill Constr., No. 06-03-030, 2009 WL 295391 (Del. C.P. Feb. 5, 2009).

45. Smith v. Erftmier, 315 N.W.2d 446 (Neb. 1982).

46. Tonkin v. Bob Eldridge Constr. Co., Inc., 808 S.W.2d 849 (Mo. Ct. App. 1991).

47. Harbor Constr. Co., Inc. v. Bd. of Sup'rs. of La. State Univ., 69 So. 3d 498 (La. Ct. App. 2011).

48. Norair Eng'g Corp. v. Saint Joseph's Hosp., Inc., 249 S.E.2d 642 (Ga. Ct. App. 1978).

49. Rhone Poulenc Rorer Pharm. Inc. v. Newman Glass Works, 112 F.3d 695 (3rd Cir. 1997); Traveler's Indem. Co. v. S. M. Wilson & Co., No. 4:04-CV-01365-ERW, 2005 WL 2234582 (E.D. Mo. Sept. 14, 2005).

This new section ties the commencement of material, equipment, and special warranties to the date of substantial completion of the project and requires that the material, equipment, and other special warranties specifically required by the contract documents (such as a 20-year roof warranty) be uniformly issued in the owner's name and commence at the same time. Because it requires warranties to be issued in the name of the owner, the contractor may want to scrutinize contract terms to avoid the risk of the contractor becoming the "guarantor" of such warranties between the owner and subcontractors or material vendors. This is particularly true with regard to special material or equipment warranties that extend beyond the one-year "call back period."

§ 3.6 TAXES

> The Contractor shall pay sales, consumer, use and similar taxes for the Work provided by the Contractor that are legally enacted when bids are received or negotiations concluded, whether or not yet effective or merely scheduled to go into effect.

The contractor is obligated to pay sales, consumer, use, and similar taxes for the work, but only such taxes as legally enacted when bids are received or negotiations concluded, whether or not effective or merely scheduled to go into effect. Therefore, the burden is on the contractor who must include such taxes in its bid. Because this provision identifies two different events as the deadline of tax law enactment (1) when bids are received and (2) the conclusion of negotiations, it is possible that tax law could change in the intervening time in the event of a significant gap in time between the two events. As a result, the parties may agree to modify this provision to reference a single event. Applicable law should be reviewed throughout the relevant time period to avoid unexpected and undesirable surprises.

There is a dearth of case law dealing with the contractor's obligation to pay sales, consumer, use, and similar taxes for the work as set forth in Section 3.6. A contractor was found to be obligated to pay sales taxes under applicable state law even when the normal AIA language indicating that the contractor must pay sales and use taxes was omitted from the contract with a public entity and the bidding requirements clearly stated that prices must exclude such taxes.[50]

The contractor's obligations in Section 3.6 have been interpreted to allocate responsibility to the contractor to remit taxes, but not to absorb such taxes.[51] The court relied on the fact that the contract's cost schedule explicitly included sales tax as a line item that the contractor invoiced and the owner paid. The court then held that the owner owed sales tax to the contractor on the additional amounts awarded to the contractor. The relevant contract language was similar to that of Section 3.6, but it's not known if AIA A201 was used.

§ 3.7 PERMITS, FEES, NOTICES AND COMPLIANCE WITH LAWS

> **§ 3.7.1** Unless otherwise provided in the Contract Documents, the Contractor shall secure and pay for the building permit as well as for other permits, fees, licenses, and inspections

50. ESI Cos. v. Fulton Cnty., 609 S.E.2d 126 (Ga. Ct. App. 2004).
51. Design Trend Int'l Interiors, Ltd. v. Cathay Enters., No. CV10-01079-PHX-NVW, 2011 U.S. Dist. LEXIS 36867 (D. Ariz. Mar. 28, 2011).

by government agencies necessary for proper execution and completion of the Work that are customarily secured after execution of the Contract and legally required at the time bids are received or negotiations concluded.

This section permits the owner and contractor to allocate responsibility for permits, fees, licenses, and inspections. The "unless otherwise provided" language permits the owner and contractor to allocate responsibility for certain permits to the party who may be in control or legally responsible for a permit. For example, the owner may assume the obligation to secure a storm water permit.

Absent a different allocation of responsibility, the contractor is responsible for securing and paying for the building permit and other permits, fees, licenses, and inspections. There are three requirements to consider. First, the owner and contractor must determine what permits, fees, licenses, and inspections are necessary for the proper execution and completion of the work. These can vary by jurisdiction and type of project. Second, such items must be customarily secured after execution of the contract. Finally, they must be legally required at the time bids are received or negotiations concluded. The difference in timing between when bids are received and negotiations concluded can be significant. If so, the parties are advised to consider whether there are any new requirements at the time negotiations are concluded that will impact the allocation of responsibilities for permits, fees, licenses, and inspections. A practice tip to consider is to enumerate in the contract documents, by a matrix or otherwise, who will be responsible for each of the permits, fees, licenses, and inspections.

A Pennsylvania case, not involving AIA forms, illustrates the pitfalls to a contractor of burying its head in the sand, even when a written agreement clearly assigns responsibility for a permit to the owner.[52] A homeowner, who was responsible under a written work order to obtain a permit for an above-ground pool to be erected in her front yard, failed to obtain the permit. After the pool was erected, the owner was informed by the building inspector that the pool violated a municipal ordinance prohibiting a pool in the front of a house. The court found in favor of the contractor, but reduced its damages because of its failure to ascertain whether a permit was required or obtained. The court determined that it was evident that the contractor's method of conducting business amounted to a deliberate refusal to know and do what a responsible business engaged in the construction industry should know and do.[53]

§ 3.7.2 The Contractor shall comply with and give notices required by applicable laws, statutes, ordinances, codes, rules and regulations, and lawful orders of public authorities applicable to performance of the Work.

The obligation and risk is broadly allocated to the contractor to both comply with and give notices required by applicable laws, statutes, ordinances, codes, rules and regulations, and lawful orders of public authorities applicable to performance of the work. For brevity, we will refer to these laws, statutes, ordinances, codes, rules and regulations, and lawful orders as the legal requirements. The obligation to comply with and give notices required by legal requirements is an ongoing obligation. The

52. Contractor Indus. v. Zerr, 359 A.2d 803 (Pa. Super. Ct. 1976).
53. *Id.* at 808.

contractor should be mindful of the risk of new, or changes to existing, legal requirements enacted after it bids or negotiations are concluded. Accordingly, similar to the tax liability in Section 3.6, contractors may seek to have a specific definition of the applicable legal requirements. Also, "applicable to the performance of the Work" may be construed to mean that legal requirements applicable to design are not intended to be covered, but rather only those legal requirements that are applicable to the contractor's performance of the work.

A municipal ordinance applicable to a construction contract has been held by operation of law to be an implied term of that contract.[54] The express obligations of Section 3.7.2 can therefore shift unanticipated obligations to an unwary contractor.

§ 3.7.3 If the Contractor performs Work knowing it to be contrary to applicable laws, statutes, ordinances, codes, rules and regulations, or lawful orders of public authorities, the Contractor shall assume appropriate responsibility for such Work and shall bear the costs attributable to correction.

This provision allocates the risk of complying with applicable legal requirements known to the contractor. Knowingly performing work contrary to applicable laws, statutes, ordinances, codes, rules and regulations, or lawful orders of public authorities allocates the responsibility for such work and the costs attributable to correction to the contractor. The elements of this Section are: (1) *knowing* that the performance of the work is contrary, (2) to *applicable* laws, statutes, ordinances, codes, rules and regulations, or lawful orders of public authorities, (3) assuming *appropriate* responsibility for such noncompliant work, and (4) bearing the costs *attributable* to correction of the noncompliant work.

An earlier similar version of Section 3.7.3 was cited in an action for summary judgment involving construction of a home in Connecticut.[55] The action involved a determination of whether the architect was responsible for frozen pipes because of a deficient design or whether the contractor was responsible because routing of the pipes was a "means and methods" responsibility. The decision turned on the statutory definition of "vendor" and the court concluded that the contractor was the vendor responsible for ensuring compliance with the municipal code.[56] What's not clear is whether the contractor "knowingly" performed the work contrary to the applicable municipal code or whether knowledge was imputed to the contractor.

The question for contractors is how familiar must they be with the applicable laws, statutes, ordinances, codes, rules and regulations, or lawful orders of public authorities. For example, in the Connecticut case discussed above, if the architect's design had specified pipe insulation or routing contrary to the municipal code, then holding the contractor responsible for ensuring compliance with the municipal code would seem to shift design responsibility to the contractor. Moreover, such a holding would be contrary to the *Spearin* doctrine providing that the contractor may rely on the specifications.[57] Also unknowable from the Connecticut case is what appropriate responsibility and costs attributable to correction were allocated to the contractor. Contractors who want to avoid such a result should make it clear that they are not assuming responsibility for laws pertaining to design.

---

54. Gutowski v. Crystal Homes, Inc., 167 N.E.2d 422 (Ill. App. Ct. 1960). The court held that a real estate developer violated its construction contract by building a home in violation of a municipal zoning ordinance.

55. Great N. Ins. Co. v. RLJ Plumbing & Heating, Inc., No. 3:04-CV-899(JCH), 2006 U.S. Dist. LEXIS 39314 (D. Conn. May 23, 2006).

56. *Id.* at *8.

57. United States v. Spearin, 248 U.S. 132 (1918).

§ 3.7.4 CONCEALED OR UNKNOWN CONDITIONS

> If the Contractor encounters conditions at the site that are (1) subsurface or otherwise concealed physical conditions that differ materially from those indicated in the Contract Documents or (2) unknown physical conditions of an unusual nature that differ materially from those ordinarily found to exist and generally recognized as inherent in construction activities of the character provided for in the Contract Documents, the Contractor shall promptly provide notice to the Owner and the Architect before conditions are disturbed and in no event later than ~~21~~ 14 days after first observance of the conditions. The Architect will promptly investigate such conditions and, if the Architect determines that they differ materially and cause an increase or decrease in the Contractor's cost of, or time required for, performance of any part of the Work, will recommend that an equitable adjustment be made in the Contract Sum or Contract Time, or both. If the Architect determines that the conditions at the site are not materially different from those indicated in the Contract Documents and that no change in the terms of the Contract is justified, the Architect shall promptly notify the Owner and Contractor ~~in writing~~, stating the reasons. If either party disputes the Architect's determination or recommendation, that party may ~~proceed~~ submit a Claim as provided in Article 15.

Minor changes have been made to Section 3.7.4, but notably the time for giving notice has been reduced to 14 days from 21 days. The decreased time duration was based on feedback from various industry participants that the 21-day reporting time for contractors to report suspected differing site conditions was unnecessarily long. Another change is to clearly require that the architect will recommend that an equitable adjustment be made in the contract sum or contract time or both. Finally, it has been clarified that the cross-reference to Article 15 is for the purpose of submitting a claim.

Section 3.7.4 allocates the risk of concealed or unknown conditions to the owner. There are two specified types, which are commonly known as Type I and Type II differing site conditions (DSC).[58] The purpose of a DSC clause is economic efficiency. The differing site conditions clause is designed to ensure low, competent bids[59] by eliminating the need for contractors to increase the bid price to absorb the risk of DSCs. Bidders need not weigh the cost and ease of performing their own geotechnical investigation against the risk of encountering an adverse subsurface, and adjust the contingency accordingly to cover the risk.[60]

TYPE I DSC

Type I DSC must be (i) subsurface or otherwise concealed, (ii) physical condition, and (iii) that differs materially from those indicated in the contract documents.

A Type I DSC stands or falls upon what is indicated in the contract documents.[61]

Whether geotechnical or soil reports and other site investigative reports are contract documents will often be a determining factor as to whether physical

58. Central Allied Enters., Inc. v. Adjutant Gen.'s Dep't, No. 10AP-701, 2011 Ohio App. LEXIS 4049 (Sept. 27, 2011).
59. Asphalt Roads & Materials Co., Inc. v. Commonwealth, 512 S.E.2d 804, 806 (Va. 1999).
60. Foster Constr. C.A. & Williams Bros. Co. v. United States, 435 F.2d 873 (Ct. Cl. 1970).
61. Stuyvesant Dredging Co. v. United States, 11 Cl. Ct. 853, 858, 1987 U.S. Cl. Ct. LEXIS 37, 33 Cont. Cas. Fed. (CCH) P75,152 (1987). *See also* P.J. Maffei Bldg. Wrecking Corp. v. United States, 732 F.2d 913, 916 (Fed. Cir. 1984); *Foster Constr. C.A. & Williams Bros. Co.*, 435 F.2d at 881.

conditions are materially different. A DSC claim on a Texas project was rejected by the Fifth Circuit Court of Appeals because soil reports were not part of the contract documents.[62] The district court had reasoned that all of the parties relied on the accuracy of the soil report and interpreted disclaimers of its accuracy as ensuring "that any inaccuracies in the soil report would be inadequate grounds to rescind or excuse nonperformance."[63]

As the Fifth Circuit suggests, contract documents that do not identify conditions at the project site will make it difficult to prevail on a Type I DSC claim. However, failure to disclose material facts has been held to constitute misrepresentation of site conditions for which recovery could be allowed.[64] Similarly, misrepresentation of site conditions has been found when an owner specified only about half of the amount its internal estimates concluded would be needed to obtain acceptable subsoil conditions.[65]

Courts have developed a six-part test to determine whether a Type I DSC exists.[66] The six elements are: (1) the contract documents must have affirmatively indicated the subsurface conditions, (2) contractor must have acted as a reasonably prudent contractor in interpreting the contract documents, (3) contractor must have reasonably relied on the indications of subsurface conditions in the contract, (4) the subsurface conditions actually encountered must have differed materially from those indicated in the contract, (5) the actual subsurface conditions encountered must have been reasonably unforeseeable, and (6) contractor's claimed increased costs must have been solely attributable to such materially different subsurface conditions.[67]

TYPE II DSC

Type II DSC must be (i) an unknown condition, (ii) a physical condition, and (iii) of an unusual nature, that (iv) differs materially from those ordinarily found to exist and generally recognized as inherent in construction activities of the character provided for in the Contract Documents. An example of a Type II DSC is encountering an artesian condition in an area where such a condition is not previously known to be present. Sustained elevated water levels in a lake caused by the U.S. Army Corps of Engineers' exercise of its control over a dam, which dictated the lake's water level, were held to form a Type II DSC in a State Department of Transportation contract, not involving the A201 General Conditions.[68]

Unlike a Type I DSC, the information from which the Type II DSC may be determined is not restricted to the contract documents. Regardless of the type of DSC involved, the procedure is for the contractor to notify the owner and the architect. The architect will then investigate and issue a decision.

§ 3.7.5 If, in the course of the Work, the Contractor encounters human remains or recognizes the existence of burial markers, archaeological sites or wetlands not indicated in the Contract Documents, the Contractor shall immediately suspend any operations

62. Millgard Corp. v. McKee/Mays, 49 F.3d 1070 (5th Cir. 1995).
63. *Id.* at 1072.
64. PT&L Constr. Co., Inc. v. State, 531 A.2d 1330 (N.J. 1987).
65. County Asphalt, Inc. v. State, 337 N.Y.S.2d 415 (App. Div. 1972).
66. *See* Central Allied Enters., Inc. v. Adjutant Gen's Dep't, No. 10AP-701, 2011 Ohio App. LEXIS 4049, at *23 (Sept. 27, 2011); Fruin-Colnon Corp. v. Niagara Frontier Transp. Auth., 585 N.Y.S.2d 248 (App. Div. 1992).
67. *Central Allied Enters., Inc.*, 2011 Ohio App. LEXIS 4049, at 23; *Fruin-Colnon Corp.*, 858 N.Y.S.2d 248.
68. Commonwealth v. AMEC Civil LLC, 699 S.E.2d 499 (Va. 2010).

that would affect them and shall notify the Owner and Architect. Upon receipt of such notice, the Owner shall promptly take any action necessary to obtain governmental authorization required to resume the operations. The Contractor shall continue to suspend such operations until otherwise instructed by the Owner but shall continue with all other operations that do not affect those remains or features. Requests for adjustments in the Contract Sum and Contract Time arising from the existence of such remains or features may be made as provided in Article 15.

As with differing site conditions under Section 3.7.4, the key to this section is what the contract documents indicate. The contractor is obligated to suspend operations if in the course of the work, it encounters or recognizes human remains or recognizes the existence of burial markers, archaeological sites, or wetlands not indicated in the contract documents. Implicit is that the contract documents may indicate the presence of human remains, burial markers, archaeological sites, or wetlands, in which case the contractor should comply with applicable laws and the contract documents.

A West Virginia case, not apparently involving AIA documents, illustrates the operation of this section.[69] At several locations along a pipeline being relocated, the contractor used a small bulldozer to pull a truck loaded with pipe through a wooded section of the pipeline. The bulldozer was driven, with its blade raised off the ground, through an area containing grave sites. This wooded area contained graves that allegedly were not indicated on any map, reserved or otherwise identified in a deed, identified by any obvious sign, included on a list of grave sites maintained by any state agency, or otherwise reasonably identifiable as an area containing graves. The defendants contended that the area was significantly overgrown with vegetation and contained mostly unmarked grave sites. Some actual grave markers were buried in forest debris. When the operator of the bulldozer realized that the bulldozer had passed through the grave site area, he immediately blocked off the area and relocated the route to connect with the pipeline at a different location.[70] Had Section 3.7.5 been in this contract, the actions of the contractor seem compliant with the Section 3.7.5 requirement to suspend operations.

The term "archaeological sites" is not defined in the A201. One definition that may be useful is:

> The term "archaeological resource" means any material remains of past human life or activities which are of archaeological interest, as determined under uniform regulations promulgated pursuant to this chapter. Such regulations containing such determination shall include, but not be limited to: pottery, basketry, bottles, weapons, weapon projectiles, tools, structures or portions of structures, pit houses, rock paintings, rock carvings, intaglios, graves, human skeletal materials, or any portion or piece of any of the foregoing items. Nonfossilized and fossilized paleontological specimens, or any portion or piece thereof, shall not be considered archaeological resources, under the regulations under this paragraph, unless found in an archaeological context. No item shall be treated as an archaeological resource under regulations under this paragraph unless such item is at least 100 years of age.[71]

69. Hairston v. Gen. Pipeline Constr., Inc., 704 S.E.2d 663 (W. Va. 2010).
70. *Id.* at 667.
71. 16 U.S.C.S. § 470bb (West 1988).

§ 3.8 ALLOWANCES

> **§ 3.8.1** The Contractor shall include in the Contract Sum all allowances stated in the Contract Documents. Items covered by allowances shall be supplied for such amounts and by such persons or entities as the Owner may direct, but the Contractor shall not be required to employ persons or entities to whom the Contractor has reasonable objection.

The contract sum is to include all allowances stated in the contract documents.[72] The architect generally specifies the allowance items of work for which dollar value allowances are required; however, the contractor may take the initiative to identify items that are not sufficiently specified to adequately establish pricing. Examples of items for which allowances may be established are carpeting, doors, and painting. However, an owner and contractor can get creative in identifying allowance items. In a contract litigated in Rhode Island, allowances were included for roofing, HVAC, elevators, and winter conditions.[73] A North Carolina school contract established an allowance for "unsuitable soils."[74]

> **§ 3.8.2** Unless otherwise provided in the Contract Documents,
>
> **.1** allowances shall cover the cost to the Contractor of materials and equipment delivered at the site and all required taxes, less applicable trade discounts;
>
> **.2** Contractor's costs for unloading and handling at the site, labor, installation costs, overhead, profit, and other expenses contemplated for stated allowance amounts shall be included in the Contract Sum but not in the allowances; and
>
> **.3** whenever costs are more than or less than allowances, the Contract Sum shall be adjusted accordingly by Change Order. The amount of the Change Order shall reflect (1) the difference between actual costs and the allowances under Section 3.8.2.1 and (2) changes in Contractor's costs under Section 3.8.2.2.

The amounts (i.e., the dollar value) of allowance items are usually set forth in the agreement (see, for example, section 4.4 of AIA A101–2007 or in the Guaranteed Maximum Price Amendment (GMP) for a GMP type of contract (see, for example, AIA A133–2009, Section 2.2.3.3)). Section 3.8.2 describes what is to be included in the allowance amounts, what is excluded, and how the amounts will be adjusted.

Section 3.8.2.1 provides that allowances shall cover the cost to the contractor of materials and equipment delivered at the site and all required taxes, less applicable trade discounts. This is essentially the direct cost of the item delivered to the site plus taxes. It does not include any service costs that may be quoted by a supplier or manufacturer. The contractor's costs for unloading and handling at the site, labor, installation costs, overhead, and profit are expressly excluded from the allowances. The last sentence of the section recognizes that those costs included in the contract sum may also change because of the selection or final specifications of an allowance item.

72. AIA A201–2017, § 1.1.1.
73. Allstate Interiors & Exteriors, Inc. v. Stonestreet Constr. LLC, 907 F. Supp. 2d 216 (Dist. R.I. 2012).
74. John W. Daniel & Co. v. Durham Pub. Schs. Bd. of Educ., No. 1:07-CV-61, 2008 U.S. Dist. LEXIS 119244 (M.D.N.C. June 6, 2008).

A contract for a performing arts center in Kentucky contained an allowance for a chandelier system.[75] The contract included AIA A201–1997, which contained allowance language similar to the 2017 edition. The chandelier system was subcontracted to an electrical subcontractor whose scope of work included a lifting and lowering device as part of the chandelier system. Fabrication of the original chandelier design was difficult, so the architect provided a new design that resulted in a larger chandelier that was too heavy for the lifting and lowering device that had already been fabricated. The subcontractor asserted claims for delays, disruptions and acceleration, and recovery of unpaid retainage, which were denied. The case illustrates whether increased costs caused by delay in either the fabrication or the selection or final design of an allowance item fall within the purview of Section 3.8.2.

Of significance in Section 3.8.2.3 is that the method of adjusting the contract sum for costs relating to allowances shall be by change order. That was the procedure used in *Allstate Interiors and Exteriors, Inc.*[76] In contrast, a contractor on a project in New Jersey did not present costs of allowance items until the end of the project.[77] It's not known if an AIA contract was used, but the case is instructive that a contract's change order provisions should be followed with respect to allowances covered by Section 3.8.2. In the New Jersey case, the owner refused to pay and the contractor was saved by the court finding that *quantum meruit* recovery was available to the contractor, despite the apparent existence of a written agreement.

> **§ 3.8.3** Materials and equipment under an allowance shall be selected by the Owner with reasonable promptness.

This subsection stipulates an owner obligation to select materials and equipment under an allowance with reasonable promptness. Reasonable promptness is not defined and could be the subject of dispute. Delay to the work would be one measure; however, is some delay reasonable? As a practical matter, it is suggested that the owner, contractor, and architect agree on the timing for selecting allowance items, which timing should be consistent with the project schedule.

§ 3.9 SUPERINTENDENT

> **§ 3.9.1** The Contractor shall employ a competent superintendent and necessary assistants who shall be in attendance at the Project site during performance of the Work. The superintendent shall represent the Contractor, and communications given to the superintendent shall be as binding as if given to the Contractor.

This obligation reinforces the contractor's responsibility for all construction means, methods, techniques, sequences, and procedures, as required by Section 3.3.1. In determining that sufficient evidence existed to conclude that scaffolding presented a safety issue, the Appellate Court of Illinois tacitly adopted the position that a "competent" superintendent need not necessarily have substantive knowledge of

75. Dynalectric Co. v. Whittenberg Constr. Co., No. 5:06-CV-00208-JHM, 2010 U.S. Dist. LEXIS 110136 (W.D. Ky. Oct. 13, 2010).
76. *Allstate Interiors & Exteriors, Inc.*, 907 F. Supp 2d 216.
77. Ippolito v. Jeffrey Angelo Constr. Co., No. A-4149-07T2, 2009 N.J. Super. Unpub. LEXIS 1219 (Super. Ct. App. Div. May 20, 2009).

every subcontractor's trade, but should be able to identify unsafe conditions in subcontractors' work.[78]

PRACTICE TIP

The obligation to supervise and provide competent oversight of the work should be flowed to each subcontractor to ensure checks and balances for project coordination and safety.

§ 3.9.2 The Contractor, as soon as practicable after award of the Contract, shall ~~furnish in writing to~~notify the Owner ~~through the~~and Architect of the name and qualifications of a proposed superintendent. ~~The~~ Within 14 days of receipt of the information, the Architect may ~~reply within 14 days to~~notify the Contractor ~~in writing~~, stating ~~(1)~~ whether the Owner or the Architect (1) has reasonable objection to the proposed superintendent or (2) ~~that the Architect~~ requires additional time ~~to~~for review. Failure of the Architect to ~~reply~~ provide notice within the 14-day period shall constitute notice of no reasonable objection.

The owner and architect may object to the proposed superintendent; however, such objection must be reasonable and timely. The modification to the A201 from the 2007 version altered both the contractor's and the owner's notice obligations. First, the contractor must now notify both the owner and the architect regarding the qualifications of the proposed superintendent. Next, if the architect or owner has an objection to the proposed superintendent, the architect must notify the contractor.

PRACTICE TIP

As long as the objection is made reasonably early in the project life cycle, it may be advisable, for purposes of the owner/contractor relationship, to replace a superintendent to whom the architect or owner objects, even if the objection is unreasonable or outside the 14-day period. However, the contractor should be equitably compensated with an adjustment of time and/or money for any delays incurred by the replacement of a superintendent.

§ 3.9.3 The Contractor shall not employ a proposed superintendent to whom the Owner or Architect has made reasonable and timely objection. The Contractor shall not change the superintendent without the Owner's consent, which shall not unreasonably be withheld or delayed.

To the extent it is necessary to change the superintendent, the owner's consent must be obtained.

§ 3.10 CONTRACTOR'S CONSTRUCTION AND SUBMITTAL SCHEDULES

§ 3.10.1 The Contractor, promptly after being awarded the Contract, shall ~~prepare and~~ submit for the Owner's and Architect's information a Contractor's construction schedule

78. Diaz v. Legat Architects, Inc., 920 N.E.2d 582 (Ill. App. Ct. 2009).

for the Work. The schedule shall contain detail appropriate for the Project, including (1) the date of commencement of the Work, interim schedule milestone dates, and the date of Substantial Completion; (2) an apportionment of the Work by construction activity; and (3) the time required for completion of each portion of the Work. The schedule shall provide for the orderly progression of the Work to completion and shall not exceed time limits current under the Contract Documents~~,~~. The schedule shall be revised at appropriate intervals as required by the conditions of the Work and Project~~, shall be related to the entire Project to the extent required by the Contract Documents, and shall provide for expeditious and practicable execution of the Work~~.

The new language in this provision requires that the contractor's schedule contain the additional enumerated details including interim project milestone dates and an apportionment of the work by construction activity (which, when combined with associated costs, constitutes the schedule of values referenced in Article 9 of the A201). In addition to the new scheduling details, owners may want to consider adding further requirements such as use of a specific type of project scheduling software and a prohibition against "front loading" of the schedule of values by subcontractors. The purpose of this change is to balance the owner's need for a more detailed schedule with the ability to use AIA Document A201–2017 on projects of varying complexity. With regard to a schedule prepared for a cost-plus (with or without a guaranteed maximum price) delivery system, the contractor may wish to specify that it is not bound to preliminary schedules that are presented to the owner prior to finalization of a GMP proposal or control estimate.

§ 3.10.2 The Contractor ~~shall prepare a submittal schedule~~, promptly after being awarded the Contract and thereafter as necessary to maintain a current submittal schedule, ~~and~~ shall submit ~~the~~ a submittal schedule~~(s)~~ for the Architect's approval. The Architect's approval shall not be unreasonably ~~be~~ delayed or withheld. The submittal schedule shall (1) be coordinated with the Contractor's construction schedule, and (2) allow the Architect reasonable time to review submittals. If the Contractor fails to submit a submittal schedule, or fails to provide submittals in accordance with the approved submittal schedule, the Contractor shall not be entitled to any increase in Contract Sum or extension of Contract Time based on the time required for review of submittals.

While the contractor is obligated to comply with the submittal schedule or face waiver of potential delay claims, the architect's only obligation in turning around reviews and approvals is to make certain that they are not unreasonably delayed or withheld.[79] However, the term unreasonable is not defined with regard to delay. From the contractor's and the owner's perspective, it may be best for the parties to expressly define unreasonable delay as delay in the progress of the work.

§ 3.10.3 The Contractor shall perform the Work in general accordance with the most recent schedules submitted to the Owner and Architect.

79. Please see also A201–2017 at section 4.2.7. "The Architect's action will be taken in accordance with the submittal schedule approved by the Architect or, in the absence of an approved submittal schedule, with reasonable promptness. . . ."

The obligation to perform in accordance with the "most recent" schedule may result in a contractor assuming the financial burden of proceeding with changed work contained in the most recently submitted schedule even though the owner refuses to approve pending change orders and accompanying requests for extensions of contract time. In contracts that provide for liquidated damages in the event of project delays, this issue takes on added significance, Owners and contractors alike should scrutinize change order provisions in conjunction with this section to minimize the potential for such scenarios.

§ 3.11 DOCUMENTS AND SAMPLES AT THE SITE

> The Contractor shall ~~maintain~~<u>make available,</u> at the <u>Project</u> site ~~for~~, the ~~Owner one copy of the Drawings, Specifications, Addenda.~~<u>Contract Documents, including</u> Change Orders, <u>Construction Change Directives,</u> and other Modifications, in good order and marked currently to indicate field changes and selections made during construction, and ~~one copy of~~<u>the</u> approved Shop Drawings, Product Data, Samples, and similar required submittals. These shall be <u>in electronic form or paper copy,</u> available to the Architect and ~~shall be~~<u>Owner, and</u> delivered to the Architect for submittal to the Owner upon completion of the Work as a record of the Work as constructed.

Section 3.11 requires certain documents and samples to be made available at the site. The owner may not maintain a construction office at the site, so access to the contractor's files may be important during site visits. Some minor modifications to the section have been made from the 2007 version. The obligation to "maintain" is replaced by "make available." The broader defined term "Contract Documents" replaces the former enumerated list of "Drawings, Specifications, Addenda" and the term "Construction Change Directives" has been added to the list of documents that the contractor must make available. Such documents are in addition to change orders, other modifications, approved shop drawings, product data, samples, and similar required submittals. Perhaps the most significant change is these documents and samples may now be available in electronic form or paper copy, presumably at the contractor's discretion, unless otherwise agreed.

The contractor is required to make such documents and samples available in good order and marked currently to indicate field changes and selections made during construction. Such marked documents are commonly called the "record set." The record set has historically been marked by hand by the contractor to show minor field revisions not affecting the design, the contract time, or the contract sum. For example, minor rerouting of piping to maintain clearances. The wording change "in electronic form or paper copy" indicates the move to the digital age at construction sites and the increasingly common practice to use building information modeling (BIM).

§ 3.12 SHOP DRAWINGS, PRODUCT DATA AND SAMPLES

> **§ 3.12.1** Shop Drawings are drawings, diagrams, schedules, and other data specially prepared for the Work by the Contractor or a Subcontractor, Sub-subcontractor, manufacturer, supplier, or distributor to illustrate some portion of the Work.

This provision defines the categories of documents included in the term "Shop Drawings." The continued inclusion of the undefined term "schedules" within the definition of "Shop Drawings" raises a potential ambiguity that should be addressed for the benefit of the parties to the agreement as well as the architect. This is because Section 3.12.4, which states that "Shop Drawings . . . are not Contract Documents," necessarily means that the "schedules" referenced in this section are not contract documents. However, the contractor's construction and submittal schedules referenced in Section 3.10.1 are included among the contract documents as a matter of common industry practice. Similarly, owners may prefer to elevate the contractor's schedule documents to contract documents and specify that the contractor's adherence to the schedule is a material term of the contract. To avoid misunderstandings and mitigate the risk of disputes, it would be wise for the parties to the agreement to clarify that the "schedules" referenced in this section are distinct from the contractor's construction and submittal schedules.

> **§ 3.12.2** Product Data are illustrations, standard schedules, performance charts, instructions, brochures, diagrams, and other information furnished by the Contractor to illustrate materials or equipment for some portion of the Work.

> **§ 3.12.3** Samples are physical examples that illustrate materials, equipment, or workmanship, and establish standards by which the Work will be judged.

> **§ 3.12.4** Shop Drawings, Product Data, Samples, and similar submittals are not Contract Documents. Their purpose is to demonstrate ~~the way by which~~how the Contractor proposes to conform to the information given and the design concept expressed in the Contract Documents for those portions of the Work for which the Contract Documents require submittals. Review by the Architect is subject to the limitations of Section 4.2.7. Informational submittals upon which the Architect is not expected to take responsive action may be so identified in the Contract Documents. Submittals that are not required by the Contract Documents may be returned by the Architect without action.

As addressed in the comment on Section 3.12.1, the continued inclusion of the undefined term "schedules" within the definition of "Shop Drawings" raises a potential ambiguity that should be addressed for the benefit of the parties to the agreement as well as the architect.

> **§ 3.12.5** The Contractor shall review for compliance with the Contract Documents, approve, and submit to the Architect, Shop Drawings, Product Data, Samples, and similar submittals required by the Contract Documents, in accordance with the submittal schedule approved by the Architect or, in the absence of an approved submittal schedule, with reasonable promptness and in such sequence as to cause no delay in the Work or in the activities of the Owner or of ~~separate contractors~~Separate Contractors.

This provision does not require the architect to approve submittals in a timely manner to avoid delaying the contractor's work. But, to the contrary, the contractor is affirmatively obligated to review and approve submittals so as to cause no delay in the activities of the owner or separate contractors. Since the mutual waiver of consequential damages in Section 15.1.6 is only between the owner and the contractor, the contractor's failure to timely review and approve submittals may give rise to delay claims (including alleged consequential damages) from separate contractors.

> **§ 3.12.6** By submitting Shop Drawings, Product Data, Samples, and similar submittals, the Contractor represents to the Owner and Architect that the Contractor has (1) reviewed and approved them, (2) determined and verified materials, field measurements and field construction criteria related thereto, or will do so, and (3) checked and coordinated the information contained within such submittals with the requirements of the Work and of the Contract Documents.

This provision contains representations from the contractor regarding the accuracy of the submittals in relation to field criteria. If the contractor fails to verify materials, field measurements, and field construction criteria prior to submitting shop drawings, product data, and samples, the architect's subsequent review and approval of such submittals will not insulate the contractor from liability in the event of errors in the submittals.[80]

> **§ 3.12.7** The Contractor shall perform no portion of the Work for which the Contract Documents require submittal and review of Shop Drawings, Product Data, Samples, or similar submittals, until the respective submittal has been approved by the Architect.

This provision is a cross-reference to the architect's contract administrative obligations from Article 4 and, at the same time, provides an additional mechanism in furtherance of the owner's and architect's control over the change order process. By requiring architect approval prior to performing the work, this procedure ensures that any concerns regarding accuracy and completeness of submittals or scope of work issues are addressed before the contractor starts the relevant portion of the work. At least one court has held that this language, while requiring contractor submittals to be approved by the architect before performing work, is ambiguous as to the timing of such submittals (i.e., whether submittals must be approved prior to materials being ordered).[81]

> **§ 3.12.8** The Work shall be in accordance with approved submittals except that the Contractor shall not be relieved of responsibility for deviations from the requirements of the Contract Documents by the Architect's approval of Shop Drawings, Product Data, Samples, or similar submittals, unless the Contractor has specifically ~~informed~~notified the Architect ~~in writing~~ of such deviation at the time of submittal and (1) the Architect has given written approval to the specific deviation as a minor change in the Work, or (2) a Change Order or Construction Change Directive has been issued authorizing the deviation. The Contractor shall not be relieved of responsibility for errors or omissions in Shop Drawings, Product Data, Samples, or similar submittals, by the Architect's approval thereof.

80. Modern Cont'l S. v. Fairfax Co. Water Auth., 70 Va. Cir. 172 (2006); D.C. McClain, Inc. v. Arlington Cnty., 452 S.E.2d 659 (Va. 1995).
81. Otis Elevator Co. v. Fulcrum Constr. Co., LLC, 721 N.W.2d 157 (Wis. Ct. App. 2006).

This provision makes clear that the contractor may not rely upon the architect's approval of submittals to cure errors in shop drawings, product data, or samples for which the contractor is responsible.[82] While the contractor is responsible for errors and omissions in submittals to the architect, it is equally true that the requirement to provide submittals, in and of itself, does not shift design liability from the architect to the contractor.[83] Further, if the contractor performs the work in accordance with the contract, it will not be held liable for a subsequent breach of warranty or contract resulting from the architect's improperly designed work.[84]

§ 3.12.9 The Contractor shall direct specific attention, in writing or on resubmitted Shop Drawings, Product Data, Samples, or similar submittals, to revisions other than those requested by the Architect on previous submittals. In the absence of such ~~written~~ notice, the Architect's approval of a resubmission shall not apply to such revisions.

This provision requires a contractor to call out to the architect, whether through the traditional method of "clouding" used on hand-drafted drawings and submittals or through electronic means on BIM or other modeling programs, new revisions to documents that are being resubmitted to the architect for review and approval. Failure to call out changes in resubmitted documents will preclude the contractor from later claiming that the changes were approved by the architect.

§ 3.12.10 The Contractor shall not be required to provide professional services that constitute the practice of architecture or engineering unless such services are specifically required by the Contract Documents for a portion of the Work or unless the Contractor needs to provide such services in order to carry out the Contractor's responsibilities for construction means, methods, techniques, sequences, and procedures. The Contractor shall not be required to provide professional services in violation of applicable law.

The general tenor of this provision is that the contractor is not liable to provide professional design or engineering services unless specifically required to do so by the Contract Documents. However, the phrase "or unless the contractor needs to provide such services in order to carry out the contractor's responsibilities for construction means, methods, techniques, sequences, and procedures" carries the implication that a contractor may be subject to design errors and omissions liability for things like scaffolding, sheeting, or shoring required to complete the scope of work. Moreover, it would be in the owner's interest, as well as the contractor's, to specify design-build requirements in the contract documents so that the contractor can appropriately flow down those requirements to subcontractors and obtain adequate professional liability coverage for such services.

§ 3.12.10.1 If professional design services or certifications by a design professional related to systems, materials, or equipment are specifically required of the Contractor by the Contract Documents, the Owner and the Architect will specify all performance and

82. *Modern Cont'l S.*, 70 Va. Cir. 172; *D.C. McClain, Inc.*, 452 S.E.2d 659.

83. John W. Cowper Co., Inc. v. Buffalo Hotel Dev. Venture, 528 N.E.2d 1214 (N.Y. 1988); Gordon v. Heritage Fellowship, 784 S.E.2d 265 (Va. 2016).

84. Gordon v. Heritage Fellowship, 784 S.E.2d 265 (Va. 2016).

design criteria that such services must satisfy. The Contractor shall be entitled to rely upon the adequacy and accuracy of the performance and design criteria provided in the Contract Documents. The Contractor shall cause such services or certifications to be provided by ~~a properly~~an appropriately licensed design professional, whose signature and seal shall appear on all drawings, calculations, specifications, certifications, Shop Drawings, and other submittals prepared by such professional. Shop Drawings, and other submittals related to the Work, designed or certified by such professional, if prepared by others, shall bear such professional's written approval when submitted to the Architect. The Owner and the Architect shall be entitled to rely upon the adequacy, and accuracy ~~and completeness~~ of the services, certifications, and approvals performed or provided by such design professionals, provided the Owner and Architect have specified to the Contractor ~~all~~the performance and design criteria that such services must satisfy. Pursuant to this Section 3.12.10, the Architect will review, and approve or take other appropriate action on submittals only for the limited purpose of checking for conformance with information given and the design concept expressed in the Contract Documents. ~~The Contractor shall not be responsible for the adequacy of the performance and design criteria specified in the Contract Documents.~~

The main change is that the contractor who performs design services as part of its scope of work is expressly entitled to rely upon performance and design criteria supplied by the owner and the architect. Such performance and design criteria must now not only be adequate but must also be accurate. This language presumably extends the right to rely upon the architect's performance and design criteria to the contractor's subcontractors who perform "design assist" work with regard to mechanical, electrical, plumbing, curtain wall, and HVAC systems.

§ 3.12.10.2 If the Contract Documents require the Contractor's design professional to certify that the Work has been performed in accordance with the design criteria, the Contractor shall furnish such certifications to the Architect at the time and in the form specified by the Architect.

This is a new provision in the 2017 version. The purpose of this provision is to clarify the contractor's requirements in the event that the contract documents require the contractor to perform design services based on specific criteria provided by the architect. In that circumstance, the contractor may elect to hire its own outside design professional. If the contractor hires its own design professional, then the contractor is obligated to provide the owner and the architect with certification that the design work has been completed in accordance with the contract documents. It would be prudent for the contractor who engages subcontractors to perform design-assist work to flow down this requirement in all such subcontracts. It would also be prudent for both the contractor and the owner to require that all design-assist subcontractors maintain adequate professional liability insurance coverage as protection against their design errors and omissions.

§ 3.13 USE OF SITE

The Contractor shall confine operations at the site to areas permitted by applicable laws, statutes, ordinances, codes, rules and regulations, ~~and~~ lawful orders of public authorities, and the Contract Documents and shall not unreasonably encumber the site with materials or equipment.

The contractor's use of the permitted portions of the site must be "reasonable." There are surprisingly few cases discussing the reasonableness requirement. It would appear that reasonableness would largely be dependent on the nature and scope of the project. Where large stockpiles of equipment and materials might be appropriate for a significant project, using a project site of relatively small scope to store materials for another project would likely not be considered reasonable under Section 3.13.

PRACTICE TIP:

To the extent that specialized or oversized equipment is necessary to the work, or where equipment's proximity to the work site has a significant impact on the ability of the contractor to perform its work, it is important to make appropriate modification to this section. For example, in sand-blasting operations, the distance between the blasting equipment and the nozzle significantly impacts blast velocity, and thus, efficiency for coating removal operations. Accordingly, it would be important to clarify that certain equipment may be placed in certain locations near the work.

§ 3.14 CUTTING AND PATCHING

§ 3.14.1 The Contractor shall be responsible for cutting, fitting, or patching required to complete the Work or to make its parts fit together properly. All areas requiring cutting, fitting ~~and~~, or patching shall be restored to the condition existing prior to the cutting, fitting ~~and~~, or patching, unless otherwise required by the Contract Documents.

§ 3.14.2 The Contractor shall not damage or endanger a portion of the Work or fully or partially completed construction of the Owner or ~~separate contractors~~Separate Contractors by cutting, patching, or otherwise altering such construction, or by excavation. The Contractor shall not cut or otherwise alter ~~such~~ construction by the Owner or a ~~separate contractor~~ Separate Contractor except with written consent of the Owner and of ~~such separate contractor; such consent~~the Separate Contractor. Consent shall not be unreasonably withheld. The Contractor shall not unreasonably withhold, from the Owner or a ~~separate contractor the Contractor's~~Separate Contractor, its consent to cutting or otherwise altering the Work.

The purpose of this provision is to assist the owner in coordinating the contractor's and separate contractors' work and minimizing the risk of damage and interference. See also the discussion of the owner's obligations when performing construction work and to award separate contracts.

§ 3.15 CLEANING UP

§ 3.15.1 The Contractor shall keep the premises and surrounding area free from accumulation of waste materials ~~or~~and rubbish caused by operations under the Contract. At completion of the Work, the Contractor shall remove waste materials, rubbish, the Contractor's tools, construction equipment, machinery, and surplus materials from and about the Project.

The owner wants the site to be maintained during construction and the site left clear at the end of the project. Only a minor change has been made to Section 3.15.1 from the 2007 version. It has been clarified that the contractor's obligation to keep the premises and surrounding area clear includes avoiding the accumulation of both waste materials and rubbish.

> **§ 3.15.2** If the Contractor fails to clean up as provided in the Contract Documents, the Owner may do so and the Owner shall be entitled to reimbursement from the Contractor.

The section requires the contractor to clean up as required by the contract documents. The AIA documents do not specify cleanup other than in a general way in Section 3.15, so the other contract documents should be reviewed for any specific cleaning requirements. Some owners—health care facilities, for example—may have very stringent requirements, such as to protect against infection. Other owners may want "broom clean" conditions or better, for example, professional-grade window cleaning.

§ 3.16 ACCESS TO WORK

> The Contractor shall provide the Owner and Architect with access to the Work in preparation and progress wherever located.

Only a minor modification has been made to this Section 3.16. The work may be performed in locations other than the actual site of the project. For example, curtain wall may be fabricated in Europe for a project located in the United States. The contractor must ensure that the owner and architect have access to all such work. For this reason, it is very important that this Section 3.16 be incorporated in subcontracts and purchase orders for equipment, systems, materials, and supplies.

§ 3.17 ROYALTIES, PATENTS AND COPYRIGHTS

> The Contractor shall pay all royalties and license fees. The Contractor shall defend suits or claims for infringement of copyrights and patent rights and shall hold the Owner and Architect harmless from loss on account thereof, but shall not be responsible for ~~such~~ defense or loss when a particular design, process, or product of a particular manufacturer or manufacturers is required by the Contract Documents, or where the copyright violations are contained in Drawings, Specifications, or other documents prepared by the Owner or Architect. However, if ~~the Contractor has reason to believe that the required design, process or product is~~ an infringement of a copyright or ~~a~~ patent is discovered by, or made known to, the Contractor, the Contractor shall be responsible for ~~such~~the loss unless ~~such~~the information is promptly furnished to the Architect.

This provision generally places the responsibility for intellectual property licensing and infringement on the contractor for the work it performs. But the contractor is not responsible for intellectual property including the design, process, or products that are specifically required by the contract documents or work performed by the owner or architect. The language changes in this section of the 2017 document leave unresolved the question of how the contractor, who is not a licensed professional designer and is not the author of the plans and specifications for a project, can "discover" an infringement of a third party's copyright or trademark contained in the plans and specifications. Read together with the language in Section 3.12.10 (stating that contractor has no professional design obligations "unless the Contractor needs to provide such services in order to carry out the Contractor's responsibilities for construction means, methods, techniques, sequences, and procedures"), does this section create an affirmative obligation for the contractor to review the architect's work product for potential infringements? Although there are currently no cases directly on point (as of 2016), the language in Sections 3.12.10 and 3.12.10.1 suggests that the contractor who performs no design work itself will not be held liable for infringement of a third party's intellectual property by the contract documents. However, it is possible that the new language does give rise to an affirmative obligation on the part of the contractor, who performs "design assist" mechanical, electrical, plumbing, curtain wall, and HVAC systems work through subcontractors, to have those subcontractors engage in peer review of the plans and specifications as to such systems for potential infringement of third-party rights. If a copyright or patent infringement contained in the contract documents is "discovered" by a design assist subcontractor, this presumably means the infringement has been "made known to" the contractor, who is then obligated to report the discovery to the owner and the architect.

§ 3.18 INDEMNIFICATION

> **§ 3.18.1** To the fullest extent permitted by law, the Contractor shall indemnify and hold harmless the Owner, Architect, Architect's consultants, and agents and employees of any of them from and against claims, damages, losses, and expenses, including but not limited to attorneys' fees, arising out of or resulting from performance of the Work, provided that such claim, damage, loss, or expense is attributable to bodily injury, sickness, disease or death, or to injury to or destruction of tangible property (other than the Work itself), but only to the extent caused by the negligent acts or omissions of the Contractor, a Subcontractor, anyone directly or indirectly employed by them, or anyone for whose acts they may be liable, regardless of whether or not such claim, damage, loss, or expense is caused in part by a party indemnified hereunder. Such obligation shall not be construed to negate, abridge, or reduce other rights or obligations of indemnity that would otherwise exist as to a party or person described in this Section 3.18.

While there have been no substantive changes to Section 3.18.1, there have been a number of significant decisions interpreting this provision. The contractor's indemnity obligation is limited to damages "attributable to bodily injury" or "injury to or destruction of tangible property," "but only to the extent caused by the Contractor, a Subcontractor," or someone for whose acts they may be liable. Interpreting similar language, the Supreme Court of Pennsylvania held that by agreeing to such language, the indemnitees were to be: indemnified for damages *"only to the extent that"* the damages were caused by the negligence of [the indemnitor] and its sub-subcontractors, employees and anyone for whom it may be liable, the parties communicated their intent to limit any indemnification to that portion of damages attributed to the negligence of [the indemnitor] and those under its supervision.[85]

85. Greer v. City of Phila., 795 A.2d 376 (Pa. 2002) (emphasis added).

The Court held that the language did not "evince an intent to provide indemnification for damages due to the negligence of other unspecified parties, including an indemnitee."[86]

The Court's holding limiting the contractor's indemnity obligation to loss to the extent caused by the negligence of the contractor is largely congruent with the language of many standard commercial general liability insurance policies.

PRACTICE TIP

It is not uncommon for owners to broaden the indemnity by modifying Section 3.18.1 to include losses beyond negligently caused bodily injury or damage to tangible property. The contractor should be aware that this may create gaps in insurance coverage that may increase its exposure to risk.

§ 3.18.2 In claims against any person or entity indemnified under this Section 3.18 by an employee of the Contractor, a Subcontractor, anyone directly or indirectly employed by them, or anyone for whose acts they may be liable, the indemnification obligation under Section 3.18.1 shall not be limited by a limitation on amount or type of damages, compensation, or benefits payable by or for the Contractor or a Subcontractor under workers' compensation acts, disability benefit acts, or other employee benefit acts.

This provision seeks to restrain the contractor from arguing that its indemnity obligation is limited by state law restrictions by requiring the contractor to waive such limitations of liability.

86. *Id.* at 379. The Court further held that the provision "that the indemnity clause [would] apply 'regardless of whether or not such claim . . . [was] caused in part by a party indemnified hereunder'" as a clarification "that any contributory negligence by [the indemnitees would] not bar their indemnification for damages due to [the indemnitor's] negligence." *Id.* at 380. *See also* United Rentals Highway Techs., Inc. v. Wells Cargo, Inc., 289 P.3d 221 (Nev. 2012); MT Builders v. Fisher Roofing, 179 P.3d 178 (Ariz. Ct. App. 2008); Braegelmann v. Horizon Dev. Co., 371 N.W.2d 644 (Minn. Ct. App. 1985).

CHAPTER 4:
Article 4 Architect

Kelli E. Goss
Paul E. Davis

ARTICLE 4 ARCHITECT

This article sets forth the duties, obligations, and limitations of authority of the architect on the project. The changes made to this article in 2017 were minor. In 2007, the AIA made significant changes to this article, starting with renaming the article from "Administration of the Contract" to "Architect." In the 2007 A201 Deskbook published by the Forum on the Construction Industry in 2008, the authors stated: "[t]he change in the title of Article 4 . . . not only reflects that the provisions in the Article encompass more than administration of the Project, but it also signals the more confined role the Architect plays as an administrator of the construction process in the 2007 edition."[1] The changes made in 2017 are consistent with the direction of the changes made in 2007. For example, section 4.2.4 has been revised to allow for more direct communication directly between the project participants rather than having the bulk of such communication go through the architect. Even the basic requirement in section 4.1.1 that the owner retain a licensed architect has been deleted from this article. That requirement has been moved to Article 2, "Owner," as it is an obligation of the owner; in its place in the new section 4.1.1 is a simple statement that the architect is the person or entity retained by the owner.

§ 4.1 GENERAL

~~**§ 4.1.1** The Owner shall retain an architect lawfully licensed to practice architecture or an entity lawfully practicing architecture in the jurisdiction where the Project is located. That person or entity is identified as the Architect in the Agreement and is referred to throughout the Contract Documents as if singular in number.~~

~~§~~ **§ 4.1.1** The Architect is the person or entity retained by the Owner pursuant to Section 2.3.2 and identified as such in the Agreement.

Section 4.1.1 has been rewritten to reference a new section 2.3.2, in Article 2, "Owner," that states exactly what this section 4.1.1 previously stated: "The Owner shall retain an architect lawfully licensed to practice architecture or an entity lawfully practicing architecture in the jurisdiction where the Project is located. . . ."

Conceptually, moving the text of the 2010 edition section 4.1.1 to Article 2 is appropriate because the requirement to obtain the services of a licensed architect is a contractual obligation of the owner and therefore should be identified as one of the owner's duties. There are no changes to the substantive requirements that the architect or architectural firm be licensed and lawfully practicing in the jurisdiction where the project is located.

Owners should be aware that contractually, under the A201, the architect is required to be licensed,[2] and this applies to any substitution of the architect. The licensure and registration requirements are designed to protect the public.[3]

1. Charles M. Sink, A. Holt Gwyn, James Duffy O'Connor, Dean B. Thomson, The 2007 A201 Deskbook 40 (ABA Forum on the Construction Industry 2008).

2. Most states have statutes regulating the practice of architecture (or engineering). *See, e.g.*, N.C. Gen. Stat. § 83A-1 (2016); Va. Code Ann. § 54.1-400 (2016); N.H. Rev. Stat. Ann. § 310-A:28 (2016); *see also* Francis v. Rehman, 110 A.3d 615 (D.C. 2015) (citing D.C.Code § 47-2853.63 (2012) (providing that "unless licensed to practice architecture . . . no person shall engage, directly or indirectly, in the practice of architecture in the District.")).

3. Sturdza v. United Arab Emirates, 11 A.3d 251, 254–55 (D.C. 2011) (noting that "the law of the District of Columbia has imposed a licensure requirement on the practice of architecture in this jurisdiction in order 'to safeguard life, health, and property, and to promote the public welfare.'").

§ 4.1.2 Duties, responsibilities, and limitations of authority of the Architect as set forth in the Contract Documents shall not be restricted, modified, or extended without written consent of the Owner, Contractor, and Architect. Consent shall not be unreasonably withheld.

The 2017 edition added two serial commas to section 4.1.2. This section requires a modification of the architect's duties, responsibilities, and scope of authority to be in writing and requires the written consent of the architect, owner, and contractor. Where written consent has been required in other sections of the AIA A201, e.g., arbitration, courts have upheld the written consent requirement.[4] The written order requirement was first addressed by the United States Supreme Court in *Plumley v. United States*,[5] upholding the requirement that modifications to the construction contract be in writing and denying the contractor recovery for extra work performed where no formal written contract modification was processed even though the parties informally agreed to the work.[6] Around the mid-20th century, however, courts began to break from strict compliance with the written modification requirement of contracts.[7] In the absence of a statute preventing oral modification of a contract, a written contract may be orally modified, even if its express terms prohibit modification except in writing.[8] Courts now may uphold oral modifications to the contract[9] and waivers of written consent requirements by actions of the parties.[10]

This section of the A201 also requires the consent of the contractor to a change of the contractual duties of the architect. The consent of the contractor is required because a modification of the architect's duties or responsibilities may directly or indirectly affect the contractor. For example, a modification of the architect's duties in processing applications for payment will almost certainly affect the contractor's method of operation. The mere fact that there has been a change in the architect's duties may also impact the contractor's schedule. It is advisable to keep all parties informed of such modification of duties. The contractor's consent may be effective without written consent and merely upon actions; however, as required by this section, the better practice is to have written consent of all required parties under the contract.

4. *See, e.g., Ex parte* Stallings & Sons, Inc. v. Sherlock, Smith & Adams, Inc., 670 So. 2d 861 (Ala. 1995) (noting that where the contract required written consent of the architect for joinder in arbitration, the court refused to adopt the doctrine of equitable estoppel against the architect where there was no written consent); Raffa Assoc., Inc. v. Boca Raton Resort & Club, 616 So. 2d 1096 (Fla. Dist. Ct. App. 1993) (excluding a claim from arbitration for lack of written consent).

5. Plumley v. United States, 226 U.S. 545 (1913).

6. *Plumley v United States* was cited throughout much of the 20th century by proponents of strict contract enforcement for the proposition that changes to the work are invalid without written order, whether the work was directed or not. *See* 4 Philip L. Bruner, Patrick J. O'Connor, Jr., Bruner & O'Connor on Construction Law § 4:38 (2002).

7. *See* Bruner & O'Connor, *supra* note 6, § 4:39; W.H. Armstrong & Co. v. United States, 98 Ct. Cl. 519,1943 WL 4254 (1943) (awarding contractor an equitable adjustment for oral changes to the work not formalized in written change order).

8. Restatement of Contracts § 407 & cmt. a (1932) (updated October 2016) ("Whenever two persons contract, no limitation self-imposed can destroy their power to contract again.").

9. First Pennsylvania Mortg. Trust v. Dorchester Sav. Bank, 481 N.E.2d 1132, 1138–39 (Mass. 1985) ("It is a settled 'principle that the mode of performance required by a written contract may be varied by a subsequent oral agreement based upon a valid consideration.'" (quoting Siegel v. Knott, 55 N.E.2d 889, 890 (Mass. 1944); and citing Wesley v. Marsman, 1004, 471 N.E.2d 51, 52 (1984))).

10. *In re* DSC Industries, Inc. v. Robert Sagot, Inc., 94 B.R. 42 (E.D. Pa. 1988) (noting that under New Jersey common law, a party's conduct may be viewed as waiver of a contractual requirement that changes to the contract be in writing); Inland Constr. Co. v. Cameron Park II, Ltd., LLC, 640 S.E.2d 415 (N.C. Ct. App. 2007) (citing J.R. Graham & Son, Inc. v. Board of Education, 212 S.E.2d 542, 544–45 (N.C. Ct. App. 1975) ("This court has held that provisions of a written contract may be modified or waived by a subsequent parol agreement, or by conduct which naturally and justly leads the other party to believe the provisions of the contract are modified or waived. This principle has been sustained even where the instrument provides for any modification of the contract to be in writing.")).

~~**§ 4.1.3** If the employment of the Architect is terminated, the Owner shall employ a successor architect as to whom the Contractor has no reasonable objection and whose status under the Contract Documents shall be that of the Architect.~~

This section has been moved to Article 2, "Owner," section 2.3.3 and follows the owner's duties to retain a licensed architect.

§ 4.2 ADMINISTRATION OF THE CONTRACT

§ 4.2.1 The Architect will provide administration of the Contract as described in the Contract Documents and will be an Owner's representative during construction until the date the Architect issues the final Certificate for Payment. The Architect will have authority to act on behalf of the Owner only to the extent provided in the Contract Documents.

An architect may be in a role of independent contractor or an agent to the owner, and this role may change throughout the various phases of a project.[11] The legal duties and obligations of an independent contractor are different from those of an agent. An independent contractor is hired to perform a specific duty and cannot generally bind the owner to any obligations,[12] whereas an agent acts on behalf of the owner and in the owner's interest.[13] This distinction is important because unlike an agent, an independent contractor cannot bind the owner.

The court in *Blue Cross & Blue Shield v. W.R. Grace & Co.*[14] held that an architect is an independent contractor to the owner during the design phase and an agent of the owner during the architect's construction administration phase. During the design phase, an architect is preparing plans and specifications and is not controlled by the owner and not subject to the owner's control with regard to the means by which the plans and specifications are produced. During the construction phase, however, the architect has a different relationship with the owner. During the construction phase, the architect is acting as the owner's agent and can bind the owner in its decisions.[15]

The architect's authority to act on behalf of the owner is not without limitations. The authority granted under section 4.2.1 is limited to the terms of the contract.[16] For example, the architect cannot change the contract sum or contract time without the owner's approval as required by numerous other sections of the A201.[17]

11. Blue Cross & Blue Shield v. W.R. Grace & Co., 781 F. Supp. 420 (D.S.C. 1991) (holding that an architect was both an agent of the owner and an independent contractor, depending upon the function being performed).

12. Tri-City Constr. Co. v. A.C. Kirkwood & Assocs., 738 S.W.2d 925 (Mo. Ct. App. 1987) (noting that an independent contractor is one that contracts with another to do something for him but is not controlled by the other, and is not subject to the other's control).

13. *Id.* at 931 ("Agency is a fiduciary relationship which results from the consent by one person, the principal, to another, the agent, for the agent to act on the principal's behalf and subject to the principal's control.").

14. 781 F. Supp. 420 (D.S.C. 1991).

15. *See* Hines v. Farr, 112 S.E.2d 33, 37 (S.C. 1960) ("It has been held that an architect, in the performance of his supervisory functions with respect to a building under construction, ordinarily acts as the agent and representative of the person for whom the work is being done.").

16. *See* Marble Falls Hous. Auth. v. McKinley, 373 S.W.2d 292 (Tex. 1971) (holding that where the contract contained a clause stating that the owner would pay the architect for the service of arranging for the work to proceed should the contractor default, and where the architect's inspector then directed a subcontractor to perform additional work, the inspector did not have apparent authority to bind the owner).

17. For example, change orders, construction change directives, and applications for payment all require owner approval.

Section 4.2.1 terminates the architect's authority to act on behalf of the owner on the date the architect issues the final certificate for payment. This limitation should be given some thought as it may create issues post-construction. For example, the owner may want the architect to be involved during the one-year period for correction of work following substantial completion found in section 12.2.2 (often erroneously referred to as the "one-year call-back warranty").[18] The owner should consider the architect's role during this one-year period and revise this clause accordingly based on how active a role the owner expects the architect to have.

> **§ 4.2.2** The Architect will visit the site at intervals appropriate to the stage of construction, or as otherwise agreed with the Owner, to become generally familiar with the progress and quality of the portion of the Work completed, and to determine in general if the Work observed is being performed in a manner indicating that the Work, when fully completed, will be in accordance with the Contract Documents. However, the Architect will not be required to make exhaustive or continuous on-site inspections to check the quality or quantity of the Work. The Architect will not have control over, charge of, or responsibility for the construction means, methods, techniques, sequences or procedures, or for the safety precautions and programs in connection with the Work, since these are solely the Contractor's rights and responsibilities under the Contract Documents~~, except as provided in Section 3.3.1~~.

The only change to this section from the 2007 version was to delete the last clause. The deleted text was a reference to section 3.3.1, which addresses the contractor's supervisory and construction responsibilities.

The architect's responsibility under this section is to review the work periodically[19] and determine whether the completed work complies with the contract documents, and if the uncompleted work, when completed, will conform to the contract documents. Section 4.2.2 is important for the protection of the architect. The architect is not responsible for *inspecting* the work, nor for the means and methods of construction, nor for safety programs in connection with the work. These are solely the responsibility of the contractor.[20] The language of section 4.2.2 has been interpreted by one court as not requiring the architect to supervise the work.[21]

The intent of the language of this section is to avoid imposing any duty upon the architect to observe means and methods of construction.[22] As late as in the 1997 edition of the A201, the architect had an express contractual duty to guard the owner against defects and deficiencies in the work.[23] The case law at that time

18. The B101 requires the architect to make a visit to the site up to one year after substantial completion for a walk-through with the owner.

19. Sams Hotel Group, LLC v. Environs, Inc., No. 1:09-cv-00930-TWP-TAB, 2012 WL 3139765, 2012 U.S. Dist. LEXIS 107459 (S.D. Ind. Aug. 1, 2012) (holding that where the contract required the architect to make three site visits during construction to become familiar with the general progress of the work, and the owner alleged that the architect's failure to inspect the site at the appropriate times was a breach of the owner-architect contract, the architect did not breach the contract because, absent special agreements relating to inspection obligations, architects are not insurers of a contractor's work).

20. Brown v. Gamble Constr. Co., Inc., 537 S.W.2d 685, 687 (Mo. Ct. App.1976) (holding that "architects are under no duty to supervise construction unless they expressly agree to do so").

21. South Burlington Sch. Dist. v. Calcagni-Frazier-Zajchowski Architects, Inc., 410 A.2d 1359 (Vt. 1980) (noting that the language of the contract in no way related to supervision of the construction project).

22. Baumeister v. Automated Prods., Inc., 690 N.W.2d 1 (Wis. 2004) (holding that under the terms of the agreement the architect was relieved of liability with regard to construction means and methods); Meridian at Windchime, Inc. v. Earth Tech, Inc., 960 N.E.2d 344 (Mass. App. Ct. 2012) (holding that where the contract language provided that the engineer shall not supervise, direct, or have control over the contractor's work, the engineer had no authority or responsibility for the means and methods of construction selected by the contractor).

23. Section 4.2.2 of the 1997 A201 provided that the architect "will visit the site at intervals appropriate to the stage of the Contractor's operations (1) to become generally familiar with and to keep the Owner informed about the progress and quality of the portion of the Work completed, (2) to endeavor to *guard* the Owner against defects and deficiencies in the Work." (emphasis added). See also similar text in section 2.2.3 in the 1976 A201 and section 4.2.2 in the 1987 A201.

interpreting comparable contract language often held the architect liable for failing to protect the owner from damages caused by the contractor's deficient work.[24] In a case involving similar contract language to section 4.2.2 of the 1997 A201, the architect approved a wall that was in an obvious defective condition and the court held that evidence existed to show that the architect breached its contractual duty to guard the owner against defects and deficiencies in the construction work based on the architect's site observations.[25]

The 2007 edition of the A201 deleted the architect's obligation to "endeavor to guard" the owner against defects and deficiencies in the contractor's work. As noted above, the 2017 edition did not change section 4.2.2. Although the language of this section attempts to insulate the architect from liability for defective construction, be aware that courts have imposed liability on the architect where the architect knew of, or reasonably should have known of, a defective condition.[26] Where a school district contracted with an architect for periodic site visits and the architect became aware of defects in the work and failed to notify the owner such that steps could have been taken to correct the work, the architect's liability was based on a breach of the contract rather than as a guarantor of the contractor's work.[27]

It is good practice for the architect to prepare written reports of each site visit, indicating the date of the visit, who was present on site that day, what work was being performed during the architect's visit, and any comments as to the quality of the completed work. However, it is recommended that the architect refrain from commenting on the contractor's means and methods of construction. A copy of the architect's reports of site visits should be sent to the owner and contractor promptly after each site visit.

> **§ 4.2.3** On the basis of the site visits, the Architect will keep the Owner reasonably informed about the progress and quality of the portion of the Work completed, and promptly report to the Owner (1) known deviations from the Contract Documents ~~and~~, (2) known deviations from the most recent construction schedule submitted by the Contractor, and (~~2~~3) defects and deficiencies observed in the Work. The Architect will not be responsible for the Contractor's failure to perform the Work in accordance with the requirements of the Contract Documents. The Architect will not have control over or charge of, and will not be responsible for acts or omissions of, the Contractor, Subcontractors, or their agents or employees, or any other persons or entities performing portions of the Work.

24. *See* Comptroller v. King, 232 S.E.2d 895 (Va. 1977) (where the language of the contract required the architect to provide administration and supervision of construction, and where the architect failed to notify the owner of defects observed, or that reasonably should have been observed, during construction, the architect could be liable to the owner); Brown v. McBro Planning & Dev. Co., 660 F. Supp. 1333 (D.V.I. 1987) (where the architect learned early on that there was a problem with the construction and failed to have the contractor correct the problem and continued to certify applications for payment, the architect was liable to the owner for damages); Roland A. Wilson v. Forty-O-Four Grand Corp., 246 N.W.2d 922 (Iowa 1976) (where the architect discovered upon a water test that windows leaked prior to final completion, and notified the contractor to caulk the windows, but at final inspection the architect did not perform the water test again to determine if the windows still leaked and then approved final payment, the court found the architect had breached a contractual duty to determine whether the contractor's work was properly done before approving final payment); Campbell County Bd. of Educ. v. Brownlee-Kesterson, 677 S.W.2d 457 (Tenn. Ct. App. 1984) (finding that the architects failed to guard the owner against defects and deficiencies in the work of the contractor when they knew or should have known them to be defective).

25. Dan Cowling & Assocs. v. Board of Educ., 618 S.W.2d 158 (Ark. 1981) (architect's contract provided that "[o]n the basis of his on-site observations as an architect, he shall endeavor to *guard* the Owner against defects and deficiencies in the Work of the Contractor") (emphasis added).

26. *E.g.*, First Nat. Bank of Akron v. Cann, 503 F. Supp. 419 (N.D. Ohio 1980) ("That exhaustive, continuous on-site inspections were not required, however, does not allow the architect to close his eyes on the construction site, refrain from engaging in any inspection procedure whatsoever, and then disclaim liability for construction defects that even the most perfunctory monitoring would have prevented.").

27. Board of Educ. v. Sargent, Webster, 539 N.Y.S.2d 814 (App. Div. 1989).

Only minor changes were made to this section from the 2007 edition. The word "promptly" was added to the architect's obligation to notify the owner of deviations. The second change repeats the "known deviations" limitation from subpart 1 for consistency and clarified the architect's obligation to notify the owner of schedule deviations. The architect's obligation to report to the owner deviations from the Contract Documents remains unchanged.

Section 4.2.3 reinforces section 4.2.2, and both sections attempt to protect the architect by emphasizing that the contractor is solely responsible for means, methods, and safety on the construction site. Pursuant to this section, the architect will not have any control over the contractor's work.[28] This is an important limitation of liability for the architect. In worker personal injury cases, for example, the language of section 4.2.3 has been held to absolve architects of liability.[29]

In cases involving property damage and building defects, there appear to be two lines of cases. One line of cases holds that the language "the architect will not be responsible for the contractor's failure to perform the work in accordance with the requirements of the Contract Documents" in conjunction with the language "the architect . . . will not be responsible for the acts or omissions of the contractor" is an exculpatory clause absolving the architect from liability as a matter of law.[30] In contrast, the second line of cases holds that where an architect knows, or possibly should have known, of a defect or construction not in compliance with the plans and specifications, the architect can be liable for its failure to inspect reasonably.[31] Where property damage is involved, section 4.2.3 has relieved architects of liability

---

28. Baumeister v. Automated Prods., Inc., 690 N.W.2d 1 (Wis. 2004) (holding that under the terms of the agreement the architect was relieved of liability for construction means and methods); Meridian at Windchime, Inc. v. Earth Tech, Inc., 960 N.E.2d 344 (Mass. App. Ct. 2012) (holding that where the contract language provided that the engineer shall not supervise, direct, or have control over the contractor's work, the engineer had no authority or responsibility for the means and methods of construction selected by the contractor).

29. Hanna v. Huer, Johns, Neel, Rivers & Webb, 662 P.2d 243 (Kan. 1983) (holding that the architect's contract with the owner did not include responsibility for safety procedures on the jobsite, and the architect did not assume those responsibilities outside the duties imposed by the contract; further, the contract provided that the general contractor would assume such duties and therefore the architect was not responsible for the injuries suffered by the plaintiffs); Case v. Midwest Mechanical Contractors, 876 S.W.2d 51 (Mo. Ct. App. 1994) (holding that where the architect was responsible for reviewing the safety program, but the responsibility for the safety of those on the site was specifically placed on the contractor, the architect was not liable for the injured worker); Romero v. Parkhill, Smith & Cooper, Inc., 881 S.W.2d 522 (Tex. App. 1994) (holding that under the contract, engineering firm had no control over job site safety and was not responsible for injuries suffered by employee of masonry subcontractor); Kelly v. Northwest Community Hosp., 384 N.E.2d 102 (Ill. App. Ct. 1978) (holding that absent a provision in the contract, an architect cannot be charged with a duty to provide safety precautions at the job site); Fisher v. M. Spinelli & Sons Co., No. 97-4016, 1999 WL 165674, 1999 Mass. Super. LEXIS 60 (Mass. Super. Ct. Feb. 9, 1999) (finding that pursuant to the terms of the owner-architect agreement, the architect was not responsible for safety in connection with the work; safety was the sole responsibility of the contractor and therefore the architect was not liable for injuries sustained by the subcontractor's employee); Hobson v. Waggoner Eng'g, Inc. 878 So. 2d 68 (Miss. Ct. App. 2003) (holding that the engineer's responsibility was to assure that the contractor delivered work consistent with the plans and specifications, and that the contractor alone had responsibility for means, methods, and safety on the project); Black + Vernooy Architects v. Smith, 346 S.W.3d 877 (Tex. App. 2011) (finding no third-party beneficiary liability of the architect where guests of the homeowner were injured when a balcony failed).

30. *See* Moundsview Indep. Sch. Dist. No. 621 v. Buetow & Assocs., 253 N.W.2d 836 (Minn. 1977); Mayor & City Council of the City of Columbus v. Clark-Dietz & Associates-Engineers, Inc., 550 F. Supp. 610 (N.D. Miss. 1982); Shepard v. City of Palatka, 414 So. 2d 1077 (Fla. Dist. Ct. App. 1981); Vonasek v. Hirsch & Stevens, Inc., 221 N.W.2d 815 (Wis. 1974); J&J Electric, Inc. v. Gilbert H. Moen Co., 516 P.2d 217 (Wash. Ct. App. 1973). *Contra* Hunt v. Ellisor & Tanner, Inc., 739 S.W.2d 933 (Tex. App. 1987) (refusing to follow the decision of the Minnesota Supreme Court in Moundsview Indep. Sch. Dist. No. 621, 253 N.W.2d 836; and Nicholson v. Turner/Cargile, 669 N.E.2d 529 (Ohio Ct. App. 1995) (holding that the provision of the contract shielded the engineering firm from liability to the employee of the subcontractor for any damages arising from the negligent construction performed by the subcontractor)).

31. Watson, Watson, Rutland/Architects, Inc. v. Montgomery County Bd. of Educ., 559 So. 2d 168 (Ala. 1990); Board of Educ. of Hudson City School Dist. v. Sargent, Webster, Crenshaw & Folley, 539 N.Y.S.2d 814 (App. Div. 1989); Young v. City of Plaquemine, 818 So. 2d 892 (La. Ct. App. 2002) (reversing grant of summary judgment to the architect where subcontractor painters were injured due to lead paint and the architect had a duty in the absence of test results to have tests performed especially since the architect was renovating an old building).

for contractor's negligence.[32] However, where a similar exculpatory provision was included in a contract excusing the architect from responsibility for means, methods, sequencing, and safety of the project, at least one court has held that such language does "not immunize [the architect] from liability flowing from a breach of its own duties to the [owner]."[33]

Even where exhaustive, continuous on-site inspections are not required under the contract, architects need to be aware that there is a legal duty to notify the owner of a known defect. "[A]n architect cannot close his eyes on the construction site and refuse to engage in any inspection procedure whatsoever and then disclaim liability for construction defects that even the most perfunctory monitoring would have been [sic] prevented."[34]

Architects should also be aware that if they allow the contractor to deviate from the plans and specifications and that deviation results in an injury to a third party, section 4.2.3 may not provide protection.[35] Further, if the architect is shown to have joint control of the work with the contractor, the architect may be liable to a third party for injuries. In *Mellon v. Shriners Hosp. for Children*,[36] the architect was responsible for designing grates to cover an opening in the sidewalk. The architect was also responsible for designing how the grates were to be seated in the sidewalk and for the design of the height of the surrounding masonry. The grates were not installed as designed. They were installed higher than the surrounding masonry, therefore creating a trip hazard. The contract contained language similar to that in sections 4.2.1, 4.2.2, 4.2.3, and 4.2.6 of the A201. A person was injured by tripping over the grate extending above the walking surface. The court denied the architect's motion for summary judgment and held that the plaintiff showed enough involvement and control of the work by the architect to impose on the architect a reasonable duty of care to prevent injuries to persons on the site such as the plaintiff. The court held "if a party retains the right of control or joint control of the work, that party must exercise that control with reasonable care for the safety of others."[37]

§ 4.2.4 COMMUNICATIONS ~~FACILITATING CONTRACT ADMINISTRATION~~

~~Except as otherwise provided in the Contract Documents or when direct communications have been specially authorized, the~~ The Owner and Contractor shall ~~endeavor~~include the Architect in all communications that relate to ~~communicate with each other through~~or affect the Architect's services or professional responsibilities. The Owner shall

32. *Moundsview Indep. Sch. Dist. No. 621*, 253 N.W.2d 836 (finding the architect not liable for the contractor's failure to follow plans and specifications that resulted in a portion of the roof blowing off, even though the architect made 90 site visits but did not observe the omission of the connections by the contractor); Weill Constr. Co. v. Thibodeaux, 491 So. 2d 166 (La. Ct. App. 1986) (holding the architect is not liable for damages caused to the project under a contract provision modeled after the standard AIA clause, where the architect was required to make periodic site visits and had no duty to supervise construction); Jewish Bd. Of Guardians v. Grumman Allied Indus., 464 N.Y.S.2d 778 (App. Div. 1983) (holding that the general contractor had a duty to protect prefab units from storm damage during construction and that the architect had no responsibility for this damage and no duty to supervise as the architect was not responsible for means and methods).

33. Leeward Constr., Inc. v. Sullivan West Central School Dist., No. 05-CV-8384, 2010 WL 1993842 (S.D.N.Y. Mar. 10, 2010) (citing Diocese of Rochester, New York v. R–Monde Contractors, Inc., 562 N.Y.S.2d 593, 596 (Sup. Ct. 1989)).

34. Watson, Watson, Rutland/Architects, Inc. v. Montgomery County Bd. of Educ., 559 So. 2d 168, 174 (Ala. 1990).

35. Campbell v. The Daimler Group, Inc., 686 N.E.2d 337 (Ohio Ct. App. 1996) (holding that the trial court erred in granting summary judgment to the engineering firm where the engineer approved a substitution to the anchor bolts and a third party was fatally injured).

36. No. 98643, 2007 WL 1977968 (Mass. Super. Ct. June 26, 2007).

37. *Id.* at *2.

promptly notify the Architect ~~about matters arising out~~ of ~~or~~ the substance of any direct communications between the Owner and the Contractor otherwise relating to the ~~Contract~~ Project. Communications by and with the Architect's consultants shall be through the Architect. Communications by and with Subcontractors and ~~material~~ suppliers shall be through the Contractor. Communications by and with ~~separate contractors~~Separate Contractors shall be through the Owner. The Contract Documents may specify other communication protocols.

This section has been modified significantly. The goal of these changes is to allow direct communications between the contractor and the owner while balancing the need of the architect to be aware of communications affecting the architect or the contract documents. This section retains communications to the architect's consultants through the architect, and similarly, communications to the subcontractors through the contractor.

Section 4.2.4 was revised in recognition that certain situations will exist during the life of a project in which direct communication between the owner and contractor makes sense. These revisions permit communication between the owner and the contractor as long as the owner and contractor include the architect in any communication that affects the architect's services or responsibilities. For other communication between the owner and the contractor, the owner is responsible for notifying the architect of the substance of such communication.

Orderly communication can be challenging to create and maintain on projects. Depending on the diligence of the parties accurately identifying the communications that affect the architect's services or responsibilities, it would seem that communication on the project could easily result in errors and miscommunication between the parties. The 2007 edition of section 4.2.4 directed the chain of communications through the parties most responsible for the subject of the communication, such as contractor to subcontractor, and architect (as the owner's agent during construction) to contractor.

It is advisable that all parties to the contract agree on matters specific to the project where the owner may have direct communication with the contractor, and from whom any direction given to the contractor should come. The 2017 edition recognizes this and added the last sentence to this section, that the parties may specify other communication protocols in the contract documents. As a best practice, the owner should carefully evaluate the role of the architect and ensure that communication protocols incorporated into the contract documents are designed consistent with the architect's role in contract administration.

§ 4.2.5 Based on the Architect's evaluations of the Contractor's Applications for Payment, the Architect will review and certify the amounts due the Contractor and will issue Certificates for Payment in such amounts.

Owners typically rely on the architect in processing contractor payment requests to identify any problems in progress payments; for example, billing ahead of the schedule of values, or billing for materials off-site. Owners rely on architects to resolve these potential issues with the contractor prior to recommending payment of the application for payment by the owner.

Courts recognize a duty of care on the part of architects certifying applications for payment.[38] Courts have held the following:

- Architect held liable for negligent certification of work performed by a contractor.[39]
- Architect liable in negligence for prematurely releasing retainage funds.[40]
- Architect has a duty to exercise reasonable care to see that work is complete before certifying applications for payment.[41]
- Where architect negligently certifies completion of defective or incomplete work the architect breaches the duty to exercise reasonable care.[42]
- Architect has a duty to ensure that general contractor has secured a payment bond on a public project before certifying payments.[43]
- Architect owes a duty of care to the general contractor's surety; and an architect's negligence in certifying applications for payment to the detriment of the surety are a breach of that duty.[44]
- Architect may be liable in negligence to the contractor's surety for approving pay applications in excess of amounts due the contractor.[45]

It is important to recognize that the architect's duties under this section are to use reasonable care in protecting the owner against payment to the contractor for work not performed or materials not delivered, but that those obligations do not make the architect a guarantor of the contractor's work.[46] This requires the architect to take an active role in observing the progress on the project site sufficient to evaluate whether the application for payment correctly reflects the amount of work actually completed during the pay period.

> **§ 4.2.6** The Architect has authority to reject Work that does not conform to the Contract Documents. Whenever the Architect considers it necessary or advisable, the Architect will have authority to require inspection or testing of the Work in accordance with Sections 13.~~5~~4.2 and 13.~~5~~4.3, whether or not ~~such~~the Work is fabricated, installed or completed. However, neither this authority of the Architect nor a decision made in good faith either to exercise or not to exercise such authority shall give rise to a duty or responsibility of the Architect to the Contractor, Subcontractors, ~~material and equipment~~ suppliers, their agents or employees, or other persons or entities performing portions of the Work.

Only minor changes have been made to this section with regard to references to other sections of the A201, and the qualifying words "material and equipment" have been deleted from "suppliers."

38. Aetna Ins. Co. v. Hellmuth, Obata, and Kassabaum, Inc., 392 F.2d. 472 (8th Cir. 1968) (holding that where the surety ended up paying for duplicate work due to the negligence of the architect in its duties on the construction site, the architect was liable to the surety regardless of lack of privity).

39. Peerless Insurance Co. v. Cerny & Assoc., Inc., 199 F. Supp. 951 (D. Minn. 1961).

40. State *ex rel.* Nat'l Surety Corp. v. Malvaney, 72 So. 2d 424 (Miss. 1954).

41. Willner v. Woodward, 109 S.E.2d 132 (Va. 1959); Auto-Owners Ins. Co. v. Mid-American Piping, Inc., No. 4:07-cv-00394, 2007 WL 2885309 (E.D. Mo. Sept. 27, 2007); Lyndon Property Ins. Co. v. Duke Levy & Assocs., LLC, 475 F.3d 268 (5th Cir. 2007).

42. City of Mound Bayou v. Roy Collins Constr. Co., Inc., 499 So. 2d 1354 (Miss. 1986).

43. Boren v. Thompson & Assocs., 999 P.2d 438 (Okla. 2000).

44. *In re* Designed Ventures, Inc., 132 B.R. 677 (D.R.I. 1991); Sweeney Company of Maryland v. Engineers–Constructors, Inc., 823 F.2d 805, 808 (4th Cir.1987) ("The certification of progress payments cannot be approached as a 'rubberstamp' process without creating exposure to negligence claims."); Peerless Insurance Co. v. Cerny & Associates, 199 F. Supp. 951 (D. Minn. 1961) (holding an architect liable to surety for negligent release of retainage).

45. Westerhold v. Carroll, 419 S.W.2d 73 (Mo. 1967) (if an architect wrongfully certifies a contractor's application for payment in an amount over which the construction has been completed, it puts the surety at risk if the contractor should default because the contract balance will not be sufficient to complete the project).

46. Roland A. Wilson v. Forty-O-Four Grand Corp., 246 N.W.2d 922 (Iowa 1976); Lee County v. Southern Water Contractors, Inc., 298 So. 2d 518 (Fla. Dist. Ct. App. 1974).

This section vests in the architect the authority to reject nonconforming work. The architect's authority to reject nonconforming work does not equate to authority to stop the work.[47] Only the owner has the right to stop work.[48] The architect is in the best position to determine if work conforms to the contract documents. Some jurisdictions hold that if the architect is designated as the judge of acceptable work, then the parties are bound by the architect's decision, in the absence of fraud or gross mistake.[49] While this section states that the architect has *authority* to reject nonconforming work, architects should be aware that some jurisdictions may consider this a *duty* of the architect to reject nonconforming work.[50] The architect's duty to reject nonconforming work under the limited responsibilities for administration of the contract set forth in section 4.2.2—to make periodic site visits to become generally familiar with the progress of the work—is limited to known defects in the construction.[51]

The architect may reject nonconforming work at any time as the project proceeds. There is no time limitation in this section. A failure of the architect to reject nonconforming work is not a waiver or acceptance of the nonconforming work.[52]

§ 4.2.7 The Architect will review and approve, or take other appropriate action upon, the Contractor's submittals such as Shop Drawings, Product Data, and Samples, but only for the limited purpose of checking for conformance with information given and the design concept expressed in the Contract Documents. The Architect's action will be taken in accordance with the submittal schedule approved by the Architect or, in the absence of an approved submittal schedule, with reasonable promptness while allowing sufficient time in the Architect's professional judgment to permit adequate review. Review

---

47. Fruzyna v. Walter C. Carlson Assocs., 398 N.E.2d 60, 66 (Ill. App. Ct. 1979) ("The right to stop work and the power to reject nonconforming work are not equivalent"). *See also* McKean v. Yates Eng'g Corp., 210 So. 3d 1037 (Miss. Ct. App. 2015) (noting that the architect's authority to reject nonconforming work does not include the authority to stop work), *aff'd*, 200 So. 3d 431 (Miss. 2016).

48. *See* AIA A201 § 2.4; Fruzyna v. Walter C. Carlson Assocs., 398 N.E.2d 60, 66 (Ill. App. Ct. 1979); *McKean*, 210 So. 3d 1037 (Miss. Ct. App. 2015).

49. Finish Line v. J.F. Pate & Assoc's. Contr., Inc., 90 So. 3d 749 (Ala. Civ. App. 2012) (holding that where the architect rejected the tile, the tile supplier could not recover the cost of the tile supplied for the project in the absence of a showing of fraud, bad faith, or gross mistake by the architect in rejecting it); Welborn Plumbing & Heating Co. v. Randolph County Bd. of Educ., 150 S.E.2d 65 (N.C. 1966). *See also* Bieman & Rowell Co. v. Donohoe Cos., 556 S.E.2d 1 (N.C. Ct. App. 2001) (citing Bolton Corp. v. T.A. Loving Co., 380 S.E.2d 796 (N.C. Ct. App. 1989) (quoting Barnes Constr. Co. v. Washington Twp., 184 N.E.2d 763 (Ind. Ct. App. 1962) ("When an architect is vested with the authority to render judgment on a contractor's performance, the determination is prima facie correct, and the other parties have the burden of proving fraud or mistake."))); Tribble & Stephens Co. v. RGM Constructors, L.P., 154 S.W.3d 639 (Tex. App. 2004) (upholding subcontract clause stating that final approval of the work is subject to the architect's approval and the subcontractor accepted that risk). *But see* Neighbors Constr. Co., Inc. v. Woodland Park at Soldier Creek, LLC, 284 P.3d 1057 (Kan. Ct App. 2012) (holding that the express language of the contract failed to show that the architect's decision was final as to certifying applications for payment unlike the express language as to the architect's determination of matters relating to aesthetics being final)

50. *See* Fruzyna v. Walter C. Carlson Assocs., 398 N.E.2d 60, 66 (Ill. App. Ct. 1979); *McKean*, 210 So. 3d 1037 ("the architect had the duty to reject nonconforming work"); HBS Contractors, Inc. v. Nat'l Fire Ins. Co. of Hartford, 501 S.E.2d 372 (N.C. Ct. App. 1998).

51. Conopco, Inc. v. Allen & Hoshall, No.01-2784 Ma/A, 2003 WL 25682114, *2 (W.D. Tenn. Aug. 22, 2003) ("The Architect shall visit the site at intervals appropriate to the stage of construction when requested to or as otherwise agreed by the Owner and Architect in writing to become generally familiar with the quality of the Work completed and to determine in general if the Work is being performed in a manner indicating that the Work when completed will be in accordance with Contract Documents. However, the Architect shall not be required to make exhaustive or continuous on-site inspections to check the quality or quantity of the Work."); Watson, Watson, Rutland/Architects, Inc. v. Montgomery County Bd. of Educ., 559 So. 2d 168, 169 (Ala. 1990). ("The Architect shall make periodic visits to the site and as hereinafter defined to familiarize himself generally with the progress and quality of the Work and to determine in general if the Work is proceeding in accordance with the Contract Documents. On the basis of his on-site observations as an Architect, he shall endeavor to guard the Owner against defects and deficiencies in the work of the Contractor. The Architect shall not be required to make continuous on-site inspections to check the quality of the Work. . . .").

52. City of Kimberling City v. Leo Journagan Constr. Co., Inc. v. 337 S.W.3d 48 (Mo. Ct. App. 2011) (holding that where the contract documents gave the architect the authority to reject the work and make minor variations in the work, they are not a waiver of any right the owner possessed); Stevens Constr. Corp. v. Carolina Corp., 217 N.W.2d 291 (Wis. 1974) (concluding that where the architect did not detect latent or hidden defects under an exercise of reasonable diligence in inspecting the project, there was no waiver of the construction defects).

> of such submittals is not conducted for the purpose of determining the accuracy and completeness of other details such as dimensions and quantities, or for substantiating instructions for installation or performance of equipment or systems, all of which remain the responsibility of the Contractor as required by the Contract Documents. The Architect's review of the Contractor's submittals shall not relieve the Contractor of the obligations under Sections 3.3, 3.5, and 3.12. The Architect's review shall not constitute approval of safety precautions or~~, unless otherwise specifically stated by the Architect,~~ of any construction means, methods, techniques, sequences, or procedures. The Architect's approval of a specific item shall not indicate approval of an assembly of which the item is a component.

Submittals have long been a part of construction projects. They typically include drawings, diagrams, layouts, descriptive literature, and performance and test data furnished by the contractor to explain in detail specific portions of the work required by the contract.[53] Submittals and shop drawings are submitted to the architect or engineer for review.[54] Shop drawings and other submittals are not part of the contract documents as defined in section 1.1.1 of the A201.[55]

Shop drawings and other submittals serve at least two purposes: (1) they are submitted by the contractor for approval as to specific details concerning the construction, and (2) they are submitted to the architect for approval with conformance to the design concept.[56]

Generally the architect reviews and approves, or takes other action upon, the shop drawings and submittals for compliance with the design intent of the contract documents.[57] This is a limited review. The contractor is solely responsible for compliance with field measurements and field construction criteria. However, the Eighth Circuit has held that the architect may be liable for negligent review of shop drawings where injury occurs to third persons.[58] The Missouri Court of Appeals has also held the engineer negligent for failure to review shop drawings, resulting in the collapse of two suspended walkways in the Hyatt Regency in Kansas City.[59]

The architect's approval of shop drawings under A201 does not relieve the contractor of responsibility for errors and omissions.[60] Based on field measurements,

53. W.M. Schlosser Co., Inc. v. United States, 767 F.2d 870 (Fed. Cir. 1985); Bates & Assoc., Inc. v. Romei, 426 S.E.2d 919 (Ga. Ct. App. 1993) ("A shop drawing is basically the mechanical process by which the design instructions of the structural engineer are transcribed for use by the fabricating personnel in the steel fabricator's shop."). *See* AIA A201 § 3.12.1 (shop drawings are typically prepared by the subcontractor, submitted to the contractor for his review with compliance with the Contract Documents, approved by the contractor and submitted to the architect).

54. See Section 3.12 above for further discussion of submittals. *See also* Degnon Contracting Co. v. City of New York, 196 N.Y.S. 63 (App. Div. 1922).

55. *See also* AIA A201 § 3.12.4.

56. Waggoner v. W & W Steel Co., 657 P.2d 147 (Okla. 1982).

57. Lutz Engineering Co., Inc. v. Industrial Louvers, Inc., 585 A.2d 631 (R.I. 1991) (concluding that the architect owed no duty of care to the subcontractor and therefore was not negligent in reviewing shop drawings where the architect's contract with the owner required the architect to review shop drawings for general conformance to the design concept).

58. Jaeger v. Henningson, Durham & Richardson, Inc., 714 F.2d 773 (8th Cir. 1983) (affirming architect was liable for injuries to third persons for negligent review of shop drawings where contract documents called for steel stair pans to be 14-gauge and landing pans to be 10-gauge and shop drawings were submitted with 14-gauge for both stair pans and landing pans; and the architect-approved landing pans collapsed on third persons causing injury). *But see* Waggoner v. W & W Steel Co., 657 P.2d 147 (Okla. 1982) (holding that where the contract required the contractor to be responsible for all construction means and required shop drawings to be submitted to the contractor for verification of all field dimensions, and contractor approved them, and architect was required to review and approve for design intent, architect was not liable for injuries to workers).

59. Duncan v. Missouri Board of Architects, 744 S.W.2d 524 (Mo. Ct. App. 1988).

60. Samuel J. Creswell Iron Works, Inc. v. Housing Authority of City of Camden, N.J., 449 F.2d 557 (3d Cir. 1971); D.C. McClain, Inc. v. Arlington County, 452 S.E.2d 659 (Va. 1995). *See* AIA A201 § 3.12.8 ("Contractor shall not be relieved of responsibility for deviations from the requirements of the Contract Documents by the Architect's approval of Shop Drawings").

a contractor submitted shop drawings to the contracting officer on a government contract and the contracting officer approved the drawings. There was a conflict in the structure based on the approved shop drawings. The court held that the government's approval did not relieve the contractor from responsibility for errors and omissions in the shop drawings.[61]

The architect's approval of shop drawings also does not result in a modification of the contract.[62] In the event that the architect's review does not catch a deviation in the shop drawings from the contract documents, the contractor will be responsible for the costs to repair or replace the nonconforming work. Where the architect approved shop drawings for doors showing particle core doors where the contract documents called for solid core doors, and the fire marshal rejected the doors, the court held that the architect had no authority to change the contract and the approval of shop drawings was not a modification of the contract, thereby denying the contractor's claim for extra costs.[63]

The timeliness of submittal reviews can threaten the project schedule. Where an architect caused unreasonable delay in review of shop drawings resulting in delay on the project, the contractor was granted a time extension.[64]

While timely response to submittals is important, if a contractor submits shop drawings that are not required by the contract documents, the architect may return them without action.[65]

> **§ 4.2.8** The Architect will prepare Change Orders and Construction Change Directives, and may ~~authorize~~order minor changes in the Work as provided in Section 7.4. The Architect will investigate and make determinations and recommendations regarding concealed and unknown conditions as provided in Section 3.7.4.

Change order (commonly referred to as CO) and construction change directive (commonly referred to as CCD) are defined in the A201 in sections 7.2.1 and 7.3.1, respectively. The issuance of either document formalizes changes to the contractor's work. The change order process involves mutual agreement between the owner and the contractor concerning scope changes, including any applicable cost and schedule adjustments. The change order process may involve the contractor receiving bids for work from subcontractors through a request for change order proposal (RCOP), and once such RCOP is approved a change order would be generated.[66] A construction change directive differs in that it is a unilateral communication from the architect to the contractor directing the contractor to change the scope of the work under the contract.[67]

---

61. McDevitt Mech. Contractors, Inc. v. United States, 21 Cl. Ct. 616, 620 (1990) (stating that "no 'reasonable and prudent' contractor in McDevitt's situation would have understood the government's approval of the drawings to absolve plaintiff from the obligation of determining the proper grades and elevations where the inlet pipe and the overhead crane would meet.").

62. Samuel J. Creswell Iron Works, Inc. v. Housing Authority of City of Camden, N.J., 449 F.2d 557 (3d Cir. 1971).

63. Fauss Constr., Inc. v. City of Hooper, 249 N.W.2d 478 (Neb. 1977).

64. Fortec Constructors v. United States, 8 Cl. Ct. 490 (1985) (granting contractor a 101-day time extension for the government's delay in disapproval of original shop drawings per the contract documents and correction of a design defect); Sterling Millwrights, Inc. v. United States, 26 Cl. Ct. 49 (1992) (holding that the government caused months of delay by its failure to review shop drawings within the period of time set forth in the contract and by further ordering an extension of the review process for shop drawings).

65. *See* AIA A201 § 3.12.4.

66. Building Constr. Enterprises, Inc. v. Public Building Com'n of Johnson County, 296 P.3d 1139 (Kan. Ct. App. 2013).

67. *Id.* ("A CCD is a communication from the project architect that asks the general contractor to alter or change the original scope of work under the contract.").

Under section 4.2.8, the architect will prepare change orders[68] and construction change directives.[69] It is often customary, however, for the contractor to prepare and submit change order proposals to the architect for review and approval and recommendation to the owner. If the change order proposal is accepted, then the scope, cost, or time extension of the change order is added to the original contract and the contractor is permitted to bill for the change order work through its application for payment.

Construction change directives are typically used where there is not an agreement between the owner, architect, and contractor as to the adjustment of either the contract amount or contract time, or both, for the change to the work.[70] Construction change directives are often invoiced on a time and material basis and thereafter converted to change orders.

The architect alone may order minor changes in the work that are consistent with the design intent and do not affect the cost or time of completion of the project.[71] These minor changes are usually issued in an Architect's Supplemental Instruction, more commonly known as ASI.[72] It is advisable that the architect verify with the contractor that the minor change in the work will not affect either the contract amount or the contract time, as the architect cannot approve changes in contract amount or contract time without the owner's approval.[73]

It is important to note here that while the architect may administratively prepare the change order or construction change directive documents, the architect does not have authority to issue these without the owner's approval.[74]

Section 4.2.8 also places on the architect the responsibility to investigate concealed or unknown conditions discovered at the project site by the contractor. The contractor is responsible for notifying the architect and owner of the concealed or unknown condition.[75] Once notified, the architect is responsible for promptly investigating the condition and making recommendations to the owner.[76]

> **§ 4.2.9** The Architect will conduct inspections to determine the date or dates of Substantial Completion and the date of final completion; issue Certificates of Substantial Completion pursuant to Section 9.8; receive and forward to the Owner, for the Owner's review and records, written warranties and related documents required by the Contract and assembled by the Contractor pursuant to Section 9.10; and issue a final Certificate for Payment pursuant to Section 9.10.

As for inspection versus observation, under the terms of the A201, the architect conducts an "inspection" of the project on only two occasions; all other site visits are "observations." The distinction between inspection and observation is important

68. *See* AIA Document G701—Change Order.
69. *See* AIA Document G714—Construction Change Directive.
70. *See* A201 § 7.3.1; typically used in the absence of total agreement on the terms of a change order.
71. A201 § 7.4.
72. *See* AIA Document G710—Architect's Supplemental Instructions.
73. A201 § 7.1.2.
74. *See* A201 §§ 7.2.1, 7.3.1.
75. A201 § 3.7.4. *See also* Robert W. Carlstrom Co., Inc. v. German Evangelical Lutheran St. Paul's Congregation of Unaltered Augsburg Confession at Jordan, 662 N.W.2d 168 (Minn. Ct. App. 2003) (discussing the notice requirements from the contractor to the architect and the architect to the owner under the 1987 AIA A201 contract).
76. A201 § 3.7.4.

in defining the architect's role on the construction site.[77] In one case where the contract terms were similar to AIA A201 sections 4.2.2, 4.2.3, and 4.2.9, but also included: "[a]ll areas to receive concrete shall be inspected and approved by the Architect-Engineer prior to any pouring of concrete," the court found the architect liable because he did not inspect the construction properly.[78]

The date of substantial completion is significant on a project. This date typically establishes when responsibilities transfer to the owner from the contractor for security, maintenance, heat, utilities, damage to the work, insurance, and potential assessment of liquidated damages,[79] and fixes the time within which the contractor will complete all items on the punch list.[80] Substantial completion is a term of art in the construction industry. Substantial completion does not mean substantial compliance with the contract or a lack of defects in the work.[81] Substantial completion occurs when the architect certifies to the owner that construction is sufficiently complete so the owner may occupy or utilize the constructed structure (or portion of the structure) for its intended purpose. "Substantial completion does not require exact performance of every detail; rather, the test applied by the architect 'is one of function and is concerned with whether the owner can make use of the work as intended.'"[82]

"Architects' certificates are and for many years have been commonly used in the construction trades. And they may not be impeached absent 'evidence . . . of such a character as to show that the action of the architect was grossly and palpably perverse, oppressive and unjust'—evidence that is 'clear and convincing and point[s] with reasonable certainty to mistake, fraud, or collusion.'"[83] In other words, absent bad faith, fraud, mistake, arbitrariness, or caprice, the architect's issuance of the certificate of substantial completion is conclusive that the work was performed according to the terms of the contract.[84]

Substantial completion also sets a bright line date on which the limitations periods begin to run in most jurisdictions.[85] Many states have enacted statutes of repose that commence at the date of substantial completion.[86]

§ 4.2.10 If the Owner and Architect agree, the Architect will provide one or more ~~project~~Project representatives to assist in carrying out the Architect's responsibilities at the site. ~~The~~The Owner shall notify the Contractor of any change in the duties, responsibilities and limitations of authority of ~~such project~~the Project representatives ~~shall be as set forth in an exhibit to be incorporated in the Contract Documents~~.

This section has been modified to provide that the owner shall notify the contractor of any changes in the duties, responsibilities, and limitations of authority of the project representative. Previously this section provided that the duties, responsibilities, and limitations of authority of the project representative were to be set forth

77. *See* § 4.2.2 ("However, the Architect will not be required to make exhaustive or continuous on-site inspections to check the quality or quantity of the Work.").

78. Dickerson Constr. Co. v. Process Eng'g, 341 So. 2d 646 (Miss. 1977).

79. 2300 Pennsylvania Ave., LLC v. Harkins Builders, Inc., 513 F. App'x 273 (4th Cir. 2013).

80. Trinity Church v. Lawson-Bell, 925 A.2d 720, 730 (N.J. Super. Ct. App. Div. 2007) ("The punch list is a 'final list of small items requiring completion, or finishing, corrective or remedial work.'").

81. *Id.*

82. KBE Bldg. Corp. v. Construction Services of NC, Inc., 2015 WL 5813848, *19 (Md. Oct. 5, 2015).

83. Bauer v. Klement, 458 N.W.2d 389 (Wis. 1990) (quoting J.G. Wagner Co. v. Cawker, 88 N.W. 599, 603 (Wis. 1902)).

84. *See* Blecick v. School Dist. No. 18, 406 P.2d 750 (1965) (for the proposition that, absent bad faith, the architect is not liable to the contractor for failing to issue a certificate of substantial completion).

85. *Trinity Church*, 925 A.2d 720.

86. Russo Farms, Inc. v. Vineland Bd. of Ed., 675 A.2d 1077 (N.J. 1996).

in an exhibit incorporated into the contract documents. The edits to this section have alleviated the need to prepare an exhibit setting forth the duties of the project representative because the owner need only notify the contractor of any "changes" to those duties that may be authorized by the owner over the life of the project. However, this change creates the need for timely notice by the owner regarding any future change to the project representative's duties.

A good practice tip is to specify what the duties of the project representative are and whether the project representative has the same authority as the architect on the project.

If the owner and architect agree to provide a full-time field representative, the field representative must be careful only to observe the progress on the project and not become a "clerk-of-the-works" and assume liability for means, methods, or safety.[87]

§ 4.2.11 The Architect will interpret and decide matters concerning performance under, and requirements of, the Contract Documents on written request of either the Owner or Contractor. The Architect's response to such requests will be made in writing within any time limits agreed upon or otherwise with reasonable promptness.

The architect is the interpreter of the requirements of the contract documents and the judge of the performance thereunder. Any and all provisions of the contract documents are subject to interpretation and clarification by the architect upon request of the owner or contractor. The architect is to render interpretations as necessary for the proper execution of the work.

The parties may agree to a time limit within which the architect will make the interpretation; however, if no time is set in the contract documents, the architect must respond in writing within a reasonable time. An architect's failure to respond or its negligence in interpreting the contract documents may lead to the architect's liability to the contractor.[88] When the architect is acting as the interpreter of the contract documents, one court has found that the architect was acting independently and was not an agent of the owner for purposes of vicarious liability.[89]

The architect's role in interpreting contract documents has been held to end at substantial completion[90] or upon the termination of the project.[91]

§ 4.2.12 Interpretations and decisions of the Architect will be consistent with the intent of, and reasonably inferable from, the Contract Documents and will be in writing or in the form of drawings. When making such interpretations and decisions, the Architect will endeavor to secure faithful performance by both Owner and Contractor, will not show partiality to either, and will not be liable for results of interpretations or decisions rendered in good faith.

87. See sections 4.2.2 and 4.2.3 for the architect's responsibilities on the project site.

88. Prichard Bros., Inc. v. Grady Co., 436 N.W.2d 460 (Minn. Ct. App. 1989) (stating that the damages were for delays caused by the architect's negligent interpretation of specifications and response to shop drawings).

89. *Id.*

90. County of Rockland v. Primiano Constr. Co., 409 N.E.2d 951 (N.Y. 1980) (concluding that where the parties agreed that disputes relating to the execution of the work or the interpretation of the contract documents were to be referred to the architect for initial decision, and a dispute arose two years after substantial completion, the court held that the architect's jurisdiction to decide the matter was limited to the operational phases of construction).

91. Liebhafsky v. Comstruct Assocs., 466 N.E.2d 844 (N.Y. 1984) (holding that the architect's role as mediator of disputes relating to the execution of the work or the interpretation of the contract documents terminated once the architect was no longer responsible for supervising the contractor's performance, whether the contractor's services have ended because of substantial completion or the contractor was terminated under the contract).

An architect's interpretation of the contract documents must be consistent with the intent of the documents. Where an architect interpreted the contract documents to require a contractor to paint duct work in an unfinished area where there was no such designation on the drawings, the court held that the architect's decision was clearly erroneous and contrary to the terms of the contract.[92] The court found that the architect in this instance ignored the terms of the contract in his interpretation and was acting in excess of his authority.[93] Provided an architect's interpretation of the contract documents is plausible and cannot be characterized as having no rational basis, courts may uphold the interpretation. If the architect is in error in his interpretation of the contract documents, courts are not likely to overturn that decision unless the error is so gross as to amount to bad faith or fraud.[94] In one case where there was an apparent ambiguity in the contract documents, the court held that the architect's interpretation of the ambiguities did not create a new contract as the contractor argued. Instead the court noted: "[c]ontract terms are not changed when the architect decides in the course of the job what a specification means or whether one specification trumps another."[95]

An architect's interpretation of the contract documents must be an impartial decision. The decision must not favor the owner or the contractor. Contractors may believe the architect is partial to the owner in these interpretations and owners may resent the architect siding with the contractor, but the architect's decision must be impartial[96] and reasonably inferable from the contract documents. The architect's decision on interpretation of the contract documents will be given great weight by the courts.[97]

§ 4.2.13 The Architect's decisions on matters relating to aesthetic effect will be final if consistent with the intent expressed in the Contract Documents.

Instruments of service are defined in the A201 as the expression of the tangible and intangible creative work of the architect.[98] The architect is deemed to be the author of this creative work.[99] Section 4.2.13 bestows upon the architect final decision on all matters relating to aesthetics. Aesthetics is not defined in the contract, and therefore, courts have had to define this term, usually in the context of determining whether certain claims were final as decided by the architect or were subject to

92. John W. Johnson, Inc. v. J.A. Jones Constr. Co., 369 F. Supp. 484 (E.D. Va. 1973) (noting that the architect cannot give a decision contrary to trade usage or custom that is part of the contract).

93. *Id.*

94. Ballou v. Basic Constr. Co., 407 F.2d 1137 (4th Cir. 1969) ("The architect's decision to enforce literal compliance with the contract may have reflected poor judgment or excessive rigidity, but there is no evidence or even suggestion that they were motivated by malice or bad faith."); Prichard Bros., Inc. v. Grady Co., 436 N.W.2d 460 (Minn. Ct. App. 1989) (holding that where the architect may have been negligent in his interpretation of specifications and review of shop drawings, his errors of judgment did not rise to the level of bad faith); M&L Bldg. Corp. v. Housing Auth., 646 A.2d 244 (Conn. App. Ct. 1994) (stating that where the architect makes a decision that is not arbitrary or made in bad faith, the court will not substitute its judgment for that of the architect).

95. Ostrow Electr. Co. v. J.L. Marshall & Sons, Inc., 798 N.E.2d 310, 314 (Mass. App. Ct. 2003) (stating that the architect's resolution of the severe ambiguity in the contract documents as to who was responsible for providing speaker backboxes could not support an inference of bad faith).

96. Meco Systems, Inc. v. Dancing Bear Entertainment, Inc., 948 S.W.2d 185 (Mo. Ct. App. 1997) (holding that summary judgment in favor of the architect was inappropriate where issues concerning the architect's partiality and lack of good faith were raised by the contractor and the architect's selective decision-making raised genuine fact issues about its impartiality and good faith).

97. Bolton Corp. v. T.A. Loving Co., 380 S.E.2d 796, 803 (N.C. Ct. App. 1989) ("the architect's decision is prima facie correct, and the burden is upon the other parties to show fraud or mistake"); James A. Cummings, Inc. v. Young, 589 So. 2d 950, 954 (Fla. Dist. Ct. App. 1992) ("When parties to a contract agree by its express terms to be bound to the determination made by an architect, that agreement is binding upon the parties. In the absence of fraud, or such mistake as would amount to fraud, the determination made by the architect shall be final.") (citing Willcox v. Stephenson, 11 So. 2d 659 (Fla. 1892)).

98. AIA A201 § 1.1.7.

99. AIA A201 § 1.5.1.

arbitration. "'Aesthetic effect' is a term of art, not a legal term."[100] The Supreme Court of Mississippi found no evidence of a prevailing industry standard or commonly accepted trade usage of the term "aesthetic effect," and although the term is used in the construction industry, it should be defined on a case-by-case basis.[101] The Arkansas Supreme Court defined "aesthetic" as "pertaining to a sense of the beautiful or to the science of aesthetics."[102]

In another case, the project owner sued the contractor alleging it used substandard materials and exhibited poor workmanship in constructing interior walls. The contractor moved to compel arbitration. The contract's dispute resolution provisions excluded from arbitration any claim involving aesthetics. The court held that the claims were about the finish of the interior walls and the appearance of the building, which were claims involving aesthetic effect, and therefore not arbitrable.[103]

Where claims are a mix of some aesthetic effect and some not, courts may analyze each claim on a case-by-case basis.[104] A Maryland court was faced with a scenario where the architect rejected brickwork on a project as being aesthetically undesirable and, because the contractor refused to replace the brick, the architect refused to certify the contractor's application for payment.[105] The contractor filed a mechanic's lien after the architect's refusal to certify the application for payment. The trial court held that the architect's decision to withhold certification was not subject to judicial review. The Maryland Court of Appeals held that the architect's decision to withhold payment was in part economic and not purely artistic. The architect's decision to reject the brickwork on the grounds of aesthetic effect was not subject to review. The architect had further determined that the entire brickwork would have to be replaced because the contracted-for brick was no longer in production. The architect's determinations of how to correct the defect and the reasonable costs thereof were not final and were reviewable.[106]

Decisions on matters entrusted to the architect are deemed final and conclusive only when it appears from the express terms of the contract that the parties intended the architect's decision to have that effect.[107] Courts have upheld

---

100. Harrison County Commercial Lot, LLC v. H. Gordon Myrick, Inc., 107 So. 3d 943, 959 (Miss. 2013).

101. *Id.*

102. May Constr. Co., Inc. v. Benton Sch. Dist., 895 S.W.2d 521, 523 (Ark. 1995) (citing RANDOM HOUSE DICTIONARY OF THE ENGLISH LANGUAGE (2d ed.1987)).

103. Flandreau Public School District v. G.A. Johnson Constr., Inc., 701 N.W.2d 430 (S.D. 2005). *But see* Noroton Heights Fire Dept. v. Vuono Lione, Inc., 362 A.2d 982 (Conn. 1976) (holding that where the contract terms provided that any decision by the architect regarding aesthetic effect was final, and all other claims were subject to arbitration, and where the specifications required six pounds of green color to be added to each bag of concrete and a dispute arose over the color of the concrete floor, the court held that the whether the contractor fulfilled its obligations under the specifications including the color achieved in the concrete was subject to arbitration, while leaving any question as to the artistic effect to the architect).

104. Harrison County Commercial Lot, LLC v. H. Gordon Myrick, Inc., 107 So. 3d 943 (Miss. 2013) (identifying various claims as aesthetic effect (breach of contract actions for performing work in a sloppy, defective and unworkmanlike manner, punch list items related to aesthetics—paint, texture of finish, negligence claims for negligent construction and unworkmanlike manner) or not aesthetic effect (claims arising out of liens for nonpayment, breach of contract claims for unauthorized changes, failing to complete the work in a timely manner, submitting inflated or incorrect payment applications, negligence claims for deviating from bid documents, negligent misrepresentation claims for submitting inaccurate applications for payment, liquidated damages claims because they relate to timeliness and not aesthetics)).

105. NCS Contractors, Inc. v. Borders, 564 A.2d 408 (Md. 1989) (noting that the contract contained language similar to the language of AIA A201 § 4.2.13). *See also* Bolton Corp. v. T.A. Loving Co., 380 S.E.2d 796 (N.C. Ct. App. 1989) (interpreting a similar provision of the contract and noting that "[f]inality is accorded only to decisions relating to artistic effect and technical matters.").

106. *NCS Contractors, Inc.*, 564 A.2d 408.

107. Neighbors Constr. Co., Inc. v. Woodland Park at Soldier Creek, LLC, 284 P.3d 1057 (Kan. Ct. App. 2012) (holding that the express language of the contract failed to show that the architect's decision was final as to certifying applications for payment unlike the express language as to the architect's determination of matters relating to aesthetics being final); O.K. Johnson Elec., Inc. v. Hess-Martin Corp., 464 P.2d 206 (Kan. Ct. App. 1970) (citing Mercantile Trust Co. v. Hensey, 205 U.S. 298 (1907)); Meco Sys. v. Dancing Bear Entertainment, Inc., 948 S.W.2d 185 (Mo. Ct. App. 1997).

the rights of contracting parties to agree that decisions of an expert, such as an architect, are final.[108]

§ 4.2.14 The Architect will review and respond to requests for information about the Contract Documents. The Architect's response to such requests will be made in writing within any time limits agreed upon or otherwise with reasonable promptness. If appropriate, the Architect will prepare and issue supplemental Drawings and Specifications in response to the requests for information.

The architect is responsible for responding to the contractor's requests for information, commonly referred to as RFIs. Architects must timely respond to RFIs as delay in response may result in delay on the construction site. If delays to the project result in a lawsuit, it is not uncommon for an architect's untimely response or incomplete response to RFIs to be alleged as causes of delay on the project.[109] Requests for information and the architect's responses to them should be in writing.[110] Requests for information are used to clarify questions with the design drawings and specifications. They are used to resolve conflicts and ambiguities in the plans or other contract documents. They are not a process for making revisions to the project.[111] Responses to RFIs may be supplemented by drawings and specifications if necessary to clarify any conflicts or discrepancies in the plans and specifications that were identified by the RFI.

While often responses to RFIs are a clarification for the contractor with respect to the drawings or specifications, it is not unusual that responses to RFIs may result in the need for a change order for a change in the contract amount or the contract time. Architects should be mindful that many contracts include time limitations on contractor's notices for claims to adjust the contract amount or the contract time. These notice requirements are typically concurrent with, or shortly thereafter, the event causing the change in cost or change in time. If a contractor submits requests for a change order or time extension at the end of the project, the contractual notice requirement may not have been met, and the contractor may have waived its request.[112]

---

108. Finish Line v. J.F. Pate & Assoc's. Contr., Inc., 90 So. 3d 749 (Ala. Civ. App. 2012); Biemann & Rowell Co. v. Donohoe Cos., 243, 556 S.E.2d 1, 4 (N.C. Ct. App. 2001) ("[w]hen an architect is vested with the authority to render judgment on a contractor's performance, the determination is prima facie correct, and the other parties have the burden of proving fraud or mistake." (citing *Bolton Corp.*, 380 S.E.2d at 803 (quoting Barnes Constr. Co. v. Washington Twp., 184 N.E.2d 763, 764–65 (Ind. Ct. App. 1962)))), Acmat Corp. v. Daniel O'Connell's Sons, Inc., 455 N.E.2d 652, 656 (Mass. App. Ct. 1983) ("The decision of the architect is final and binding on the parties where he is given that power by the contract, as here, and where his exercise of that power is not arbitrary and capricious."); Saydel Community School Dist. v. Denis Della Vedova, Inc., 735 N.W.2d 202 (Iowa Ct. App. 2007) (upholding the lower court's decision that the architect's decision was final as to removing and replacing the woodwork because it was consistent with the intent expressed in the contract that the architect had final decision on issues of aesthetic effect and the decision was not made in bad faith).

109. *See e.g.*, Miller Bldg. Corp. v. NBBJ North Carolina, Inc., 497 S.E.2d 433 (N.C. Ct. App. 1998) (alleging that the designer did not timely respond to RFIs); Mepco Services, Inc. v. Saddleback Valley Unified School Dist., 117 Cal. Rptr. 3d 494 (Ct. App. 2010) (alleging that the failure of the designer to respond to RFIs caused delay to the project); Guy M. Cooper, Inc. v. East Penn School District, 903 A.2d 608 (Pa. Commw. Ct. 2006) (alleging construction delays caused in part by architect's failure to timely respond to RFIs); First Millennium Constr., LLC v. Parish of Jefferson, 70 So. 3d 942 (La. Ct. App. 2011) (alleging that the architect ignored and failed to adequately respond to RFIs regarding numerous discrepancies in the drawings, ultimately causing the contractor delays and liability for liquidated damages); Diocese of Metuchen v. Prisco & Edwards, AIA, 864 A.2d 1168 (N.J. Super. Ct. App. Div. 2005) (alleging project delayed by inadequate responses to RFIs).

110. *See* AIA Document G716–Request for Information ("RFI").

111. *In re* Electric Machinery Enterprises, Inc., 416 B.R. 801 (M.D. Fla. 2009).

112. Starks Mechanical, Inc. v. New Albany-Floyd County Consol. School Corp., 854 N.E.2d 936 (Ind. Ct. App. 2006).

CHAPTER 5:
Article 5 Subcontractors

Robert J. Dietz
David L. Fine

ARTICLE 5 SUBCONTRACTORS

Article 5 defines the terms "Subcontractor" and "Sub-subcontractor" (see Section 5.1), sets forth the procedure for the award of subcontracts, including the rights of the owner and architect to participate in that process (see Section 5.2), addresses "flow down" requirements (see Section 5.3), and provides the owner the right to take assignment of subcontracts in the event of a default or termination (see Section 5.4).

§ 5.1 DEFINITIONS

> **§ 5.1.1** A Subcontractor is a person or entity who has a direct contract with the Contractor to perform a portion of the Work at the site. The term "Subcontractor" is referred to throughout the Contract Documents as if singular in number and means a Subcontractor or an authorized representative of the Subcontractor. The term "Subcontractor" does not include a ~~separate contractor~~Separate Contractor or the subcontractors of a ~~separate contractor~~Separate Contractor.

Section 5.1.1 was revised to clarify that a "Separate Contractor"[1] is excluded from the definition of a "Subcontractor."

Section 5.1.1 defines the term "Subcontractor" as used throughout the contract documents. Section 5.1.1 continues to limit the definition of a "Subcontractor" to one who has a direct contract with the general contractor and performs work at the project site. A "Subcontractor" does not include material suppliers or those other persons or entities who work off-site.[2] Section 5.1.1 further confirms that a subcontractor is separate and distinct from a contractor as defined in Section 3.1.1.[3]

> **§ 5.1.2** A Sub-subcontractor is a person or entity who has a direct or indirect contract with a Subcontractor to perform a portion of the Work at the site. The term "Sub-subcontractor" is referred to throughout the Contract Documents as if singular in number and means a Sub-subcontractor or an authorized representative of the Sub-subcontractor.

Section 5.1.2 defines the term "Sub-subcontractor" as used throughout the contract documents. As with Section 5.1.1, Section 5.1.2 limits sub-subcontractors to those who actually perform work at the project site, excluding from its definition material suppliers and those who work off-site.

---

1. The term "Separate Contractor" is defined in Section 6.1.1.

2. *See* Aetna Cas. & Sur. v. Canam Steel Corp., 794 P.2d 1077 (Colo. Ct. App. 1990) (excluding steel supplier from subrogation waiver provision of Article 11.3.7); Waco Scaffolding Co. v. National Union Fire Ins. Co. of Pittsburg, PA, No. 74656, 1999 WL 980629 (Ohio Ct. App. Oct. 28, 1999) ("With regard to construction contracts, '[s]ubcontractor has a well[-]defined meaning and as used in its technical sense it means one who takes from the principal contractor a specific part of the work, and does not include laborers or materialmen.'"); Higgins Erectors & Haulers, Inc. v. Niagara Frontier Transp. Auth., 529 N.Y.S.2d 654, 655 (App. Div. 1988) (same); Luczak Bros. v. Generes, 451 N.E.2d 1267, 1277 (Ill. App. Ct. 1983) (comparing definition of Contractor and Subcontractor under Section 5.1.1); Baldwin Co. v. Weyland Mach. Shop, Inc., 685 S.W.2d 537, 538 (Ark. Ct. App. 1985) (holding that steel supplier does not fall within the definition of subcontractor in § 5.1.1); Vulcraft v. Midtown Bus. Park, Ltd., 800 P.2d 195 (N.M. 1990) (analyzing whether supplier of materials was subcontractor or materialman).

3. *See* Indiana Erectors, Inc. v. Trustees of Indiana Univ., 686 N.E.2d 878, 881 (Ind. Ct. App. 1997).

§ 5.2 AWARD OF SUBCONTRACTS AND OTHER CONTRACTS FOR PORTIONS OF THE WORK

> **§ 5.2.1** Unless otherwise stated in the Contract Documents ~~or the bidding requirements~~, the Contractor, as soon as practicable after award of the Contract, shall ~~furnish in writing to~~notify the Owner ~~through the~~and Architect ~~the names~~ of the persons or entities ~~(~~proposed for each principal portion of the Work, including those who are to furnish materials or equipment fabricated to a special design~~) proposed for each principal portion~~. Within 14 days of receipt of the ~~Work. The~~information, the Architect may ~~reply within 14 days to~~ notify the Contractor ~~in writing stating (1)~~ whether the Owner or the Architect (1) has reasonable objection to any such proposed person or entity or (2) ~~that the Architect~~ requires additional time for review. Failure of the ~~Owner or~~ Architect to ~~reply~~provide notice within the 14-day period shall constitute notice of no reasonable objection.

Section 5.2.1 addresses the contractor's obligation to provide the owner, through the architect, the names of all persons or entities proposed for each principal portion of the work, whether they are subcontractors, sub-subcontractors, or others, including those furnishing materials or equipment fabricated to a special design. Prior to the 2007 modification, the architect, after receiving a written list of proposed subcontractors, was required to "promptly" object. This was replaced with a 14-day turnaround period for the owner or architect to object on reasonable grounds, or state that it requires "additional time for review." The 2017 modification keeps the 14-day period intact, but clarifies that the period commences upon the architect's "receipt of the information." As explained in the analysis of Section 1.6 in Chapter 1, the 2017 modification to the A201 changes the definition of notice.

An owner has a material interest to ensure its contractor selects qualified and reliable subcontractors.[4] Courts vary significantly when determining the extent of the materiality of the term requiring that the contractor submit subcontractor selections, and the consequences for the failure to comply with that term.[5]

Under this section, the architect and the owner evaluate the suitability of contractors and subcontractors. In doing so, the architect may enjoy a "qualified privilege." *Kecko Piping Co., Inc. v. Town of Monroe*[6] is illustrative. As part of its contractual duties in connection with a school renovation project, the architect was to assist the town in obtaining bids or negotiated proposals and in awarding and preparing construction contracts.[7] The low bidder included a proposal from a

---

4. *See* Williard, Inc. v. Powertherm Corp., 444 A.2d 93, 97 (Pa. 1982) (stating the construction "contract is replete with provisions manifesting the owner's concern with the selection, supervision and welfare of the subcontractors").

5. *See* Gaglioti Contracting, Inc. v. City of Hoboken, 704 A.2d 1301 (N.J. Super. Ct. App. Div. 1997) (failure to include list of subcontractors at time of bid was a violation of the state's procurement statute and constituted a "material defect" constraining the City to reject the bid); Chicago College of Osteopathic Medicine v. George A. Fuller Co., 719 F.2d 1335 (7th Cir. 1983) (question of fact whether owner waived requirement for contractor to furnish list of subcontractors within 45 days, where owner proceeded with project for two years and subsequently terminated contractor without identifying failure to list subcontractors as cause of default); Hanson Southwest Corp. v. Dal-Mac Constr. Co., 554 S.W.2d 712 (Tex. Civ. App. 1977) (approval of major subcontractors was not a "condition precedent" to payment from owner). The *Hanson* provision reads as follows: "Prior to letting of any major subcontract or the granting of any purchase order, the Contractor will submit the same for Owner's review. . . . Owner may recommend a substitution for any subcontractor or purchase order, which recommendation shall not be binding upon the Contractor." Thus, the *Hanson* provision differs significantly from § 5.2 in that the AIA version does not give the Contractor discretion to disregard the owner's objection. *See* § 5.2.2 ("[t]he Contractor *shall not* contract with a proposed person or entity to whom the Owner or Architect has made reasonable and timely objection") (emphasis added). Query whether the *Hanson* court would have found differently had the operable contract included the then-current version of the A201, i.e., is a contractor's compliance with § 5.2.2 a "condition precedent" to payment from the owner?

6. 374 A.2d 179 (Conn. 1977).

7. *Id.* at 180.

mechanical subcontractor as part of its bid package.[8] Shortly after the bid opening, the architect wrote to the mechanical subcontractor requesting certain information so the architect could assess the subcontractor's qualifications.[9] The architect never received the requested information. The architect also had evidence the subcontractor was overextended with other work. In fact, the contractor itself expressed to the owner and architect concern that this subcontractor would be unable to perform.[10] As a result, the architect advised the town it would be in the town's best interest to find another subcontractor. The subcontractor filed an action against the architect for tortious interference with contract, among other counts. After a bench trial, the court found for the architect, which judgment was affirmed on appeal. The court noted an absence of any fraud, misrepresentation, or intimidation on the part of the architect. The architect did not exceed the limits of its "qualified privilege."[11]

Kecko Piping serves as a cautionary tale for individuals and entities exercising the powers set forth in Section 5.2. While architects may enjoy a qualified privilege in advising owners as to the suitability of subcontractors, that advice could subject the architect to liability if not made with prudence. Owners too could be exposed if they rely unreasonably on an architect's advice. On the other hand, subcontractors must acknowledge the right of reasonable refusal authorized by Section 5.2, and should exercise discretion when determining whether to pursue a claim in the event a subcontractor is rejected. Indeed, under Section 5.2, an owner has the right to reject a proposed subcontractor *even where the subcontractor was prequalified by the owner to bid the project.* In *North American Mechanical, Inc. v. Diocese of Madison*,[12] the architect prepared a list of prequalified general contractors and subcontractors that were invited to bid on a major renovation project.[13] The subcontractors submitted their bids to the general contractors, who incorporated the subcontractors' bids into their bids. The project manual provided that the owner could raise reasonable objections to particular subcontractors even after the general construction contract was signed.[14] After the contract was awarded, the owner objected to the HVAC subcontractor whose bid had been carried by the low general contractor, on the grounds that the HVAC subcontractor was nonunion. The rejected subcontractor filed suit, alleging misrepresentation and promissory estoppel. The subcontractor argued that the prequalified designation constituted a representation that the owner would not later object. The court held that Section 5.2 gave the owner the opportunity to exercise its right to disqualify a subcontractor after the subcontract was chosen by the general contractor, and that the owner had no duty to disclose its preference for union contractors during the bid process.[15]

> **§ 5.2.2** The Contractor shall not contract with a proposed person or entity to whom the Owner or Architect has made reasonable and timely objection. The Contractor shall not be required to contract with anyone to whom the Contractor has made reasonable objection.

Section 5.2.2 affords both the contractor and the owner, through the architect, the right to make a reasonable objection to a proposed subcontractor. However, if

8. *Id.* at 181.
9. *Id.*
10. *Id.* at 181.
11. *Id.* at 181–82.
12. 600 N.W.2d 55 (Wis. Ct. App. 1999) (unpublished opinion).
13. *Id.* at *1.
14. *Id.*
15. *Id.* at *4–6.

the bidding documents require the contractor to work with a specific subcontractor, the fact of the contractor's bid is deemed an implicit waiver by the contractor to later pose an objection under § 5.2.2.[16]

> **§ 5.2.3** If the Owner or Architect has reasonable objection to a person or entity proposed by the Contractor, the Contractor shall propose another to whom the Owner or Architect has no reasonable objection. If the proposed but rejected Subcontractor was reasonably capable of performing the Work, the Contract Sum and Contract Time shall be increased or decreased by the difference, if any, occasioned by such change, and an appropriate Change Order shall be issued before commencement of the substitute Subcontractor's Work. However, no increase in the Contract Sum or Contract Time shall be allowed for such change unless the Contractor has acted promptly and responsively in submitting names as required.

Section 5.2.3 affords the owner the right to make a reasonable objection to a subcontractor proposed by the contractor. If the owner or architect makes such an objection, the contractor must propose a new subcontractor. As long as the original proposed subcontractor was "reasonably capable" of performing the work, any resulting change in the contract sum or contract time is memorialized through a change order, to be issued prior to the commencement of the substitute subcontractor's work.

As a drafting tip, it benefits both the owner and the contractor to add language making the replacement of the subcontractor conditioned upon a properly executed change order providing for the change in time and price.

This section allows the contractor to recover increased costs incurred if the new subcontractor's bid is higher than the original subcontractor's bid. Conversely, if the new bid is lower, the owner will be entitled to a credit.

Note that the contractor has the burden of acting "promptly and responsively" in submitting the names of proposed subcontractors, as well as proving the original subcontractor was "reasonably capable" of performing the work. The contractor must therefore present timely and reliable evidence of any cost increase, or else waive the right to an adjustment under this section.

Oftentimes owners are at a disadvantage when it comes to the subcontractor selection process, with little or no ability to measure the quality and value of proposed subcontractors before exercising veto rights, particularly less sophisticated owners proceeding without the benefit of an owner's representative, clerk of the works, or some other similar protective agent. For this reason, owners may be well served to revise this section by adding requirements for certain objectives metrics that will help define whether the subcontractor was "reasonably capable" of performing the work—for example, evidence of the subcontractor's past performance, and/or tying the proposed subcontract price to the schedule of values.

> **§ 5.2.4** The Contractor shall not substitute a Subcontractor, person, or entity for one previously selected if the Owner or Architect makes reasonable objection to such substitution.

16. *See* Norair Eng'g v. St. Joseph's Hosp., 249 S.E.2d 642 (Ga. Ct. App. 1978) (completion contractor deemed responsible for subcontractors selected by owner where contractor had actual knowledge when it bid the contract that it was to employ, to the extent requested, all major subcontractor and suppliers to the original contractor).

Section 5.2.4 was not substantively revised. A contractor is prohibited from unilaterally replacing subcontractors without notice to and/or permission from the owner and architect. This prohibition adds certainty and reduces the risk of claims on both sides.

§ 5.3 SUBCONTRACTUAL RELATIONS

> By appropriate written agreement~~, written where legally required for validity~~, the Contractor shall require each Subcontractor, to the extent of the Work to be performed by the Subcontractor, to be bound to the Contractor by terms of the Contract Documents, and to assume toward the Contractor all the obligations and responsibilities, including the responsibility for safety of the Subcontractor's Work~~, which~~ that the Contractor, by these Contract Documents, assumes toward the Owner and Architect. Each subcontract agreement shall preserve and protect the rights of the Owner and Architect under the Contract Documents with respect to the Work to be performed by the Subcontractor so that subcontracting thereof will not prejudice such rights, and shall allow to the Subcontractor, unless specifically provided otherwise in the subcontract agreement, the benefit of all rights, remedies, and redress against the Contractor that the Contractor, by the Contract Documents, has against the Owner. Where appropriate, the Contractor shall require each Subcontractor to enter into similar agreements with Sub-subcontractors. The Contractor shall make available to each proposed Subcontractor, prior to the execution of the subcontract agreement, copies of the Contract Documents to which the Subcontractor will be bound, and, upon written request of the Subcontractor, identify to the Subcontractor terms and conditions of the proposed subcontract agreement that may be at variance with the Contract Documents. Subcontractors will similarly make copies of applicable portions of such documents available to their respective proposed Sub-subcontractors.

There were no significant changes made to Section 5.3.

Section 5.3 continues to obligate the contractor to bind its subcontractors to the terms of the contract documents by means of a clause commonly referred to as a "flow down," "pass through," or "conduit" clause. The purpose of Section 5.3 is to ensure that the subcontractor's obligations under its subcontract mirror the contractor's obligations to the owner in the contract documents, thereby reducing the potential of differing or inconsistent obligations on the part of the contractor, both upstream and downstream. This section does not provide the owner with the right to deal directly with, or control the work of, the subcontractor. Likewise, the section does not create third-party beneficiary rights for a subcontractor against the owner or any duty on the part of the contractor to its subcontractors.[17]

Most reported decisions interpreting Section 5.3 concern attempts to enforce a term in the contract documents in a dispute between the contractor and a subcontractor when the terms of the subcontract at issue differ from the contract documents. Generally, courts hold that Section 5.3 does not bind subcontractors to the terms of the contract documents; rather, Section 5.3 creates an obligation on the part of the contractor to include a "flow down" provision in the particular

17. *See* Andrews v. DT Constr., Inc., 205 S.W.3d 4 (Tex. App. 2006) (refusing to find that § 5.3 creates duty on part of contractor to subcontractor's employees for safety).

subcontract.[18] Therefore, Section 5.3 has been held to constitute a promise by the contractor to bind its subcontractors to the terms of the contract documents, and the contractor's failure to do so is merely a breach of the contract documents by the contractor.[19]

For instance, in *Tribble & Stephens Co. v. RGM Constructors, L.P.*, the prime contractor attempted to enforce a provision in the prime contract against a subcontractor, which required the subcontractor to submit its claim to the project's architect prior to initiation of arbitration or litigation.[20] Although the subcontract contained no such requirement, the prime contractor maintained that certain provisions in the subcontract bound the prime contractor to the subcontractor in the same manner as the owner was bound to the prime contractor and that Section 5.3 of the prime contract incorporated the prime contract's claims process into the subcontract.[21] Relying substantially on the AIA A201 Commentary, the court held that Section 5.3 was not self-executing, but rather placed the burden upon the prime contractor to include appropriate provisions in its subcontracts to ensure that subcontractors assume the same responsibilities to the prime contractor that the prime contractor assumed to the owner.[22] Since the subcontract did not contain such reciprocal obligations, the court held that the subcontractor need not comply with the prime contract's claims process.[23]

While most courts have held that Section 5.3, operating alone, does not itself incorporate the contract documents in the applicable subcontract, some courts have held that, when combined with an ambiguous "flow down" provision in the subcontract, Section 5.3 provides evidence that the parties intended for the subcontractor to be bound by the terms of the contract documents.[24] In contrast, where the subcontract provisions directly conflict with the terms of the contract documents, courts have held that Section 5.3 demonstrates a specific intention *not* to bind the subcontractor to the "rights, remedies, and redress" contained in the contract documents. For example, in *Falcon Steel Co. v. Weber Engineering Co.*, the court focused on the phrase "unless specifically provided otherwise in the subcontract agreement" contained in Section 5.3 as an exclusion from the flow-down of remedies from the prime contract to subcontracts.[25] The *Falcon Steel* Court relied upon this specific exclusion as evidence of an agreement between the owner and prime contractor that subcontractors could bargain for remedies against the contractor different from those between the prime contractor and owner.[26]

As a practice pointer, to ensure compliance, AIA Document A401 (Standard Form of Agreement Between Contractor and Subcontractor) contains a "flow down" provision that satisfies the requirements of Section 5.3.[27] In the event that

18. *See* Paul Ryan Assocs. v. Hawaiiana Painting & Maint., Inc., No. A136052, 2013 WL 1828928, at *6 (Cal. Ct. App. Apr. 30, 2013) ("Paragraph [5.3] is expressly conditioned on there being an 'appropriate agreement,' ostensibly separate from the General Contract. . . ."); Tribble & Stephens Co. v. RGM Constructors, L.P., 154 S.W.3d 639, 664 (Tex. App. 2004) (noting that § 5.3 is not self-executing, but rather places burden on contractor to obtain appropriate written agreement from subcontractor); St. Augustine Pools, Inc. v. James M. Barker, Inc., 687 So. 2d 957, 958 (Fla. Dist. Ct. App. 1997) (finding § 5.3 does not require arbitration per clause contained in prime contract where subcontract required litigation); Luczak Bros. v. Generes, 451 N.E.2d 1267 (Ill. App. Ct. 1983) (holding that A201 does not constitute waiver of lien rights).

19. Jonathan J. Sweet, Sweet on Construction Industry Contracts: Major AIA Documents § 17.05 (4th ed. 2006); Paul Ryan Assocs. v. Welch Marble & Tile, Inc., No. CGC-10-504430, 2013 WL 5214038, at *4 (Cal. Ct. App. Sep. 17, 2013) (finding that separate agreement between contractor and subcontractor was necessary for vitality of Section 5.3).

20. 154 S.W.3d at 664.

21. *Id.*

22. *Id.*

23. *Id.*

24. *See, e.g.*, Maxum Foundations, Inc. v. Salus Corp., 779 F.2d 974, 980 (4th Cir. 1985); Turner Constr. Co. v. Midwest Curtainwalls, Inc., 543 N.E.2d 249, 251 (Ill. App. Ct. 1989).

25. 517 A.2d 281, 285 (Del. Ch. 1986).

26. *Id.*

27. *See* AIA Document A201–2007 Commentary § 5.3, at www.aia.org.

the contractor utilizes another subcontract form, it must take care to coordinate the non-AIA forms with the AIA Document A201.[28] The contractor may include terms and conditions in its subcontract form that vary from those in the contract documents; however, such terms and conditions cannot prejudice the rights of the owner and architect.[29] Unless specifically provided otherwise, the contractor must also permit the subcontractors the same rights, remedies, and redress against the contractor that the contractor has against the owner under the contract documents.[30]

Finally, Section 5.3 obligates the contractor to make available copies of the contract documents to the subcontractors. A subcontractor's failure to request copies of the contract documents does not absolve the subcontractor from conditions placed upon it by the contract documents, if such conditions are properly incorporated into the subcontract by a "flow down" clause.[31] Further, upon written request, the contractor must identify those terms in the subcontract that vary from the terms of the contract documents. Although there are no reported cases interpreting this obligation, one would expect that a contractor's failure to comply with a written request could be used adversely against the contractor, possibly operating as a defense to any right of the contractor to enforce the contract documents against the subcontractor.

§ 5.4 CONTINGENT ASSIGNMENT OF SUBCONTRACTS

§ 5.4.1 Each subcontract agreement for a portion of the Work is assigned by the Contractor to the Owner, provided that

.1 assignment is effective only after termination of the Contract by the Owner for cause pursuant to Section 14.2 and only for those subcontract agreements that the Owner accepts by notifying the Subcontractor and Contractor ~~in writing~~; and

.2 assignment is subject to the prior rights of the surety, if any, obligated under bond relating to the Contract.

When the Owner accepts the assignment of a subcontract agreement, the Owner assumes the Contractor's rights and obligations under the subcontract.

Section 5.4.1 exists largely for the owner's convenience in the event of a termination of the contractor for cause. If a surety is involved, the surety is likely to exercise its rights under the performance bond to complete the project, thus eliminating the need for the owner to take assignment of subcontracts. However, if there is no surety, or if the surety leaves it to the owner to complete the work, the owner must have the ability to finish the project with the subcontractors and suppliers

28 *See* MPACT Constr. Grp., LLC v. Superior Concrete Constructors, Inc., 802 N.E.2d 901, 909 (Ind. 2004) ("This indicates that if the general contractor uses subcontract forms other than those provided by the AIA . . . it must in its own contract include a provision requiring the subcontractors to assume the same responsibilities that it assumes toward the owner."); AIA Document A201–2007 Commentary § 5.3, at www.aia.org.

29. *See* AIA Document A201–2007 Commentary § 5.3, at www.aia.org.

30. *See* TechnoSteel, LLC v. Beers Constr. Co., 271 F.3d 151, 162 (4th Cir. 2001) (finding that § 5.3, where subcontractor incorporates general contract, permits subcontractor to enforce owner/general contractor remedies only where subcontract does not specifically conflict). *But see* Massachusetts Elec. Sys., Inc. v. R.W. Granger & Sons, Inc., 594 N.E.2d 545 (Mass. App. Ct. 1992) (binding subcontractor to arbitration clause in prime contract that was incorporated into subcontract).

31. *See* E & F Constr. Co. v. Rissil Constr. Assocs., Inc., 435 A.2d 343, 345 (Conn. 1980) (stating that party cannot "escape his contractual obligations by saying . . . that he did not read what was expressly incorporated as specific provisions of the contract into which he entered").

that have been working on the project, that have existing contracts and purchase orders for labor and material, and that have a working familiarity with the property and the project.

AIA Document A401 (Standard Form of Agreement Between Contractor and Subcontractor) incorporates by reference the A201 General Conditions. Therefore, each subcontractor has already agreed to such assignment.[32] The 2007 modifications added the new sentence at the end of Section 5.4.1, clarifying the rights and obligations of the owner upon accepting assignment of a subcontract. This language helped all of the parties concerned with the termination and assignment process by clarifying that the owner effectively steps into the shoes of the contractor insofar as its rights and obligations toward the assumed subcontractors. For example, if the terminated contractor owed a balance to a subcontractor, the owner that assumes the subcontract now owes that balance.

In the event of an owner's termination of the contractor for cause, the owner must notify the subcontractor and contractor in order to effect an assignment of a subcontract. This change is consistent with the other revisions in Article 5 concerning the change to the definition of notice in Section 1.6.

The owner has the option under Section 5.4.1 to pick and choose which subcontracts to assume. This decision should not be taken lightly. An owner is well-advised to perform due diligence upon termination of its contractor, including a review of each subcontract and an assessment of the status of payments to, and performance by, each subcontractor. Factors an owner should consider include whether the subcontractor or supplier has been performing, and whether the subcontractor or supplier will complete the project for the previously agreed price. An owner must balance its desire to keep prices intact with an appreciation for the fact that there are often price increases as a result of delays due to the contractor's termination.

In contract negotiations, the contractor should be satisfied with the standard form language, having the owner assume all of the contractor's rights and obligations under the subcontract. The owner may wish to insert limiting language, for example, confirmation that the owner shall not be liable for matters that took place prior to the date of termination, or for change orders that were not authorized or approved by the owner. The owner may also wish to make the assignment effective in the event of a termination for convenience, as well as for cause, since the owner's interest in facilitating a smooth transition exists equally no matter the reason for termination.

§ 5.4.2 Upon such assignment, if the Work has been suspended for more than 30 days, the Subcontractor's compensation shall be equitably adjusted for increases in cost resulting from the suspension.

Section 5.4.2 accounts for the fact there is usually some delay in a project when the contractor is terminated for cause. An owner may wish to delete or revise this section, increasing the period of time before a cost adjustment is required in the event of an assignment of a subcontract.

32. WERNER SABO, LEGAL GUIDE TO AIA DOCUMENTS § 4.36, Contingent Assignment of Subcontract: ¶ 5.4 (5th ed. 2008).

> **§ 5.4.3** Upon ~~such~~ assignment to the Owner under this Section 5.4, the Owner may further assign the subcontract to a successor contractor or other entity. If the Owner assigns the subcontract to a successor contractor or other entity, the Owner shall nevertheless remain legally responsible for all of the successor contractor's obligations under the subcontract.

Section 5.4.3 was not substantively revised.

When a default occurs resulting in termination of a contractor, it is common for the owner to eventually retain a successor contractor to complete the project. Section 5.4.3 addresses the owner's responsibility toward the assumed subcontracts, even after those subcontracts are assigned to the successor contractor. This was a new provision in the 2007 modifications, again aimed at clarifying the rights and obligations of the parties in the event an owner (or lender) takes over a project following a termination for cause. Owners should be aware that this provision essentially makes the owner (or lender) a *guarantor* to fulfill the subcontracts, even when a new contractor is retained to finish a project.

CHAPTER 6:
Article 6 Construction by Owner or by Separate Contractors

Kimberly A. Davison
Keith Heath

ARTICLE 6 CONSTRUCTION BY OWNER OR BY SEPARATE CONTRACTORS

The owner may determine that it is in its best interest to perform work on a project outside of the contractor's scope utilizing either its own forces or the services of a separate contractor or contractors. Although this situation is inherent in "multi-prime" construction projects, it may also arise on larger and more complex commercial construction projects where the owner uses separate agreements to secure work that may be outside of the contractor's scope of work or expertise. While the owner has the freedom to contract with separate contractors for the benefit of itself and the project, the owner also assumes the responsibility and liability for coordination of the work to ensure that the project schedule and the interests of the various parties are not adversely impacted. To the extent that the contractor is adversely impacted by the owner's decision to utilize its own forces or separate contractors, the contractor may be entitled to assert a claim.

Article 6 anticipates these circumstances; provides a framework for the relationship between the contractor and such separate contractors; and establishes the respective rights, duties, and obligations of the respective parties when the owner's own forces or separate contractors are performing work on the project. While previous versions of the A201 did not contain a clear definition for the term separate contractors, this has been resolved in subparagraph 6.1.1 of the A201–2017, where the term "Separate Contractor(s)" is defined as "other contractors retained by the Owner under separate agreements." This definition is carried throughout Article 6 and the rest of the A201–2017. Otherwise, there were few substantive changes in A201–2017, Article 6.

§ 6.1 OWNER'S RIGHT TO PERFORM CONSTRUCTION AND TO AWARD SEPARATE CONTRACTS

~~§ 6.1.1~~**§ 6.1.1** The term "Separate Contractor(s)" shall mean other contractors retained by the Owner under separate agreements. The Owner reserves the right to perform construction or operations related to the Project with the Owner's own forces, and ~~to award separate contracts in connection~~ with ~~other portions of the Project or other construction or operations on the site~~Separate Contractors retained under Conditions of the Contract ~~identical or~~ substantially similar to ~~these~~those of this Contract, including those ~~portions~~ provisions of the Conditions of the Contract related to insurance and waiver of subrogation~~. If the Contractor claims that delay or additional cost is involved because of such action by the Owner, the Contractor shall make such Claim as provided in Article 15~~.

So long as the work performed by the owner's own forces or a separate contractor is not part of the contractor's work[1] and the language of such separate agreements is not inconsistent with the terms of the general contract, the owner is given broad rights to perform work with its own forces or with a separate contractor.

1. The definition of "Work" in Subparagraph 1.1.3 of A201–2017 remains unchanged from the 2007 version of the A201. "Work" is defined as "the construction means and services required by the Contract Documents, whether completed or partially completed, and includes all other labor, materials, equipment, and services provided or to be provided by the Contractor to fulfill the Contractor's obligations. The Work may constitute the whole or a part of the Project."

In *Eldeco, Inc. v. Charleston County School District*, 642 S.E.2d 726 (S.C. 2007), the plaintiff, an electrical subcontractor, filed suit against the owner and the general contractor for damages arising from the contractor's award of additional work to a separate subcontractor. The plaintiff contended that the contractor breached their contract when, on the instructions of the owner, it awarded electrical work not included in the original work to another subcontractor. The Supreme Court of South Carolina upheld the trial court's determination that, since the additional work was not identified in the Contract Documents and the contractor had no right or obligation to perform the additional work, it was not included in the subparagraph 1.1.3 definition of "Work." Consequently, the owner was free to direct the contractor to enter into an agreement with the separate electrical contractor.

Subparagraph 6.1.1 provides the primary framework for the relationship between the owner and the contractor in such circumstances.

The owner may determine that it would be in its best interest to enter into separate agreements with multiple "prime" contractors to perform various aspects of the project or different scopes of work. In this "multi-prime" scenario, the owner may choose to contract directly with a general contractor to oversee general construction operations, but may also enter into separate agreements directly with the structural, mechanical, plumbing, electrical, or other specialty trade contractors to provide other construction services on a project. Since the design only needs to be complete for the discrete trades and not for the project as a whole, this approach offers the owner flexibility when bidding the work. It may also offer the owner cost savings by reducing contractor markups that may be passed through to the owner for labor and materials secured by a single prime contractor. In addition, it places significant control of, and for that matter responsibility for, the project schedule in the owner's hands when the owner introduces work by others into the project.

Separate contractors may also be utilized on larger or more complex commercial construction projects where the owner elects to contract directly with one or more separate contractors for their particular subject matter expertise. For example, it is not uncommon for the owner to enter into separate agreements with hazardous material abatement contractors to deal with asbestos and other hazardous or toxic materials that may be encountered on a project. In such case, the owner will be obligated to coordinate the work of its separate contractor with the contractor's work, but may not have as pervasive a role in the overall management of the project as a whole.

The A201–2017 requires that the conditions of the separate agreements between the owner and its separate contractors be "substantially similar to those of this contract, including those provisions of the Conditions of the Contract related to insurance and waiver of subrogation."[2] Among other things, this constitutes a representation on the part of the owner that it or its separate contractors will have sufficient insurance to cover insured losses to persons or damage to property in an amount that is equivalent to that provided by the general contractor. This language ensures consistency among the various agreements related to the project and is intended to preclude the owner from negotiating terms that disproportionally favor itself or its separate contractors in those separate contract agreements. From the contractor's perspective, revisions to Section 6.1.1 should be kept to a minimum to ensure that these protections are maintained.

§ 6.1.2 When separate contracts are awarded for different portions of the Project or other construction or operations on the site, the term "Contractor" in the Contract Documents in each case shall mean the Contractor who executes each separate Owner-Contractor Agreement.

This section contains no changes from the A201–2007. Since it is anticipated that the separate agreements of the owner will be similar, subparagraph 6.1.2 attempts

2. The owner's separate contractors generally acquire third-party beneficiary status under the terms of the AIA A201 general conditions. As a result, the waiver of subrogation provisions set forth in subparagraph 11.3.7 may be applicable to claims against the owner's separate contractors. *See* RLI Ins. Co. v. Southern Union Co., 341 S.W.3d 821, 831 (Mo. Ct. App. 2011).

to clarify that, for purposes of the respective agreements, "Contractor" simply means the contractor executing each separate agreement with the owner.[3]

> **§ 6.1.3** The Owner shall provide for coordination of the activities of the Owner's own forces and of each ~~separate contractor~~Separate Contractor with the Work of the Contractor, who shall cooperate with them. The Contractor shall participate with ~~other separate contractors~~any Separate Contractors and the Owner in reviewing their construction schedules. The Contractor shall make any revisions to ~~the~~its construction schedule deemed necessary after a joint review and mutual agreement. The construction schedules shall then constitute the schedules to be used by the Contractor, ~~separate contractors~~Separate Contractors, and the Owner until subsequently revised.

When the owner utilizes its own forces or a separate contractor, it is inevitable that the owner's work will intersect or potentially overlap the work of the contractor. As drafted, the owner, either through its own efforts or by agreement with other parties, assumes responsibility for the coordination of the work between the contractor and the owner's own force or its separate contractors. Absent contractual provisions designating overall supervision of the project to another,[4] the owner must anticipate taking a very active role, either through its own employees or by utilizing outside consultants, contractors, or managers, to coordinate the work of the various contractors and overall the project schedule. Even if using only a single separate contractor, the owner is responsible for coordinating that separate contractor's activities with those of the contractor.

The "duty to coordinate" generally means that the owner must manage and supervise worksite activities among the prime contractors and their schedules to ensure timely completion of the work. This equates to an obligation on the part of the owner to use reasonable measures to ensure that all contractors work together to meet the project's scheduling goals.[5] In the event that there are active or passive conditions on the project that make the performance of the contractor more difficult or expensive, it will most likely result in increased project delays and construction costs. If such delays or inefficiencies are caused by the owner's failure to take reasonable measures to manage the performance of its own forces or that of its separate contractors, then the owner may have breached its contractual duty to coordinate the work and may be liable for damages to the contractor arising therefrom.[6]

Likewise, the performances of the prime contractors in a multi-prime contract tend to be closely interrelated and, consequently, a delay or breach by one prime may lead to losses by others. The failure of one contractor to make reasonable efforts to cooperate with the owner or its separate contractors to coordinate their respective work may result in damages to the owner or the separate contractor.

---

3. In *G.T. Leach Builders, LLC v. Sapphire V.P., L.P.*, 458 S.W.3d 502, 527 (Tex. 2015), an engineering contractor and an insurance broker attempted to compel arbitration with the owner based on an arbitration provision contained in the agreement between the owner and the contractor. They argued that, since they had each executed separate agreements with the owner, they were each a "Contractor" under the general contract and thus entitled to enforce the arbitration clause. The court disagreed, finding that the contract expressly provides that the "Contract Documents shall not be construed to create a contractual relationship of any kind . . . between [the owner] and a Subcontractor . . . or between any persons or entities other than [the owner] and the [contractor]." There was no language in the general contract's arbitration clause that would allow the other parties to compel arbitration against the owner.
4. Broadway Maint. Corp. v. Rutgers, State Univ., 447 A.2d 906, 912 (N.J. 1982).
5. Apac-Ga. v. DOT, 472 S.E.2d 97, 100 (Ga. Ct. App. 1996).
6. In *Amp-Rite Electric Co. v. Wheaton Sanitary District*, 580 N.E.2d 622 (Ill. App. Ct. 1991), the court held that the contractor has the right to recover its increased cost of performance from the owner for damages resulting from delay caused by the owner's default. It was the obligation of the owner to keep the work in "such a state of forwardness as to enable the contractor to perform the contract within the time limits specified."

In such case, the contractor may be liable for damages to the owner or the separate contractor arising from the same.[7]

Notwithstanding the owner's obligation to coordinate its work with the activities of the contractor, it is also recognized that the contractor and separate contractors have an obligation to reasonably cooperate with the owner's efforts.[8] Pursuant to subparagraph 3.10, the contractor is required to submit a construction schedule providing for the orderly progress of its own work after being awarded the contract. Subparagraph 6.1.3 requires the contractor and the owner's separate contractors to review their respective construction schedules to coordinate their respective work. To clarify when this obligation goes into effect, the contractor may want to consider adding "when directed to do so" at the end of the second sentence. Although this language was set out in prior versions of the A201, it was deleted in the AIA's 2007 revisions. With this modification, the owner has an affirmative duty to notify the contractor that coordination with the owner and its separate contractors is desired and necessary to ensure the orderly progress of the work. Absent this modification, the contractor may find itself liable for schedule impact claims from the owner or the owner's separate contractors even if the contractor was not aware that such coordination was necessary.

After a joint review of the project schedule, in light of the competing interests of the owner's forces and its separate contractors, the owner and the contractor may mutually agree that the contractor's construction schedule should be revised to ensure the timely and orderly progress of the work. If the parties mutually agree, then the contractor's schedule will be revised accordingly. This language is especially important for the contractor and prohibits the owner from unilaterally modifying the contractor's schedule without its consent.

With the addition of the last sentence, the A201–2017, unlike previous versions of the A201, clearly recognizes that project schedule revisions arising from coordination issues may result in an increase to the contractor's time and/or cost to perform the work. In the event that the contractor determines that a change is warranted, the contractor will be entitled to assert a claim pursuant to Article 15.

> **§ 6.1.4** Unless otherwise provided in the Contract Documents, when the Owner performs construction or operations related to the Project with the Owner's own forces or with Separate Contractors, the Owner or its Separate Contractors shall ~~be deemed to be subject to~~have the same obligations and ~~to have the same~~ rights that ~~apply to~~ the Contractor has under the Conditions of the Contract, including, without excluding others, those stated in Article 3, this Article 6, and Articles 10, 11, and 12.

When the owner performs any construction- or operations-related work on the project, either with its own forces or with that of separate contractors, it is bound to the same obligations ascribed to the contractor under the general conditions of the contract.

7. "It should be recognized at the outset that a contractor whose performance is delayed through no fault of its own has two types of relief available under the standard contract prepared by the American Institute of Architects. It first has the right to an extension of the time available for performance in accordance with Article 8.3.1. In addition to a time extension, a contractor may have the right to be compensated for the increased costs it has incurred as a result of this delay in accordance with Article 6.2." Moore Constr. Co. v. Clarksville Dep't of Electric., 707 S.W.2d 13 (Tenn. Ct. App. 1985).

8. In *Everman's Electric Co. v Evan Johnson & Sons Construction, Inc.*, 955 So. 2d 979, 986 (Miss. Ct. App. 2007), the court determined that a co-prime contractor could assert a claim against another co-prime contractor for negligently breaching its duty to coordinate its work on the project.

§ 6.2 MUTUAL RESPONSIBILITY

> **§ 6.2.1** The Contractor shall afford the Owner and ~~separate contractors~~Separate Contractors reasonable opportunity for introduction and storage of their materials and equipment and performance of their activities, and shall connect and coordinate the Contractor's construction and operations with theirs as required by the Contract Documents.

Only minor changes have been made to subparagraph 6.2.1 reflecting the defined nature of "Separate Contractor" established in subparagraph 6.1.1. In their respective dealings, the actions of the owner, the owner's separate contractors, and the contractor must be reasonable in light of the circumstances.[9] Pursuant to subparagraph 6.2.1, the contractor is obligated to provide reasonable access for the owner and the owner's separate contractors to store material and equipment on the site and to allow them to perform their respective obligations on the project. Further, the contractor must make reasonable efforts to assist the owner and its separate contractors with coordinating their respective construction schedules, roles, and activities on the project. Likewise, the demands of the owner and the owner's separate contractors on the contractor to coordinate their respective activities must also be reasonable.

> **§ 6.2.2** If part of the Contractor's Work depends for proper execution or results upon construction or operations by the Owner or a ~~separate contractor~~Separate Contractor, the Contractor shall, prior to proceeding with that portion of the Work, promptly notify the Architect of apparent discrepancies or defects in ~~such other~~the construction or operations by the Owner or Separate Contractor that would render it unsuitable for ~~such~~ proper execution and results~~.~~ of the Contractor's Work. Failure of the Contractor ~~so~~to notify the Architect of apparent discrepancies or defects prior to ~~report~~proceeding with the Work shall constitute an acknowledgment that the Owner's or ~~separate contractor's~~Separate Contractor's completed or partially completed construction is fit and proper to receive the Contractor's Work~~, except as to defects not then reasonably discoverable~~. The Contractor shall not be responsible for discrepancies or defects in the construction or operations by the Owner or Separate Contractor that are not apparent.

If the contractor's work relies on the proper execution and performance of separate work by the owner's own forces or a separate contractor, it is anticipated that the contractor will make a reasonable inspection of such separate work to ensure that it is suitable for use by the contractor and in accordance with the contract documents. Otherwise, it is assumed that the work of the separate contractor is in accordance with the contract documents and suitable for acceptance of the contractor's work. Subparagraph 6.2.2 has been revised in the A201–2017 to clarify both the scope and timing of the contractor's inspection obligations. Specifically, the contractor is obligated to notify the architect only of those apparent discrepancies or defects that directly impact the proper execution of the contractor's work or the results anticipated therefrom.

Since there is subjectivity in determining what discrepancies are truly "apparent," the contractor may want to consider limiting its obligation to report only those discrepancies and defects in the work that are "known" to it. This modification

9. *See id.*

reduces the likelihood that the contractor will be liable for defects in the work performed by a separate contractor that, although unknown to it, are arguably "apparent." Notwithstanding, in the event that the contractor is impacted by such apparent discrepancies or defects, the contractor should promptly notify the owner and the architect before proceeding with the portion of the work impacted by such apparent discrepancy or defect. Failure of the contractor to notify the architect prior to proceeding with the work constitutes the contractor's acceptance of such separate work and waives the contractor's right to any additional compensation for apparent discrepancies or defects that should have reasonably been reported.

Conversely, new language at the end of subparagraph 6.2.2 provides that the contractor does not waive any rights for defects in such separate work to the extent the defects are not apparent. If the contractor identifies such apparent discrepancies or defects in accordance with subparagraph 6.2.2 or encounters latent defects that were not reasonably discoverable prior to proceeding with the work, the contractor may be entitled to an adjustment in the contract time or the contract sum to the extent that such discrepancies or defects delay the contractor's performance or cause an increase in the cost of the work.

> **§ 6.2.3** The Contractor shall reimburse the Owner for costs the Owner incurs that are payable to a ~~separate contractor~~Separate Contractor because of the Contractor's delays, improperly timed activities or defective construction. The Owner shall be responsible to the Contractor for costs the Contractor incurs because of a ~~separate contractor's~~Separate Contractor's delays, improperly timed activities, damage to the Work or defective construction.

Only minor changes have been made to subparagraph 6.2.3 reflecting the defined nature of "Separate Contractor" established in subparagraph 6.1.1. Subparagraph 6.2.3 imposes liability on the owner and the contractor in the event that their actions on the project impact, or are impacted by, the performance of a separate contractor. In the event that a separate contractor is injured by the action of the contractor on the project, then the separate contractor will look to the owner for compensation for its damages and the owner will be entitled to seek reimbursement from the contractor for the same. Likewise, if the contractor is delayed or damaged as a result of the performance of a separate contractor, then the owner is liable for such damages to the extent that such damages are recoverable pursuant to the terms of the general contract. Unfortunately, this provision raises some concerns regarding the types of damages that may be recoverable by the owner in the event that the contractor's performance delays a separate contractor.

Specifically, pursuant to subparagraph 15.1.7, the owner and the contractor waive claims against each other for consequential damages arising out of or relating to the work. Presumably any payments from the owner to a separate contractor would be considered "direct" damages and would fall outside of the mutual waiver of consequential damages contemplated by subparagraph 15.1.7. Notwithstanding, the owner is not prohibited from agreeing to pay damages that would typically be defined as consequential to a separate contractor. Consequently, the contractor may unknowingly be agreeing to assume liability for damages that are not contemplated, and actually waived, in its agreement with the owner. Conversely, the waiver language in subparagraph 15.1.7 would still be applicable to the contractor prohibiting its recovery of consequential damages incurred as a result of such separate contractor's performance.

> **§ 6.2.4** The Contractor shall promptly remedy damage that the Contractor wrongfully causes to completed or partially completed construction or to property of the Owner or ~~separate contractors~~Separate Contractor as provided in Section 10.2.5.

Only minor changes have been made to subparagraph 6.2.4 reflecting the defined nature of "Separate Contractor" established in subparagraph 6.1.1. If the contractor's performance causes damage to the construction performed by a separate contractor or the owner or to their property, then the contractor is responsible for the costs associated with remedying such damage. The owner may wish to consider expanding this provision to provide for full indemnification of the owner and the architect in the event that the contractor causes damage contemplated by this subparagraph 6.2.4. Since the contractor's obligations only extend to damage that was wrongfully caused, the owner may want to consider deleting the word "wrongfully." This change would expand the contractor's obligations to remedy damage caused by it whether wrongful or not. Finally, it should also be noted that the architect may withhold a certificate for payment, in whole or part, to the extent reasonably necessary to protect the owner because of damage to the owner or its separate contractors pursuant to subparagraph 9.5.1.5.

> **§ 6.2.5** The Owner and each ~~separate contractor~~Separate Contractor shall have the same responsibilities for cutting and patching as are described for the Contractor in Section 3.14.

Only minor changes have been made to subparagraph 6.2.5 reflecting the defined nature of "Separate Contractor" established in subparagraph 6.1.1. The owner and its separate contractors assume the same obligations for cutting and patching their work as the contractor assumes pursuant to subparagraph 3.14. Specifically, the owner's forces or a separate contractor will be responsible for cutting, fitting, or patching required to complete their work and will not cut, patch, or alter the work of the contractor without written consent.

§ 6.3 OWNER'S RIGHT TO CLEAN UP

> If a dispute arises among the Contractor, ~~separate contractors~~Separate Contractors, and the Owner as to the responsibility under their respective contracts for maintaining the premises and surrounding area free from waste materials and rubbish, the Owner may clean up and the Architect will allocate the cost among those responsible.

Once again, minor changes have been made to subparagraph 6.3 reflecting the defined nature of "Separate Contractor" established in subparagraph 6.1.1. In the event that there is a dispute among the contractor, a separate contractor, and the owner regarding their respective obligations for cleaning up or maintaining the premises, then the owner may take it upon itself to perform such obligations with its own forces or by separate agreement, and the architect will make a determination allocating the costs among the various parties. The parties may want to consider adding a notice component to subparagraph 6.3 so that the parties will have a final opportunity to either clean up or settle the matter prior to the architect making any determination as to responsibility. Any determination by the architect will be subject to the dispute resolution procedures of Article 15.

CONCLUSION

When multiple parties with different goals, concerns, and interests converge on a construction project, it is nearly inevitable that coordination issues will develop. Although Article 6 anticipates these circumstances and does an admirable job of dealing with this potentially complex and complicated scenario, the size and scope of the project, along with the owner's needs, may warrant a careful review to ensure that the respective needs and concerns of both the owner and the contractor are being adequately addressed.

CHAPTER 7:
Article 7 Changes in the Work

Nora E. Loftus
Allison Taller Reich

ARTICLE 7 CHANGES IN THE WORK

In nearly every construction contract, changes in the work arise after the contract is executed: the owner changes its mind on a detail or feature or adds additional elements to the design; the architect makes a modification or finds an error or omission in its plans and specifications; the contractor proposes a change for efficiency, cost savings, or availability of material. The list goes on. It benefits all parties to establish procedures for changes in the work within the text of the contract rather than risk invalidating the contract.

Changes generally mean not only changes to the design, drawings, specifications, or construction itself, but they often affect the project completion time (Contract Time) or sums to be paid to the contractor performing the work (Contract Sum). As we will discuss in this section, there are several different mechanisms through which changes can be made, depending on the nature and magnitude of the change. These detailed mechanisms do not have many significant changes from the prior 2007 version of the A201, so they should be familiar to many.

Our recommendation, and best practice, is to confirm all changes in writing, even if it is a minor change that will not affect time or sum. If the change is one that will affect the Contract Time or the Contract Sum, a written construction change directive or written change order must be in place before performing the changed work, even if the magnitude of the change is not yet known.

§ 7.1 GENERAL

§ 7.1.1 Changes in the Work may be accomplished after execution of the Contract, and without invalidating the Contract, by Change Order, Construction Change Directive or order for a minor change in the Work, subject to the limitations stated in this Article 7 and elsewhere in the Contract Documents.

AIA "Document A201, Article 7, specifies the manner in which changes in the work may be accomplished by change order, construction change directive, or order for a minor change and defines each method."[1] However, changes in the work are affected by other articles within the A201, and it is important that all are considered in conjunction with this Article 7.

For example, change orders, construction change directives (CCDs), and minor changes are often ordered by the architect. Article 4 governs the actions of the architect, which gives the architect the authority to administer the contract. Further, Article 3 requires the contractor to perform the work in accordance with the Contract Documents, though the contractor may be asked to deviate from those Contract Documents by the architect. Article 8 deals with the Contract Time and adjustments thereto, while Article 9 prescribes rules for payment. The parties must take these sections into account when preparing and executing change orders, moving forward with CCDs, and deciding the best course of action to take in response to minor changes in the work.

1. Gilbert H. Moen Co. v. Spokane River Assocs., No. 15947-7-III, 1997 Wash. App. LEXIS 2123 (Dec. 23, 1997); Angotti & Reilly v. Alexander Group, Nos. A127917, A128743, 2011 Cal. App. Unpub. LEXIS 8917, at *6 (Nov. 21, 2011) ("Pursuant to Article 7 of the General Conditions, changes in the work can be accomplished without invalidating the contract by one of three means: change orders; construction change directives; or minor changes.").

> **§ 7.1.2** A Change Order shall be based upon agreement among the Owner, Contractor, and Architect~~; a~~. A Construction Change Directive requires agreement by the Owner and Architect and may or may not be agreed to by the Contractor~~; an~~. An order for a minor change in the Work may be issued by the Architect alone.

A change order requires the agreement of the owner, contractor, and architect; a CCD requires the agreement of the owner and architect, and a minor change can be made by the architect acting alone. "A change order is a written instrument prepared by the architect and signed by the owner, contractor and architect stating their agreement on the change, price adjustment if any and time adjustment if any."[2] Change orders are the "cleanest" way to accomplish changes in the Work after execution of the contract. In a change order, all parties—architect, owner, and contractor—agree to the change in the work, the change in the Contract Sum, and the change in the Contract Time. Change orders are to be reduced to writing and signed by all parties *before* work on the change begins. Failure to reduce the change to writing and have the document signed by all parties before commencing work can operate as a waiver of the contractor's right to seek an adjustment in the Contract Sum and/or Contract Time.

"A construction change directive, to be used when the parties cannot agree on the terms of a change order, is a written order prepared by the architect and signed by the owner and architect directing a change in the work and proposing a basis for adjustment of price or time or both."[3] Section 7.3 sets forth the procedure for issuing a construction change directive, which is to be used "in the absence of total agreement on the terms of a Change Order." "Among other things, this article sets forth the procedure for addressing and resolving disagreements between the contractor and owner regarding changes in work and price and also specifically obligates the contractor to maintain and provide an itemized accounting and 'appropriate supporting data.'"[4]

CCDs are used so as not to delay the project—which often leads to other damages, costs, and claims—in instances where the architect, owner, and contractor cannot agree on the corresponding change in the Contract Sum and/or Contract Time. The architect and owner agree on the change and on the proposed adjustment to the Contract Sum and/or Contract Time, ordering the contractor to proceed with the changed work and reserve its claim for adjustment of the Contract Sum and/or Contract Time. This can lead to litigation, arbitration, or formal mediation if the parties are unable to agree on appropriate adjustments. Contractors should be sure to follow the notice and timing requirements set forth in the A201 form (in § 1 as well as § 15) to preserve its claims related to CCDs.

"The architect is given authority to order minor changes not involving changes in the price or time by written order."[5] The magnitude and type of change needs to be taken into account in determining whether a change order is needed. Architects may order "minor changes" that do not affect the Contract Sum or Contract Time without a change order. This procedure can be used as an avenue through which architects could order many changes without providing change orders to contractors.

2. *Gilbert H. Moen Co.*, 1997 Wash. App. LEXIS 2123; *Angotti & Reilly*, 2011 Cal. App. Unpub. LEXIS 8917, at *6 ("A Change Order is a written instrument prepared by the Architect and signed by the Owner, Contractor and Architect, stating their agreement upon all of the following: [¶] .1 change in the Work; [¶] .2 the amount of the adjustment, if any, in the Contract Sum; and [¶] .3 the extent of the adjustment, if any, in the Contract Time.").
3. *Gilbert H. Moen Co.*, 1997 Wash. App. LEXIS 2123.
4. *Angotti & Reilly*, 2011 Cal. App. Unpub. LEXIS 8917.
5. *Id.*

However, there are times where the architect orders what it believes may be a "minor change" that does, or should, result in an adjustment of the Contract Sum or Contract Time. Under new changes to the A201 form, the contractor is required to give notice of the necessity of adjustments to Contract Sum or Contract Time before proceeding with the "minor change."

> **§ 7.1.3** Changes in the Work shall be performed under applicable provisions of the Contract Documents~~, and the~~. The Contractor shall proceed promptly with changes in the Work, unless otherwise provided in the Change Order, Construction Change Directive, or order for a minor change in the Work.

The purpose of the mechanisms in Article 7—particularly CCDs and the authority of the architect to order minor changes—is to keep the project moving without major impacts to the project schedule. However, sometimes the change order, CCD, or minor change in the work may, itself, require that the changed work be postponed for a particular period of time or may require other work to be completed or changed before the change can be implemented. Further, there are times when even this section, meant to keep the project moving forward, requires the contractor to pause, despite the project schedule. If the contractor proceeds with change order work without a fully executed change order, it may waive its right to seek a change order for that work. Similarly, under new Section 7.4, if the contractor fails to notify the architect that minor changes ordered in the Work are actually not minor and may require adjustments to the Contract Time or Contract Sum, and if the contractor proceeds with the work without obtaining the necessary adjustments, then the contractor may waive its right to seek those adjustments.

§ 7.2 CHANGE ORDERS

> **§ 7.2.1** A Change Order is a written instrument prepared by the Architect and signed by the Owner, Contractor, and Architect stating their agreement upon all of the following:
>
> **.1** The change in the Work;
> **.2** The amount of the adjustment, if any, in the Contract Sum; and
> **.3** The extent of the adjustment, if any, in the Contract Time.

Change orders should be written, signed by the owner, architect, and contractor, prior to performance of the changed work. A change order is a modification of the contract, containing all relevant information necessary to effect the change—including adjustments to the contract amount and schedule in addition to the details of the actual change—as agreed upon by all parties.[6] "It is well settled, that the terms of a written contract cannot be varied by any previously executed contract, written or parol, nor by any contemporaneous parol contract."[7] "It is equally well settled, that the terms of a written contract may be varied, modified, waived, annulled, or wholly

6. Coonrod & Walz Constr. Co. v. Motel Enterprises, Inc., 535 P.2d 971, 979–80 (Kan. 1975).

7. *Id.* (citing Todd v. Allen, 18 Kan. 543, 545 (1878); and 29 AMERICAN & ENGLISH ENCYCLOPEDIA OF LAW 829 (James Cockcroft, David S. Garland, Charles Porterfield eds., 2d ed. 1904)); Whitfield Constr. Co. v. Commercial Dev. Corp., 392 F. Supp. 982, 1000–01 (D.V.I. 1975).

set aside, by any subsequently executed contract, whether such subsequently executed contract be in writing or in parol."[8]

The change order must itself "possess all elements necessary to form a contract, including the element of consideration."[9] Even if the Contract Sum or Contract Time is not affected by the changed work, a change order noting "no net change" should be executed, in order to provide clarity and avoid confusion and disputes later on.[10]

If the Contract Time is extended in a change order, the owner cannot later assert that it was damaged by a delay of the contractor, that others were delayed by the contractor, or assess delay damages or liquidated damages.[11] Similarly, if the Contract Sum is adjusted by a change order, the contractor cannot later make a claim for additional compensation. The change order should include all costs, including any claimed overhead and profit, acceleration, additional costs of the work, additional costs for material or supervision, any increase in the project time.

Owners and contractors may not be required to pay for extra labor or materials furnished without a written change order.[12] However, the written change order requirement may be waived by the conduct or statements of the parties.[13] "Parties to an unperformed written contract may, by mutual consent, modify it by altering, excising or adding provisions, and that such modifications may be by parol agreement or inferred from the conduct of the parties."[14] "A provision in a written contract to the effect that no changes in the work should be made unless provided for in writing and signed by both parties is waived to the extent of changes made pursuant to an executed oral agreement."[15] "Waiver can be implied from conduct such as making payments for or accepting performance which does not meet contract requirements; waiver can also be expressed verbally or in writing."[16]

If an owner and a contractor proceed without a written change order—the contractor performs the work and the owner pays it—the owner may not be able to rely on the written change order requirement in its contract, unless the contractor acts in bad faith.[17] "It has generally been recognized that in the absence of express authority, an architect, as such, has no power to waive or modify a stipulation requiring a written change order for alterations or extras."[18]

In *United States ex rel. Agate Steel, Inc. v. Jaynes Corp.*, Agate argued that the contractual change order requirement was waived when the parties agreed to changes and changed price via e-mail. [19] American Steel later disputed the contract amount, despite its e-mail approving the amended contract amount and directing

---

8. *Coonrod*, 535 P.2d at 979–80 (citing *Todd*, 18 Kan. at 545).
9. Smith v. Githens, 271 S.W.2d 374, 380 (Mo. Ct. App. 1954).
10. *See* Articles 8 and 9 of the A201 regarding changes in time and sum and how to calculate adjustments. In the alternative, a "no net change" may be documented as an architect's order for a minor change in the Work. See Section 7.4.
11. J.A. Jones Constr. Co. v. Greenbriar Shopping Center, 332 F. Supp. 1336 (N.D. Ga. 1971).
12. Pan Am Eng'g Co. v. Poncho's Constr. Co., 387 So. 2d 1052, 1053 (Fla. Dist. Ct. App. 1980).
13. Harrington v. McCarthy, 420 P.2d 790, 793 (Idaho 1966); Custom Builders, Inc. v. Clemons, 367 N.E.2d 537 (Ill. App. Ct. 1977).
14. *Harrington*, 420 P.2d at 793.
15. Shimek v. Vogel, 105 N.W.2d 677 (N.D. 1960).
16. United States *ex rel.* Agate Steel, Inc. v. Jaynes Corp., No. 2:13-CV-01907-APG-NJK, 2016 U.S. Dist. LEXIS 79878, at *13 (D. Nev. June 17, 2016) (quoting Udevco, Inc. v. Wagner, 678 P.2d 679, 682 (Nev. 1984)).
17. Care Systems, Inc. v. Laramee, 547 N.Y.S.2d 471, 473 (App. Div. 1989): When an owner knowingly receives and accepts the benefits of extra work orally directed by himself and his agents, that owner is equitably bound to pay the reasonable value thereof, notwithstanding the provisions of his contract that any extra work must be supported by a written authorization signed by the owner; such conduct constitutes a waiver of that requirement. *See* La Rose v Backer, 203 N.Y.S.2d 740, 746 (App. Div. 1960) (when the owner knowingly receives and accepts the benefits of extra work orally directed by himself and his agents, the owner is equitably bound to pay the reasonable value thereof, notwithstanding the provisions of his contract that any extra work must be supported by a written authorization signed by the owner, such conduct working a waiver of that requirement), *aff'd*, 181 N.E.2d 632 (N.Y. 1962).
18. Whitfield Constr. Co. v. Commercial Dev. Corp., 392 F. Supp. 982, 1000–01 (D.V.I. 1975).
19. No. 2:13-CV-01907-APG-NJK, 2016 U.S. Dist. LEXIS 79878, at *13 (D. Nev. June 17, 2016).

Agate to proceed with the changed work.[20] The court held that it would be "highly unreasonable" to conclude that, even though American Steel agreed to the revised contract price and urged Agate to complete the changed work as soon as possible, American Steel did not agree to change the contract amount.[21]

Many contracts include an anti-waiver clause to prevent waivers of several provisions, including waiver of the change order requirement. These provisions purportedly invalidate waivers of any contract provisions, unless expressly waived in writing.[22] "Such 'anti-waiver' clauses, however, are of little effect because of the overriding principle that any contract clause may be modified or waived expressly or impliedly by persons with requisite contracting authority."[23]

Change orders are preferred over construction change directives, which are not agreed upon and may later result in claims or disputes.

This provision has no changes from the 2007 A201.

§ 7.3 CONSTRUCTION CHANGE DIRECTIVES

> **§ 7.3.1** A Construction Change Directive is a written order prepared by the Architect and signed by the Owner and Architect, directing a change in the Work prior to agreement on adjustment, if any, in the Contract Sum or Contract Time, or both. The Owner may by Construction Change Directive, without invalidating the Contract, order changes in the Work within the general scope of the Contract consisting of additions, deletions, or other revisions, the Contract Sum and Contract Time being adjusted accordingly.

When the parties cannot agree to an adjustment of the Contract Time or Contract Sum for additional work such that a change order is executed, the owner has the ability to direct changes in the work through a construction change directive. The AIA has a separate form (Form G714) that can be used to memorialize the CCD. The CCD must be signed by both the owner and architect and presented to the contractor. Although the CCD is used when the parties have not agreed on the Contract Time and Contract Sum, a CCD should include a proposed basis for adjusting the contract price that the contractor can accept by counter-signing the CCD. As seen in the next provision, Section 7.3.2, a CCD is used when the parties are not in total agreement on all terms of an adjustment of the Contract Time and Contract Sum such that it is not possible to execute a change order prior to the contractor proceeding with the work.

The CCD works to the benefit of both the owner and the contractor. It gives the owner a mechanism to direct the contractor to immediately proceed with extra work even if the parties have not agreed to an equitable adjustment for that work, and it protects the contractor from the provision that states that a contractor waives rights to additional compensation if it proceeds with extra work without a written directive to do so. Often contractors are pressured into beginning work without anything in writing because the parties can't agree on the adjustment to Contract Time and Contract Sum to complete a change order. The CCD fixes this problem, and contractors must insist that the owner execute and sign a CCD before the

20. *Id.*
21. *Id.*
22. Certified Power Sys. v. Dominion Energy Brayton Point, LLC, Nos. 118267, BRCV2008-1217, BRCV2008-01114, BRCV2008-1198, BRCV2009-292, 2011 Mass. Super. LEXIS 317, at *181 (Dec. 30, 2011) (citing Philip L. Bruner & Patrick J. O'Connor, Bruner & O'Connor on Construction Law § 4:40, at 592 (2002)).
23. *Id.*

contractor proceeds with work, or the contractor is at considerable risk of waiving its rights to be compensated for the change or additional work.

One key clause within this section is the phrase "within the general scope of the Contract." It is key because there must be some limit on an owner's ability to direct a contractor to perform additional work, even if the contractor is entitled to an equitable adjustment of the Contract Time and Contract Sum. For instance, an electrical subcontractor should not be forced to perform roof work through a CCD because that is not within the general scope of the electrical subcontracting services that were within the original contract scope. Courts that have addressed a CCD that is so far outside the scope of the original contract (either by type of work or volume) have held that the owner was deemed to have abandoned the contract.

For example, in one California case, the written contract called for compensation to be paid on a cost-plus basis with a not-to-exceed price of $15,900 (it was 1952).[24] Throughout the course of the project, the owner repeatedly made design changes and instructed the contractor to complete the changed work through a CCD. When the work was complete, including the extra work, the owner agreed to pay the contractor for the extra work, but only up to the "not-to-exceed" price of the original contract. The contractor argued that the owner abandoned the written contract by the owner's constant changes to, and interference with, the contractor's work. Although the appellate court did not expressly agree with the contractor's position, it ultimately held that the "not-to-exceed" provision was abandoned and the contractor was entitled to its cost plus all extra work.

More recently, a federal district court in Kentucky did an extensive analysis of case law as it relates to changes so great that they would constitute an owner's abandonment of the original contract, citing one case that held "there was substantial evidence that change orders and extra work imposed by the Owner upon the contractor were of such magnitude as to change the scope of the work originally contemplated under the contract, [and] the record supports the court's findings and conclusions that the parties had abandoned the original contract."[25]

The same court also did an extensive analysis of another theory relating to changes outside the contemplation of the original contract that are known as "cardinal changes." This legal theory is based on the fact that the changes provision in a contract, like the one in Article 7, does not authorize a "drastic modification beyond the scope of the contract" and that such a drastic change would constitute a breach of the contract and entitle the contractor to damages.[26] Under the theory of "cardinal change" an owner is not deemed to have abandoned the contract such that the contractor would not be required to continue performance. Instead, the contractor is still obligated to perform but is entitled to damages for the continued performance.

A contractor has to keep in mind that it *must* proceed with work if the owner issues a CCD, unless that work is outside the general scope of the contract, or the contractor will likely be found in breach of the contract and at risk for a default termination.[27] However, the contractor is only required to proceed with work when

24. Opdyke & Butler v. Silver, 245 P.2d 306 (Cal. Dist. Ct. App. 1952).

25. L.K. Comstock & Co. v. Becon Constr. Co., 932 F. Supp. 906 (E.D. Ky. 1992) (citing C. Norman Peterson Co. v. Container Corp. of America, 218 Cal. Rptr. 592, 598 (Ct. App. 1985)). *See also* Daughtery Co. v. Kimberly-Clark Corp., 92 Cal. Rptr. 120 (Ct. App. 1971) ("Performing the work prescribed by the contract and the changes does not preclude the possibility that the labor and materials furnished were outside the contemplation of the parties when the bid was made and the contract signed.").

26. *L.K. Comstock & Co.*, 932 F. Supp. at 936 (citing Air-A-Plane Corp. v. United States, 408 F.2d 1030 (Ct. Cl. 1969)).

27. All Seasons Constr., Inc. v. City of Shreveport, 742 So. 2d 626 (La. Ct. App. 1999) ("Upon receipt of a signed Construction Change Directive, the Contractor must promptly proceed with the change described.").

the CCD is signed by the owner. If the architect alone signs the CCD, the contractor is likely not bound to proceed with the work unless the language of the contract was changed to expressly state that the owner does not need to separately sign the CCD and that the architect has the authority to bind the owner.

Courts have repeatedly held that the architect does not have authority to bind the owner by issuing a CCD without the owner's signature. In one case, the court excused a contractor from performing work where only the architect had signed the CCD and not the owner, holding that the CCD that the architect signed was not binding on the contractor, that the contractor "was under no obligation to do the additional work," and that it was the owner's failure to properly execute a CCD that caused the work stoppage.[28]

This Article 7.3.1 has no changes from the prior 2007 version of the A201.

§ 7.3.2 A Construction Change Directive shall be used in the absence of total agreement on the terms of a Change Order.

This provision simply confirms that the owner has a mechanism to direct extra work before the parties agree on an equitable adjustment of either the Contract Time or the Contract Sum, or both. This is important because it works in conjunction with the requirement that a contractor is not to proceed with changes in the work without a written directive. If the parties cannot agree on the amount of an equitable adjustment, either in time or sum, or both, there must be another mechanism for the owner to direct the contractor to proceed with extra work so as to prevent any delays and prevent a contractor's ability to hold the project progress hostage while negotiating an equitable adjustment, all without the contractor waiving its right to seek additional time or money.

This provision also has no changes from the prior 2007 version of the A201.

§ 7.3.3 If the Construction Change Directive provides for an adjustment to the Contract Sum, the adjustment shall be based on one of the following methods:

.1 Mutual acceptance of a lump sum properly itemized and supported by sufficient substantiating data to permit evaluation;
.2 Unit prices stated in the Contract Documents or subsequently agreed upon;
.3 Cost to be determined in a manner agreed upon by the parties and a mutually acceptable fixed or percentage fee; or
.4 As provided in Section 7.3.~~7~~4.

This provision mandates that, when the change involves an adjustment to the Contract Sum, then the adjustment *must* be calculated by one of the four methods listed. These predetermined methods are critical to give the parties some reasonable comfort of fair compensation for additional work. An owner can state in the CCD any of the four methods of calculation. If the contractor agrees with the method at the time the CCD is issued, or at any time before the architect makes a determination

28. Lott Constructors, Inc. v. Jackson Township Bd. of Educ., No. 92-2402 (CSF), 1992 U.S. Dist. LEXIS 12891 (D.N.J. Aug. 12, 1992). *See also* Contract Applicators, Inc. v. Borough of Park Ridge, No. A-6080-10T3, 2012 N.J. Super. Unpub. LEXIS 2232 (App. Div. Oct. 2, 2012) (architect did not have apparent authority to authorize extra work when contract required owner's written authorization).

under Article 7.3.4, or if the owner agrees to use another method of calculation to which the contractor agrees, the CCD can then be converted to a change order. If the parties do not agree, Article 7.3.4 comes into play.

Again, there are no changes to this Article 7.3.3 from the 2007 version of the A201.

Prior Articles 7.3.4, 7.3.5, and 7.3.6 were simply renumbered and moved further down in the Article without any changes. Each provision will be discussed further in the remainder of this chapter.

> **§ 7.3.~~7~~4** If the Contractor does not respond promptly or disagrees with the method for adjustment in the Contract Sum, the Architect shall determine the ~~method and the~~ adjustment on the basis of reasonable expenditures and savings of those performing the Work attributable to the change, including, in case of an increase in the Contract Sum, an amount for overhead and profit as set forth in the Agreement, or if no such amount is set forth in the Agreement, a reasonable amount. In such case, and also under Section 7.3.3.3, the Contractor shall keep and present, in such form as the Architect may prescribe, an itemized accounting together with appropriate supporting data. Unless otherwise provided in the Contract Documents, costs for the purposes of this Section 7.3.~~7~~4 shall be limited to the following:
>
> **.1** Costs of labor, including ~~social security, old age and unemployment insurance~~applicable payroll taxes, fringe benefits required by agreement or custom, ~~and~~ workers' compensation insurance, and other employee costs approved by the Architect;
>
> **.2** Costs of materials, supplies, and equipment, including cost of transportation, whether incorporated or consumed;
>
> **.3** Rental costs of machinery and equipment, exclusive of hand tools, whether rented from the Contractor or others;
>
> **.4** Costs of premiums for all bonds and insurance, permit fees, and sales, use, or similar taxes, directly related to the ~~Work~~change; and
>
> **.5** ~~Additional costs~~Costs of supervision and field office personnel directly attributable to the change.

The architect has the authority to determine a proper adjustment of the Contract Sum if the contractor "does not respond promptly or disagrees with the method of adjustment of the Contract Sum." This is critical and can be fatal for a contractor because, absent prompt objection, the architect's decision is final and binding relating to a determination of the Contract Sum, unless it is based on fraud or bad faith.[29] While it is important to note that the provision is written to limit the architect's discretion so that the architect's determination is based on actual costs plus overhead and profit, a contractor will likely be bound by the architect's determination if it does not timely object.

This provision also requires the contractor to keep an accounting with supporting documentation of costs associated with the changed work from which the architect can make its determination of "reasonable expenditures."

The changes in this section to the prior 2007 version of the A201 are minor and simply clarify some of the items that the contractor is entitled to include as costs in

29. *See e.g.*, Morse/Diesel, Inc. v. Trinity Indus., 67 F.3d 435 (2d Cir. 1995); RaDec Constr. v. School Dist. No. 17, 535 N.W.2d 408 (Neb. 1995).

adjusting the Contract Sum for changed work. The section is intended to give the parties some certainty as to the additional compensation to which the contractor will be entitled if there is a change in the work.

> **§ 7.3.5** If the Contractor disagrees with the adjustment in the Contract Time, the Contractor may make a Claim in accordance with applicable provisions of Article 15.

This provision allows a contractor to pursue a claim if it disagrees with a decision regarding the adjustment of Contract Time, and was not changed from the prior 2007 version of the A201. It is worth noting here that both owners and contractors should be mindful to include adjustments to the Contract Time when appropriate. Often changes to the Contract Sum are the focus and adjustments to Contract Time get overlooked, and that has serious consequences.

It is highly recommended that the contractor add to this provision that if the contractor disagrees with an adjustment in the Contract Sum, it can also make a claim in accordance with Article 15 to avoid any misinterpretation and make it expressly clear that a contractor has a right to dispute, through the Article 15 dispute resolution process, any architect determination of an adjustment to the Contract Sum under Section 7.3.4.

> **§ 7.3.6** Upon receipt of a Construction Change Directive, the Contractor shall promptly proceed with the change in the Work involved and advise the Architect of the Contractor's agreement or disagreement with the method, if any, provided in the Construction Change Directive for determining the proposed adjustment in the Contract Sum or Contract Time.

As discussed above, it is imperative that the contractor proceed with the work set forth in the CCD, unless it is clearly outside the scope of the contract. It is equally critical that the contractor advise the architect if it disagrees with the method of adjustment of Contract Time or Contract Sum or it may be bound by the architect's decision. The contractor also needs to keep track of its costs associated with the change because it ultimately will have the burden of proof on its damages for additional work.[30]

> **§ 7.3.7** A Construction Change Directive signed by the Contractor indicates the Contractor's agreement therewith, including adjustment in Contract Sum and Contract Time or the method for determining them. Such agreement shall be effective immediately and shall be recorded as a Change Order.

A contractor can, at any time after the owner issues a CCD, sign the CCD and agree to the method or stated adjustment of the Contract Sum or Contract Time. A contractor should be careful to note that, if it signs and distributes the CCD, even if not sent to the owner or the architect, the contractor has agreed to the terms. In one case, a contractor signed a CCD but did not send a signed copy to the architect

30. C.R. Humphreys Gen. Contr., Inc. v. Tangipahoa Parish Sch. Sys., No. 2007 CA 0993, 2007 La. App. Unpub. LEXIS 343 (Dec. 21, 2007) ("The burden of proof is upon the contractor to establish by a preponderance of the evidence that he is entitled to the amount requested in a change-order request.").

or the owner. Instead, the contractor sent the signed copy to a supplier. The court determined that the signed copy, even though not sent back to either the owner or architect, bound the contractor to the proposed adjustments to the Contract Time and Contract Sum set forth in the CCD.[31]

> **§ 7.3.8** The amount of credit to be allowed by the Contractor to the Owner for a deletion or change that results in a net decrease in the Contract Sum shall be actual net cost as confirmed by the Architect. When both additions and credits covering related Work or substitutions are involved in a change, the allowance for overhead and profit shall be figured on the basis of net increase, if any, with respect to that change.

This provision is written such that a change in the work that results in a decrease of the Contract Sum will not decrease the contractor's overall overhead and profit. It was not changed from the 2007 version.

> **§ 7.3.9** Pending final determination of the total cost of a Construction Change Directive to the Owner, the Contractor may request payment for Work completed under the Construction Change Directive in Applications for Payment. The Architect will make an interim determination for purposes of monthly certification for payment for those costs and certify for payment the amount that the Architect determines, in the Architect's professional judgment, to be reasonably justified. The Architect's interim determination of cost shall adjust the Contract Sum on the same basis as a Change Order, subject to the right of either party to disagree and assert a Claim in accordance with Article 15.

This is an important provision because it permits the contractor to seek payment of costs incurred in proceeding with the changed work even if the contractor and owner have not reached a final agreement on the total compensation and the architect has not reached a determination under Article 7.3.4. There are no changes from the 2007 version of the A201.

> **§ 7.3.10** When the Owner and Contractor agree with a determination made by the Architect concerning the adjustments in the Contract Sum and Contract Time, or otherwise reach agreement upon the adjustments, such agreement shall be effective immediately and the Architect will prepare a Change Order. Change Orders may be issued for all or any part of a Construction Change Directive.

The best outcome of a CCD is to convert it into a written change order resolving all disputes between the parties relating to adjustments in Contract Time and Contract Sum. This will occur if the parties agree to the architect's determination of an equitable adjustment to the Contract Time or Contract Sum, but can also be done at any point during the process that the parties are able to reach an agreement, with or without a final determination from the architect.

This provision (unchanged from the 2007 version) also makes clear that the parties can agree to a change order for any part of work included in a CCD without

31. *All Seasons Constr., Inc.*, 742 So. 2d 626.

invalidating the CCD, or the procedures relating to the CCD, as to the remaining work. This is a nice provision that should be used as often as possible because it eliminates areas of dispute. The more the parties can agree on, the less contentious the remaining issues in dispute will likely be.

§ 7.4 MINOR CHANGES IN THE WORK

> The Architect ~~has authority to~~ may order minor changes in the Work that are consistent with the intent of the Contract Documents and do not ~~involving~~involve an adjustment in the Contract Sum or an extension of the Contract Time ~~and not inconsistent with the intent of the Contract Documents. Such~~. The Architect's order for minor changes shall be in writing. If the Contractor believes that the proposed minor change in the Work will ~~be effected by written order signed by~~affect the Contract Sum or Contract Time, the Contractor shall notify the Architect and shall ~~be binding on the Owner and Contractor.~~ not proceed to implement the change in the Work. If the Contractor performs the Work set forth in the Architect's order for a minor change without prior notice to the Architect that such change will affect the Contract Sum or Contract Time, the Contractor waives any adjustment to the Contract Sum or extension of the Contract Time.

Section 7.4 was changed quite a bit from the 2007 version to the 2017 version. The 2007 version, as written, gave the architect the "authority to order minor changes in the work that did not involve an adjustment in the Contract Sum or an extension of the Contract Time and not inconsistent with the intent of the Contract Documents." The minor changes ordered by the architect were to be "effected by written order signed by the Architect and . . . binding on the Owner and Contractor."

The 2017 version still gives the architect the authority to order minor changes that do not involve or result in an adjustment in the Contract Sum or Contract Time, and those changes must be in writing. However, unlike the 2007 version, the 2017 version requires the contractor to review the minor changes for effects on the Contract Sum and the Contract Time—which those minor changes should not affect—and, if the contractor believes the change will result in or require an adjustment of the Contract Sum or an extension of the Contract Time, the contractor is required to notify the architect. Further, the 2017 version places the obligation on the contractor not to perform the work, without prior notice, if a change in Contract Sum or Contract Time may result from the minor changes. The requirement that the contractor not proceed with the work if adjustments may be needed seems contrary to the spirit of Section 7.4, which is to keep the project moving despite minor changes in the work.

If the contractor fails to comply with these new obligations—set forth in the 2017 version—to notify the architect that the minor changes ordered by the architect will cause changes in the Contract Time or Sum, and proceeds with the minor changes, the contractor is punished. The contractor waives its right to request adjustment to the Contract Time or Contract Sum if it proceeds with the minor changes ordered by the architect without giving prior notice of the adjustments.

Both the 2007 version and 2017 version require the minor changes to be consistent with the "intent of the Contract Documents." The minor changes must be consistent with, and not substantially differ from, the project drawings, specifications, and other design documents that served as the basis for the work and contract. Substantial deviations would constitute major changes, which would require change orders.

Minor changes are thought to be those changes that are so insignificant that they do not affect, or result in any change in, the Contract Time or Contract Sum.[32]

Section 7.4 gives authority to the architect to order changes without getting prior written consent or approval from the owner; however, that authority is not unlimited. "[T]he architect is not, by virtue of his employment, the general agent of the owner for all purposes in the work he is engaged to supervise."[33] "His authority is a limited one—defined by the terms of his contract of employment or by the terms of the contract between the owner and the contractor."[34] "He has no authority to make alterations in the plans and specifications nor to bind the owner with respect thereto except as provided in the contractual documents."[35]

Contractors may be held liable for not strictly conforming to the plans, despite the alleged approval of the architect.[36] If the contractor performs work that differs from the project plans and specifications due to oral directives or field changes issued by the architect, but those changes are not in writing, the contractor bears the risk of being required to redo or correct nonconforming work. The changes to Section 7.4 put even more risk on contractors, requiring them to give notice of required adjustments to Contract Sum and Contract Time, and not proceed if an adjustment is needed.

Architects may try to avoid processing change orders by ordering field changes or orally ordering changes, even where those changes are not minor and even where the changes may require an adjustment of the Contract Sum or Contract Time. Another way architects may attempt to order changes without processing change orders is through "Architect's Supplemental Instructions" (ASI).[37]

CONCLUSION

Traditionally, changes in work cause anxiety for all parties. The owner is concerned that the contractor will use changed work as an opportunity to gain leverage or charge more than a fair price because the project is in progress and any delays or disputes will affect the final completion to the owner's detriment. The contractor is concerned that it won't be fairly compensated for the full impact of changed work because the full impact may be difficult to determine at the time the change is directed, or the parties cannot agree on what is an equitable adjustment of Contract Time or Contract Sum. Changes can also cause anxiety for the architect because one change often affects other scopes of work, which just adds to the stress that goes with changes in the work.

---

32. *See id.* at 631 ("'Minor changes,' those with no added cost and no added time, can be approved by the project architect."); Dehnert v. Arrow Sprinklers, 705 P.2d 846, 848 (Wyo. 1985) ("The Architect shall prepare Change Orders for the Owner's approval and execution in accordance with the Contract Documents, and shall have authority to order minor changes in the Work not involving an adjustment in the Contract Sum or an extension of the Contract Time which are not inconsistent with the intent of the Contract Documents.").

33. Whitfield Constr. Co. v. Commercial Dev. Corp., 392 F. Supp. 982, 1001–02 (D.V.I. 1975).

34. *Id.*

35. *Id.*

36. *Id.*

37. *See* Walsh v. West Valley Mission Community College Dist., 78 Cal. Rptr. 2d 725 (Ct. App. 1998) "ASI's are standard AIA forms, which state, 'The work shall be carried out in accordance with the following supplemental instructions issued in accordance with the contract documents without change in contract sum or contract time. Prior to proceeding in accordance with the instructions[,] indicate your acceptance of these instructions for minor change to the work as consistent with the contract documents and return a copy to the architect.'" *Id.* In *Walsh*, an expert testified that "the California Code of Regulations, title 21 language pertaining to change orders for minor changes dates back to the late 1970's. It soon became apparent that the requirement for a change order executed by contractor, architect, owner, and approved by OSA, was having the effect of bringing public works projects to a standstill. As a consequence, 'OSA came up with the stamp, and the stamp meant this is formal, official approval of this change in your contract. We recognize that it may be a while before you can get an approved signed change order to process that everybody agrees to. So we are telling you, go ahead with your work. But some day subsequently you have to . . . come in here with a change order that has everybody's signature [on] it.'"

Article 7 of the A201 helps alleviate much of this anxiety by providing detailed steps to follow in the event of a change, and these steps were not changed much from the 2007 version of the A201. Article 7 provides a mechanism for the owner to direct a contractor to make changes in the work, provided the changes are the same general type of work as the original contract, but also a mechanism for the contractor to protect its right to get fairly compensated through additional time or additional money, or both.

The most important aspect of this Article 7 is that changes in the work must be documented in writing through a construction change directive, change order, or other writing by the architect for minor changes that don't affect the Contract Time or Contract Sum. It is so easy for a contractor or owner to push the paperwork off, particularly when a change is discussed and may be documented informally in e-mails or meeting minutes. However, it is absolutely critical for the contractor to ensure that changes are documented in the proper format or it risks the ability to be compensated for additional time or additional sums.

CHAPTER 8:
Article 8 Time

Jeremy S. Baker
Emily Wurtenberger Schott

ARTICLE 8 TIME

Benjamin Franklin famously quipped "lost time is never found again." While building projects can be accelerated to recover lost time, this aphorism reveals a truth of the construction industry: proper time and schedule management can mean the difference between success and failure on many projects. Commercial projects, in particular, can be time sensitive. Owners often rely on anticipated revenue from completed projects to service construction loans. An untimely delivered project may jeopardize an owner's business or cause it to incur additional costs and lost revenue. Contractors and designers commit staff and resources based on assumptions concerning project duration and phases of work. A delayed project may cost a contractor extended general conditions and home office overhead. Project delays often lead to claims. Depending on the size and complexity of the project, avoiding delay can require attention to many interdependent schedule activities. A contract with appropriate time and schedule terms can provide a road map for timely project completion.

The 2017 version of AIA Document A201, like its predecessors, covers the basic terms relevant to project time in Article 8. Its terms address project commencement, duration, and completion. It binds the contractor to these terms, highlights their importance, and provides owners and contractors with the reason and opportunity to give them thoughtful consideration. Article 8 requires the contractor to proceed expeditiously with adequate forces, but not before insurance is in place to protect the contractor, owner, and architect. Article 8 also provides basic terms to address delays, extensions of time, and dispute resolution.

The terms of Article 8 suffice for many small and simple projects—particularly when it is read in connection with other provisions that affect Contract Time, including Section 3.10 (Contractor's Construction and Submittal Schedules). The standard AIA text of Article 8 often remains unedited even when an owner and contractor take the time to negotiate custom changes to other articles of AIA Document A201.

For larger and more complex projects, the standard AIA text of Article 8 is often supplemented with additional terms. In particular, Section 8.3 (Delays and Extensions of Time) is often modified to more explicitly specify when a contractor is entitled to additional time or money if project delay occurs. Some commentators believe the failure of AIA Document A201 to address the scheduling methodology required of the contractor is a "glaring shortcoming."[1] The 2017 version of AIA Document A201 has not been substantially revised to address this perceived shortcoming. The complexity of construction scheduling and the unique needs of each project may account for the comparatively light treatment of time and schedule in AIA Document A201. On large or complex projects it may be appropriate to supplement the standard AIA schedule and time provisions with more fulsome language to coordinate, sequence, and track critical activities.

This chapter explains each section of Article 8, reviews the changes between its 2007 and 2017 versions, addresses judicial treatment of earlier versions of Article 8, and provides commentary on terms that some may consider adding to Article 8 in order to address concepts not found in the standard AIA text. All of the potential modifications to the standard AIA text of Article 8 discussed below should be considered carefully in light of the type and size of the project and its unique requirements, circumstances, and risks.

1. 2 Philip L. Bruner, Patrick J. O'Connor, Jr., Bruner & O'Connor on Construction Law § 5:160 (2002).

§ 8.1 DEFINITIONS

> **§ 8.1.1** Unless otherwise provided, Contract Time is the period of time, including authorized adjustments, allotted in the Contract Documents for Substantial Completion of the Work.

Section 8.1.1 of AIA Document A201 defines the "Contract Time." Contract Time is the time allotted in the contract documents for "Substantial Completion of the Work." Substantial completion is the stage in the progress of the work when the work, or a designated portion thereof, is sufficiently complete in accordance with the contract documents so the owner can occupy or utilize the work for its intended use.[2] The date of commencement of the work and the date of substantial completion are usually set forth in other AIA documents that incorporate AIA Document A201 by reference.[3] Contract Time is tied to substantial completion, not final completion. Final completion, which is governed by Section 9.10 (Final Completion and Final Payment), typically occurs after the Contract Time has expired. The Contract Time is subject to adjustment as provided in the contract documents, including under Article 7 (Changes in the Work) of AIA Document A201. Failure to substantially complete the work within the Contract Time may expose the contractor to actual damages or liquidated damages, a liquidated sum sometimes designated by contract and recoverable by the owner for each day of unexcused contractor delay.

The 2017 version of AIA Document A201 makes no changes to the 2007 version of Section 8.1.1. The straightforward and standard AIA text of Section 8.1.1 is generally revised only in connection with a fulsome revision of Article 8.

Because construction disputes infrequently hinge on a legal interpretation of Section 8.1.1, minimal case law exists on the meaning of this section. However, courts have used standard rules of contract construction to hold that the meaning of Section 8.1.1 is clear and unambiguous, and that application of the plain and ordinary meaning of its terms is appropriate, allowing courts to enforce it in accordance with its terms.[4] If necessary, the court can interpret Section 8.1.1 by reference to other provisions within the four corners of the contract, but often such reference is not required.[5]

> **§ 8.1.2** The date of commencement of the Work is the date established in the Agreement.

Section 8.1.2 of AIA Document A201 provides that the date of commencement of the work is the date established elsewhere in the contract documents. The standard text in AIA Documents A101, A102 and A103 provides that, unless the parties agree otherwise, the date of commencement is the date of the Agreement itself. This approach can be modified to cause the date of commencement to be triggered by any number of events, including the owner's confirmation of financing, the owner's receipt of a valid building permit, or the contractor's receipt of an owner's written notice to proceed. In many agreements, the Contract Time starts to run on

2. *See* AIA Document A201 § 9.8.1.
3. *See, e.g.*, AIA Document A101, Article 3; AIA Document A102, Article 4; AIA Document A103, Article 4.
4. Great Am. Ins. Co. of N.Y. v. Superior Contracting Corp., No. 2:14-cv-00055-PPS-JEM , 2014 WL 5597272, at *6 (D. Ind. Nov. 4, 2014).
5. *Id.* at 6.

the commencement date established by the contract documents without regard to whether the contractor actually begins work on that date.

No changes to the 2007 version of Section 8.1.2 were made in the 2017 version of AIA Document A201. The straightforward and standard AIA text of this section is rarely revised unless as part of a wholesale rewriting of Article 8.

Construction disputes seldom hinge on a legal interpretation of Section 8.1.2, so case law analyzing this section is sparse. However, Section 8.1.2 has been subject to challenge in states where contractors are required by statute to specify commencement and completion dates in residential construction contracts.[6] In *Ippolito v. Olympic Construction, LLC*, the parties agreed contractually that the date of commencement of the work was tied to the date of commencement of the construction phase; in another contract section, the parties agreed that the construction phase commenced when the owner accepted the guaranteed maximum price proposal or the owner issued a notice to proceed.[7] Because the court could readily deduce the date of commencement of the work by reviewing the parties' contract in its entirety, it held the residential contract enforceable.[8]

§ 8.1.3 The date of Substantial Completion is the date certified by the Architect in accordance with Section 9.8.

Substantial completion is an important concept that permeates the entire AIA family of documents. Substantial completion is defined in Section 9.8 of AIA Document A201 as the point when the owner can occupy or utilize the project for its intended use. The date of substantial completion is often considered the point when the contractor's initial responsibility to insure the project, and to control and make safe the project site, is ultimately and finally vested in the property owner. Section 8.1.3 provides that the official date of substantial completion is to be set forth in a certificate prepared by the architect. Architects typically provide such certificates after they inspect the project and determine that the work is substantially complete.

No changes to the 2007 version of Section 8.1.3 were made in the 2017 version of AIA Document A201.

Courts have upheld the date of substantial completion certified by architects provided their determinations are not made in bad faith, the product of gross mistake, or arbitrary or capricious.[9] That an architect's decision reflects an error in judgment, or that a court may have ultimately reached a different conclusion as to the date of substantial completion, does not by itself establish bad faith.[10] In one particular case involving Section 8.1.3, an architect's certification of the date of substantial completion was held to prevail over a conflicting date in a written change order or an oral statement by the owner.[11] Courts have held contracting parties may agree that a certain event marks the date of substantial completion; in such cases, other events "lose their significance" with respect to achievement of that milestone.[12]

6. *See* Ippolito v. Olympic Contr., LLC, 136 A.3d 653, 663 (Conn. App. Ct. 2016).
7. *Id.* at 663–64.
8. *Id.*
9. McCarthy Bros. Constr. Co. v. Pierce, 832 F.2d 463, 468 (8th Cir. 1987).
10. *Id.*
11. MR Hotels, LLC v. LLW Architects, Inc., No. M2015-00840-COA-R9-CV, 2016 WL 4070050, at *8 (Tenn. Ct. App. July 28, 2016).
12. *See id.* at *7; Brookridge Apartments, Ltd. v. Universal Constructors, Inc., 844 S.W.2d 637, 640 (Tenn. Ct. App. 1992).

If a conflict arises between Section 8.1.3 and Section 9.8, courts have held that, because Section 8.1.3 specifically identifies what constitutes the substantial completion date, Section 8.1.3 prevails over any contradiction in Section 9.8.[13]

Owners and contractors may wish to modify Section 8.1.3 to expressly reserve their right to challenge, in any subsequent binding dispute resolution proceedings, the architect's determination as to when, and whether, substantial completion was achieved. Owners occasionally edit Section 8.1.3 to make the architect's determination on this issue merely advisory and to select another project participant, usually the owner or an owner's representative, to certify the contractually and legally operative date of substantial completion.

The traditionally significant role reserved by AIA Document A201 for architects during the construction administration phase has been the subject of recent debate. In 2007, another industry group issued a general conditions document akin to AIA Document A201 that virtually eliminates the architect's role in the construction administration phase. This may be appropriate for sophisticated owners or those who employ owners' representatives or have other means to protect themselves from construction phase risks. However, diminishing the architect's role can deprive those owners who only occasionally build the substantial benefit of AIA Document A201's traditional approach—which requires the architect to police the contractor's work, for the owner's benefit, during the construction phase.

In a similar fashion, the AIA diminished the architect's role, at least minimally, in the 2007 version of AIA Document A201. Whereas prior versions of the document empowered the architect to render initial decisions on disputes between the owner and contractor, the 2007 version called for the owner and contractor to select an initial decision maker who can be the architect or virtually anyone else. Section 8.1.3 continues to empower architects to certify the critical substantial completion date—and is an example of the important construction phase role AIA Document A201–2017 continues to reserve for architects.

Depending on the complexity and nature of the project, a contract may specify either a single substantial completion date for the entire work or multiple substantial completion dates for a project delivered in phases. The definition of substantial completion in Section 9.8.1 of AIA Document A201–2017 contemplates completion of either the entire work or, if the project is to be completed in phases, completion of "a designated portion" of the work.

In a multiphase project, it may be appropriate to add terms to Article 8 to clarify which party controls the site and is responsible for site safety as the various portions of the project achieve substantial completion. Section 8.1.3 can be modified to clarify that "occupancy of one portion of the Project by the Owner may occur while the construction of other portions of the Project are ongoing." In such cases, owners sometimes insert language to clarify that the "Contractor recognizes partial occupancy by the Owner during the construction phase" either "does not relieve the Contractor of its obligation to initiate, maintain, and supervise all safety precautions and programs" or "requires particular coordination with the Owner to protect the health and safety of the public, the Owner and its tenants, and their respective employees, and to avoid unreasonable inconvenience to and interference with their conduct of business." Conversely, a contractor who anticipates turning over "designated portions" of substantially complete work to the owner may wish to modify Section 8.1.3 in order to clarify that the contractor no longer has an obligation to ensure adequate safety precautions remain in place for portions of the work turned over to the owner.

13. *See, e.g.*, 314 W. 115th St. LLC v. Expo Dev. Corp., 2014 WL 2804118, at *18–19 (N.Y. Sup. Ct. 2014).

Where adjacent work is to be simultaneously performed by contractors separately hired by the owner, it may wish to include an obligation for each contractor to "reasonably cooperate with Separate Contractors in order to minimize disruption and delay."

> **§ 8.1.4** The term "day" as used in the Contract Documents shall mean calendar day unless otherwise specifically defined.

Section 8.1.4 clarifies that, unless the contract documents specifically otherwise provide, use of the term "day" refers to a calendar day, not a business day. Calculations of the Contract Time and any other time periods specified in the contract documents include weekends and holidays. If the owner or contractor desires the term "day" to mean a business day or to exclude holidays, the standard AIA text may be edited accordingly.

No changes to the 2007 version of Section 8.1.4 were made in the 2017 version of AIA Document A201.

Section 8.1.4 is straightforward and seemingly unambiguous, which may explain why little case law analysis of this section exists.

§ 8.2 PROGRESS AND COMPLETION

> **§ 8.2.1** Time limits stated in the Contract Documents are of the essence of the Contract. By executing the Agreement, the Contractor confirms that the Contract Time is a reasonable period for performing the Work.

The first sentence of Section 8.2.1 indicates that time is of the essence of the parties' agreement. This makes timely performance an express condition of the contract. If one party fails to perform an obligation by a date specified in the contract documents the other party may, subject to the particular facts and contract terms in question, contend a contractual breach that entitles the non-breaching party to recover damages caused by the breach.

In the second sentence of Section 8.2.1, the contractor acknowledges that it understands the Contract Time is a reasonable period in which to substantially complete the work. This sentence is an effort to prevent the contractor from subsequently contending that the Contract Time is insufficient to achieve substantial completion.

The 2017 version of Section 8.2.1 does not differ from the prior version of this Section in AIA Document A201–2007.

Whether delay leads to liability often depends on the materiality of the breach—in other words, whether untimely performance is a minor or material breach of contract. "Time is of the essence" provisions have been held by some jurisdictions to "give a minor breach as to timely performance the legal effect of a material breach."[14] Some interpret Section 8.2.1 as an attempt to elevate failure by the contractor to achieve the time limits stated in the contract documents to a material breach of contract. This provision can increase a contractor's liability exposure. An owner typically has more substantial remedies, including potential termination of the contract, for material breaches than it does for minor or immaterial breaches of contract.

14. Zancanaro v. Cross, 339 P.2d 746, 749 (Ariz. 1959).

Jurisdictions differ on the effect of "time is of the essence" provisions. Some hold that a delay will constitute a material breach of contract "where the contract by its terms makes time of the essence."[15] Courts that take this approach have held that parties "can assure a finding that 'time is of the essence' by including those, or equivalent words, within their agreement."[16] Even where a contract contains an express stipulation that "time is of the essence" some courts will inquire into the underlying circumstances or the other provisions of the contract to determine whether a delay in performance resulted in a material breach.[17]

Contractors who wish to minimize the risk of late performance sometimes seek to modify Section 8.2.1 by deleting the first sentence entirely or editing it to provide that the time limits "are terms of the Contract" rather than "are of the essence of the Contract." Even if the owner requires the project to be completed by a date certain, it may be appropriate to modify Section 8.2.1 to clarify that "time is of the essence" only with respect to major milestones designated in the contract documents. The rationale is that a contractor's failure to meet the start and end dates for every activity on the project schedule will not necessarily prevent the contractor from timely achieving substantial completion or other major project milestone dates.

Not all delays are created equal. The materiality of delay often depends on whether the delayed activity is on the "critical path" of the project schedule. Unlike many construction contracts, particularly agreements used on large or complex projects, the standard text of AIA Document A201 does not address the critical path concept. This complex concept is relevant to most properly framed contractor claims for extra time and money.

The "critical path" is sometimes defined as the "the longest path through the network of identified and logically sequenced construction activities that establishes the minimum overall project duration."[18] The extra cost and inconvenience associated with delay to non-critical path activities can be far less substantial than those associated with critical path delays. Failure by the contractor to timely complete activities on the critical path can cause cascading delays to other scheduled activities, potentially jeopardizing timely completion of the entire project.

The "Critical Path Method" or "CPM" is an analysis technique used to both manage the critical path of ongoing projects and to analyze delay claims. Most complex projects closely track and monitor the critical path with recurring schedule updates and other metrics to objectively monitor progress of the work. Construction project schedulers use the CPM to analyze interdependent schedule activities and to determine whether delay to a particular activity may jeopardize schedule milestone dates or overall project completion. This allows CPM schedulers to determine the anticipated minimum total duration of the project and whether it is on track to achieve the applicable time limits and milestones. If the project is not on track, the CPM can assist schedulers to determine the most advantageous way to "recover" the schedule in order to achieve key milestone dates. Little or no attention is typically paid to the critical path on smaller projects that lack complex interdependent activities. The critical path can often be forensically reconstructed to analyze delay claims, even on projects that are not actively managed using CPM

15. Darrell J. Didericksen & Sons, Inc. v. Magna Water & Sewer Imp. Dist., 613 P.2d 1116, 1119 (Utah 1980). *See also* Commonwealth Constr. Co. v. Cornerstone Fellowship Baptist Church, Inc., 2006 WL 2567916 (Del. Super. Ct. 2006) (holding that any delay that damages one party and is attributable to the other party must be construed as a breach).

16. Urban Archaeology Ltd. v. Dencorp Investments, Inc., 783 N.Y.S.2d 330, 335 (App. Div. 2004). *See also* 314 W. 115th St. LLC v. Expo Dev. Corp., 2014 WL 2804118, 9 (N.Y. Sup. Ct. 2014) (giving effect to an express "time is of the essence" provision).

17. *See* Asset Recovery Contracting, LLC v. Walsh Const. Co. of Illinois, 980 N.E.2d 708, 726 (Ill. App. Ct. 2012). *See also* Vermont Marble Co. v. Baltimore Contractors, Inc., 520 F. Supp. 922, 927–28 (D.D.C. 1981).

18. 5 Bruner & O'Connor, *supra* note 1, § 15:5.

techniques, if sufficient contemporaneous data exists as to the actual sequence of activities that occurred during the time periods in question.

Owners may wish to modify Article 8 to incorporate CPM concepts as both a schedule management tool and to assess a contractor's entitlement to extra time and money. Historically, owners turned over schedule management to contractors. However, owners increasingly employ their own CPM-trained schedulers and require contractors to keep the owner informed of the status of the work by providing monthly or even weekly schedule updates.

As discussed below, owners on large or complex projects often modify Section 8.3 to require the contractor to show, as a precondition of receiving extra money for delay, that unforeseeable factors beyond the contractor's control delayed its critical path activities and increased its cost in a manner that justifies extra compensation.

> **§ 8.2.2** The Contractor shall not knowingly, except by agreement or instruction of the Owner in writing, ~~prematurely~~ commence ~~operations on the site or elsewhere~~ the Work prior to the effective date of insurance required ~~by Article 11~~ to be furnished by the Contractor and Owner. ~~The date of commencement of the Work shall not be changed by the effective date of the insurance.~~

Section 8.2.2 protects the owner and others by preventing the contractor from knowingly commencing the work before the contractor's insurance coverages are effective. Most construction contracts make the contractor responsible for site safety, including the implementation of safety programs, and many require the contractor to provide insurance to protect the owner and others from claims for personal injury and property damage that may result from performance of the work. Section 8.2.2 seeks to avoid commencement of potentially dangerous site operations before required liability and other insurance is in place.

The 2007 version of Section 8.2.2 of AIA Document A201 has been amended, as follows, to create the 2017 version of this section:

(1) The adverb "prematurely" was deleted. The word was arguably superfluous because the contractor is otherwise precluded by Section 8.2.2 from commencing the work *prior to* the effective date of required insurance.
(2) The phrase "operations on the site or elsewhere" was removed and replaced with "the Work." This revision confines the contractor's insurance procurement responsibilities under Section 8.2.2 to the "Work" of the Contract Documents. The 2007 version of Section 8.2.2 was arguably broader and somewhat ambiguous because it encompassed the contractor's operations either at the project site *or elsewhere.*
(3) The reference to "Article 11" insurance requirements was deleted. The new Section 8.2.2 is thus somewhat broader than its predecessor. If insurance requirements are located elsewhere in the contract documents (the 2017 version of A201 adds a separate exhibit for insurance requirements), the contractor may not knowingly commence the work prior to the effective date of all such required insurance due to this change.
(4) The sentence previously indicating that the date of commencement will not be changed by the effective date of insurance was deleted. This sentence was arguably superfluous because Section 8.1.2 plainly identifies the date of commencement of the work.

Scant case law interpreting the prior versions of Section 8.2.2 exists.

> **§ 8.2.3** The Contractor shall proceed expeditiously with adequate forces and shall achieve Substantial Completion within the Contract Time.

Section 8.2.3 obligates the contractor to proceed expeditiously, to provide adequate staffing, and to achieve substantial completion within the Contract Time. Without more, the obligation to "proceed expeditiously with adequate forces" is subjective. Owners developing large or complex projects may wish to augment this concept by requiring contractors to maintain certain levels of manpower, to report on manpower levels actually deployed to the project site, to achieve interim milestone dates, or to make other objective showings that demonstrate it is on track to achieve substantial completion within the Contract Time.

The 2017 version of AIA Document A201 makes no changes to the 2007 version of Section 8.2.3.

Despite the arguably subjective nature of Section 8.2.3's requirement "to proceed expeditiously with adequate forces," contractors have been found to have failed to satisfy this requirement under the facts of particular projects. In one case, a procurement review panel found a contractor to have breached Section 8.2.3 when (1) it submitted over 90 letters seeking additional time or money, (2) the architect repeatedly withheld payments for nonconforming work, (3) the architect relayed concern to the contractor that the project was behind schedule, and (4) the contractor failed to accelerate performance of the work or provide additional forces.[19]

§ 8.3 DELAYS AND EXTENSIONS OF TIME

The most substantial edits to Article 8 are typically reserved for Section 8.3. This section gives comparatively light treatment to the important question of a contractor's entitlement to extra time and money if delays occur. The standard AIA text of Section 8.3.1 identifies certain causes of excusable delay that entitle contractors to extensions of time. Section 8.3.3 is a single sentence that leaves open the possibility of the owner and contractor asserting delay-related claims for monetary damages against each other. Aside from comparatively light treatment in a handful of clauses outside of Article 8, the standard text of AIA Document A201 does not specify what would justify delay-related claims for extra money.

AIA Document A201 leaves many important questions unanswered. What if the delayed activities are not on the critical path of the project schedule? What if they are on the critical path but do not cause the contractor to incur extra costs or delay project completion? What if the contractor and owner cause concurrent delay, i.e., independent delays that overlap in time? What if the project is delayed for reasons beyond the control of both the owner and contractor? What are their respective obligations to give notice of, and to mitigate, schedule delays? What circumstances entitle a contractor to additional time or money, or both?

In many cases, owners do not accept delays to non-critical path activities as grounds for extensions of the Contract Time or contractor claims for additional money. The analysis is necessarily contract specific and fact specific. However, critical path delays are generally more likely to jeopardize project deadlines and lead to extra costs than other delays.

19. *In re* Protest of Hass Construction Company, Inc.; South Carolina State University, 2000 WL 33956162 (S.C. Procurement Review Panel 2000).

One framework to determine if a contractor is entitled to additional time or money is to characterize delays as either "compensable," "excusable," or "non-excusable." Those exact terms and categories may not appear in a particular contract. However, in many cases, contract terms that address delays and extra costs may nevertheless have the effect of categorizing delay-related claims along similar lines.

A delay is sometimes considered "compensable" to the contractor when the delayed activities are on the critical path, if the delay is caused by the owner (or parties for whom the owner is legally responsible), if the contractor incurred actual costs because of the delay, and if the contractor is not responsible for causing a concurrent delay. In such cases, depending on the specific contract and particular facts in question, the contractor may be entitled to both additional time and money from the owner.

A critical path delay may be considered "excusable" but not "compensable" when the contractor and the owner (or parties for whom the owner is legally responsible) cause concurrent delay. Depending on the specific contract and particular facts in question, concurrent delay can sometimes entitle a contractor to additional time but no additional money. Delays caused by factors a contractor cannot control or predict may also be considered excusable. Contracts sometimes provide that contractors are entitled to additional time, but no additional money, for "force majeure" type delays to the critical path that are beyond the reasonable control or anticipation of both the owner and contractor.

A delay may be considered "non-excusable," entitling the contractor to neither additional time nor money from the owner, if the delay is caused by the contractor (or parties for whom the contractor is legally responsible) and if the owner (and parties for whom the owner is legally responsible) has not caused concurrent delay. Non-excusable delays can result in owner claims against the contractor, particularly when the delayed activities are on the critical path and cause the owner to incur additional costs.

The "compensable," "excusable," and "non-excusable" delay categories are just generalizations. There are many possible ways to address contractor entitlement to additional time and money if delays occur. Resolution of this issue typically turns on an intensely fact-specific and project-specific analysis.

§ 8.3 DELAYS AND EXTENSIONS OF TIME

> **§ 8.3.1** If the Contractor is delayed at any time in the commencement or progress of the Work by (1) an act or neglect of the Owner or Architect, ~~or~~ of an employee of either, ~~or of a separate contractor employed by the Owner; or~~ or of a Separate Contractor; (2) by changes ordered in the Work; ~~or~~ (3) by labor disputes, fire, unusual delay in deliveries, unavoidable casualties, ~~or~~ adverse weather conditions documented in accordance with Section 15.1.6.2, or other causes beyond the Contractor's control; ~~or~~ (4) by delay authorized by the Owner pending mediation and ~~arbitration; or~~ binding dispute resolution; or (5) by other causes that the Contractor asserts, and the Architect determines, justify delay, then the Contract Time shall be extended ~~by Change Order~~ for such reasonable time as the Architect may determine.

Section 8.3.1 identifies specific delay events that can entitle the contractor to reasonable extensions of time. This section empowers the architect to reasonably extend the Contract Time on the above-indicated grounds (another example of the large role reserved for architects in AIA Document A201). Some may characterize

Section 8.3.1 as an attempt to define delays that are "excusable" but not "compensable," therefore justifying extensions of the Contract Time but not necessarily increases to the Contract Sum. Section 8.3.1 does not, by itself, foreclose a contractor from seeking additional compensation from the owner for such delays. This possibility is explicitly left open by Section 8.3.3.

The 2007 version of Section 8.3.1 of AIA Document A201 has been amended, as follows, to create the 2017 version of this section:

(1) The phraseology is clarified by adding numerals.
(2) The phrase "separate contractor employed by the Owner" is replaced with the term "Separate Contractor," which is defined in Section 6.1.
(3) Adverse weather conditions documented under Section 15.1.6.2 are added as a potential trigger for an extension of the Contract Time. In the 2007 version of Section 8.3.1, adverse weather conditions (which the 2007 version addressed in Section 15.1.5.2) were arguably included in "other causes beyond the Contractor's control."
(4) Delays authorized by the owner pending binding dispute resolution (including mediation, arbitration, and litigation) are a potential trigger for extensions of the Contract Time. The 2007 version of Section 8.3.1 included only owner-authorized delays due to mediation and arbitration as potential triggers for extension of the Contract Time.
(5) The contractor can assert any other cause that it feels, and the architect determines, justifies delay as a means to extend the Contract Time. The prior version of Section 8.3.1 empowered the architect to unilaterally (though reasonably) determine whether any other causes of delay justify an extension of the Contract Time.
(6) The requirement that the Contract Time must be extended by written change order was deleted from this section.

Courts have held that a contractor can be entitled to an extension of the Contract Time, under some circumstances, even absent an executed change order.[20]

Interpreting earlier versions of Section 8.3.2, some courts have held, based on the specific facts in question, that the contractor's exclusive remedy for construction delays is an extension of the Contract Time—absent a finding that the owner was negligent, acted in bad faith, or breached an implied promise not to delay the work.[21] Some courts have construed earlier versions of Section 8.3.2 as an indication that the parties anticipated the possibility of delays and fully provided for such delays with an extension of time.[22]

Depending on the size and complexity of the project, owners may wish to consider modifying Section 8.3.1 to provide for an extension of the Contract Time only in cases of delay to activities on the critical path of the project schedule. Even if the project will not be actively managed by CPM techniques, the critical path can be forensically reconstructed if a delay-related claim occurs. Adding such a critical

---

20. *See* Allstate Interiors & Exteriors, Inc. v. Stonestreet Constr., LLC, 907 F. Supp. 2d 216, 237 (D.R.I. 2012) (holding that the owner's delay in providing permanent power to the project and the changes in the design of the fire protection and life safety systems were excusable delays even though the owner did not execute a change order authorizing extended general conditions); Board of Educ. of Woodford Cnty. v. D.W. Wilburn, Inc., 2010 WL 2010760, at *4 (Ky. Ct. App. May 21, 2010) (holding that the contractor was entitled to an extension of time after the contractor was delayed due to events beyond contractor's control and the decisions of the owner, even though the contractor's requested change orders were rejected by the owner); Bouchard v. Boyer, No. 543089, 1999 WL 335845, at *4 (Conn. Super. Ct. May 17, 1999) (holding that, although the contractor did not extend the Contract Time through a written change order, the contractor was entitled to recover damages when the contractor properly attempted to secure building permits but could not do so for reasons beyond its control).

21. United States *ex rel.* Straus Systems, Inc. v. Assoc. Indemnity Co., 969 F.2d 83, 85 (5th Cir. 1992).

22. Burgess Constr. Co., v. M. Morrin & Son Co., 526 F.2d 108, 114–15 (10th Cir. 1975).

path limitation can be achieved by modifying the beginning of the first sentence of Section 8.3.1 to begin "If delay to the critical path of the Project impacts Contractor's commencement or progress of the Work due to . . . ."

Owners may wish to consider adding "the Owner shall have the right to require the Contractor to substantiate all claims for time extensions with a Project schedule and CPM analysis in a form and level of detail acceptable to the Owner." Contractors may wish to counter by changing "acceptable" in that sentence to "reasonably acceptable" or by editing the last words of the sentence to read "mutually acceptable to the Owner and Contractor."

Contractors may wish to consider clarifying their potential entitlement to additional money by changing the wording of the last part of Section 8.3.1 from "the Contract Time shall be extended" to "the Contract Sum shall be adjusted and the Contract Time shall be extended." Owners who accept this modification may, in some cases, wish to consider modifying the last part of Section 8.3.1 to remove the architect as the decision maker. Owners may not want to empower architects to make decisions on monetary claims. If the architect is empowered to decide such claims, owners and contractors may wish to consider adding language to Section 8.3.1 to explicitly reserve their rights to challenge the architect's decision through the dispute resolution provisions of the Contract.

Section 8.3.1 may entitle contractors to extensions of time for "causes beyond the Contractor's control." Some argue this broad catchall provides basic fairness to contractors by granting them relief from eventualities that they cannot control. Owners, however, may wish to consider narrowing that provision to leave open the possibility of extending the Contract Time only in cases of "unforeseeable causes beyond the Contractor's control."

Contractors may wish to consider adding a more explicit description of the delay events for which they would not expect to be responsible. Such a list could include delays caused by acts of God, riots, thefts, accidents, blizzards, quarantines, blockades, governmental acts, war, rebellion, terrorism, public utility interruptions, transportation delays, shortages of fuel, labor or material, or many other conceivable causes arguably beyond the contractor's control. Owners may dispute whether all such events are beyond the contractor's control. Owners may wish to consider adding "unless such delays could have been reasonably anticipated by, or are caused, in whole or in part, by the Contractor" to Section 8.3.1.

Rather than make "labor disputes" excusable, which can result under Section 8.3.1, some owners may wish to consider imposing labor-related obligations on contractors. An owner may wish to obligate the contractor to comply with project labor agreements, if applicable. If union labor is used, the owner may wish to consider requiring the contractor to "make all necessary arrangements to reconcile, without delay, damage or cost to the Owner, and without recourse to the Owner, any labor disputes or disturbances" and "to use its best efforts to mitigate the effect of such disputes or disturbances upon the Project." If a mix of nonunion labor is used, owners may wish to consider requiring the contractor "to cause harmony to be achieved among such non-union labor and any unions having jurisdiction over the same work." Owners may also wish to consider adding something like "Any labor dispute that does not result from a breach of this Agreement shall be considered an excusable but not a compensable event of delay, entitling the Contractor to additional time but no adjustment of the Contract Sum."

§ 8.3.2 Claims relating to time shall be made in accordance with applicable provisions of Article 15.

Section 8.3.2 provides that any claims relating to time are subject to Article 15. Article 15 of AIA Document A201 includes the definition of "Claims," the procedure for giving notice of claims, and provisions specific to claims for additional time and money. For example, Section 15.1.5 (Claims for Additional Cost) requires the contractor to give notice to the owner before proceeding to execute the portion of the work that is the subject of the claim—unless it relates to an emergency endangering life or property under Section 10.4. Similarly, Section 15.1.6 (Claims for Additional Time) requires the contractor to give notice under Section 15.1.3 to the owner with an estimate of the cost and probable effect of delay on the progress of the work. Article 15 also outlines the rules relating to dispute resolution, including decisions by the initial decision maker and mediation and arbitration procedures.

No changes to the 2007 version of Section 8.3.2 are made in the 2017 version of AIA Document A201. Section 8.3.2 is merely a cross-reference to Article 15. If owners and contractors desire to revise and supplement the notice and dispute resolution procedures for claims relating to time and money, Article 15 can be edited accordingly.

In many cases, contractual requirements for giving notice and making claims are not strictly or rigidly enforced unless one party's failure to comply with such requirements prejudiced the other. However, courts occasionally construe Section 8.3.2 as definitive in its reference to Article 15 and strictly adhere to its specific requirements for giving notice and making claims. In one case, a contractor did not follow the proper process for asserting a claim for an extension of time, and the contractor's request for such extension was denied.[23]

§ 8.3.3 This Section 8.3 does not preclude recovery of damages for delay by either party under other provisions of the Contract Documents.

Depending on other applicable contract terms, Section 8.3.3 leaves open the possibility of one party seeking to recover damages for delay against the other party. The 2017 version of AIA Document A201 makes no changes to the 2007 version of Section 8.3.3.

The standard AIA text of section 8.3 has been interpreted to not preclude delay damages because this section does not provide, specifically, that an extension of time is the sole or exclusive remedy.[24]

Some may wish to revise Section 8.3.3 to substitute a "no damages for delay" clause in place of the AIA's standard text and permissive approach to recovery of damages for delay. Such a clause could provide, for example, that the contractor's sole remedy for any delay, regardless of its cause, is an extension of the Contract Time. Whether to include such a clause should be considered in light of Section 15.1.7, in which the contractor and the owner mutually waive any right to recover consequential damages from each other.

23. McCarthy Bros. Constr. Co. v. Pierce, 832 F.2d 463, 469 (8th Cir. 1987). *See also* Old Colony Constr., LLC v. Town of Southington, 113 A.3d 406 (Conn. 2015) (holding that, when a contractor did not strictly comply with the contractual procedure for an extension of time, the contractor was responsible for liquidated damages, even if the delay was otherwise excusable).

24. City of Gillette v. Hladky Constr. Inc., 196 P.3d 184, 200 (Wyo. 2008). *See also In re* Contract Controversy—Agricultural Biotechnology/Molecular Biology Complex, EllisDon Constr., Inc. v. Clemson University, 2007 WL 4898329, at *26 (S.C. Procurement Review Panel 2007) (holding that an extension of the Contract Time was proper and awarding an increase in the Contract Sum based on a per diem delay charge).

Some courts uphold "no damages for delay" clauses when the parties are sophisticated and have contracted freely.[25] Courts generally do not like to rule in a way that substitutes a new or different agreement in place of the parties' contract. Therefore, just like choosing to not delete the mutual waiver of consequential damages at Section 15.1.7, inclusion of a "no damages for delay" clause can be an effective way for parties to limit the risk of delay-related claims.

Some courts have limited the enforcement of "no damages for delay" clauses in certain circumstances. The issue is jurisdiction-specific, but some courts have recognized four judicial exceptions to the application of "no damages for delay" clauses: (1) if the delay is not of a kind contemplated by the parties, (2) if the length of the delay is unreasonable, (3) if there is evidence of bad faith, and (4) if the delay is caused by the owner's active interference (defined as the commission of some affirmative, willful act, in bad faith, that unreasonably interferes with the contractor's compliance with the terms of the contract).[26]

Section 8.3 may be revised to explicitly identify delay events that entitle the contractor to additional time or money. Such edits have many possible permutations because no two contracts or projects are exactly alike. The model clause set forth below is not appropriate for all contracts or all projects. It is merely intended to identify possible subjects to address when customizing Article 8. Owners and contractors sometimes elect to completely delete and replace all of Section 8.3 with terms similar to the following:

§ 8.3.1 "Force Majeure" means unavoidable material or labor shortages which could not have reasonably been anticipated, strikes or labor disputes (that arise for reasons other than a dispute at or directly involving the Contractor or the Project), unusually severe adverse weather conditions not common to the area where the Project is located (as determined on the basis of the past ten (10) year weather records collected by the National Oceanic and Atmospheric Administration), riots, war or insurrection, or other similar unforeseeable causes beyond Contractor's control which were not caused by the Contractor, a Subcontractor, or anyone for whom the Contractor, is legally responsible (each a "Contractor Party").

§ 8.3.2 "Force Majeure Delay" means a delay to the critical path of the Project that impacts Contractor's ability to achieve Substantial Completion within the Contract Time, but only to the extent that such delay was caused by Force Majeure and was not caused by (i) any act, omission, negligence, fraud, willful misconduct, or other misconduct of a Contractor Party; (ii) the bankruptcy, insolvency, or other financial problem of any Contractor Party; (iii) the failure of any Contractor Party to perform the Work in accordance with the terms of the Contract Documents or its contract for the Work; or (iv) Contractor's breach of this Agreement. If a delay is caused by unusually severe weather conditions, such claim shall be documented by data substantiating that weather

25. *See* C & H Electric, Inc. v. Town of Bethel, No. X07CV106015518S, 2012 WL 2899314, at *4, *11 (Conn. Super. Ct. June 15, 2012) (holding that when the contract included a "no damages for delay" clause and no "active interference" by the owner was found, the contractor's claim for damages for lost productivity was denied), *aff'd*, 96 A.3d 477 (Conn. 2014); Dynalectric Co. v. Whittenberg Constr. Co., No. 5:06-CV-00208-JHM, 2010 WL 4062787, at *3 (W.D. Ky. Oct. 15, 2010) (holding that "no damages for delay" clauses are enforceable); C & C Plumbing and Heating, LLP v. Williams Cnty., 848 N.W.2d 709 (N.D. 2014) (holding a "no damages for delay" clause is not repugnant to other provisions of the contract and includes no exception for uncontemplated delays, and that the owner was insulated from liability even though a "time is of the essence" clause was included in the contract); Gilbert H. Moen Co. v. Spokane River Assoc., 88 Wash. App. 1064 (1997) (upholding the "no damages for delay" clause and finding that the owner did not waive the express provision in writing or by conduct).

26. *See* Peter Kiewit Sons' Co. v. Iowa Southern Utilities Co., 355 F. Supp. 376, 397 (S.D. Iowa 1973) (summarizing these four judicial exceptions).

conditions were abnormal for the period of time in the area where the Project is located, could not have been reasonably anticipated by the Contractor, and had an adverse effect on the scheduled construction. Contractor shall use commercially reasonable and good faith efforts to avoid or mitigate the length and impact of any Force Majeure.

§ 8.3.3 "Compensable Delay" means a delay, other than a Force Majeure Delay, to the critical path of the Project that impacts Contractor's ability to achieve Substantial Completion within the Contract Time, but only to the extent that such delay was due to a change for which Contractor is entitled to compensation under the Contract or was caused by the fault of Owner or an entity for which Owner is responsible, or by any Separate Contractor employed by Owner and was not caused by (i) any act, omission, negligence, or willful misconduct of any Contractor Party; (ii) the bankruptcy, insolvency, or other financial problem of any Contractor Party; (iii) the failure of any Contractor Party to perform in accordance with the terms of its contract for the Work; or (iv) Contractor's breach of this Agreement.

§ 8.3.4 Within two (2) business days after becoming aware of Force Majeure or the possibility thereof, or within seven (7) business days after any event that may give rise to a Compensable Delay, Contractor shall notify Owner in writing, stating in such notice the basis for the claim of an existing or potential Force Majeure Delay or Compensable Delay, as applicable. Contractor shall advise Owner of the estimated length of the expected delay as soon as practicable after the information necessary to estimate the delay is available (which notice shall include the basis for such estimate and identify its impact on the critical path). As Contractor obtains additional information to calculate the actual delay, Contractor shall continue to update Owner, providing it with the basis used in calculating such update. As Contractor obtains information to substantiate the actual delay, Contractor shall so notify Owner, providing the information it has utilized in calculating such delay.

§ 8.3.5 Adjustment to Contract Sum and Contract Time.

.1 In the event of a Force Majeure Delay for which Contractor has provided timely notice under Section 8.3.4 above, and subject to Section 8.3.6 below, Owner shall, in its sole discretion, either (A) extend the Contract Time by Change Order to the extent of the impact of such delay on the critical path or (B) direct Contractor to accelerate the Work, in which case the Contract Sum shall be increased by the additional Cost of the Work due solely to such acceleration, plus Contractor's Fee on such costs. Contractor agrees to accept the foregoing as its sole remedy in the event of a Force Majeure Delay.

.2 In the event of a Compensable Delay for which Contractor has provided timely notice under Section 8.3.4 above, and subject to Sections 8.3.6 below, Owner shall, in its sole discretion, either (A) extend the Contract Time by Change Order to the extent of the impact of such delay on the critical path or (B) direct Contractor to accelerate the Work, in which case the Contract Sum shall be increased by the additional Cost of the Work due solely to such acceleration, plus Contractor's Fee on such costs. In addition, the Contract Sum shall be adjusted by the actual, reasonable, and unavoidable costs that Contractor would not have incurred but for the Compensable Delay, except to the extent that such costs are included in an adjustment under Section 8.3.5.1(B) above. Contractor agrees to accept the foregoing as its sole remedy in the event of a Compensable Delay.

§ 8.3.6 Notwithstanding anything to the contrary herein:

.1 The Contractor shall use commercially reasonable and good faith efforts to avoid or mitigate the effect of any delay described in Section 8.3 as a condition precedent to a claim for an adjustment of the Contract Sum or Contract Time;

.2 Contractor shall not be entitled to an increase in the Contract Sum to the extent that a Compensable Delay or a Force Majeure Delay is concurrent with a delay caused by any Contractor Party; provided, however, in the event of such concurrent delay the Contractor shall be entitled to an extension of the Contract Time;

.3 In order to receive an increase of the Contract Sum or extension of the Contract Time, the Contractor must demonstrate that there was an actual delay in the performance of the Contractor's Work that impacts the critical path and/or an actual increase in the Contractor's costs resulting from the delay after accounting for concurrent delays;

.4 Under no circumstances shall Contractor be entitled to assert a claim for additional or extended home office overhead;

.5 Under no circumstances shall Contractor be entitled to assert a claim for an increase in the Contract Sum on the basis that Contractor would have completed the Work early but for an Owner-caused delay; and

.6 The Owner shall have the right to require the Contractor to substantiate all claims for adjustments to the Contract Sum or Contract Time in a form and level of detail that is reasonably acceptable to Owner.

Owners may wish to clarify that the above language is not intended as a limitation on the owner's right to order, or the circumstances in which the owner may order, the contractor to accelerate performance of the work under the other terms of the agreement.

Owners may also wish to add that "Failure to provide timely notices as required by Section 8.3.4 shall be deemed a waiver of Contractor's right to a Change Order with respect to such Force Majeure Delay or Compensable Delay." Contractors may oppose this additional language, which might not be enforceable in binding dispute resolution proceedings. To address those concerns, that language may be qualified by adding "but only to the extent the Owner was prejudiced by Contractor's failure to provide said notice" to the end of the sentence.

In conclusion, the revised version of Article 8 in AIA Document A201–2017 is not significantly different from its AIA-Document A201–2007 predecessor. Courts will probably continue to find case law interpreting the prior versions of Article 8 as persuasive authority to interpret the 2017 version of Article 8. Like its predecessor, the new version of Article 8 is virtuous because it covers the basic terms relevant to contract time. On small and simple projects, or those in which considerations of time and schedule are not paramount, the standard AIA text of Article 8 may suffice. In many cases, owners and contractors should consider supplementing the standard AIA text of Article 8 with additional and more specific terms. The potential Article 8 modifications suggested above, while not appropriate or necessary for every contract (and may require changes for consistency to be carefully made throughout AIA Document A201–2017 and other Contract Documents), identify subjects for contractors and owners to consider when customizing Article 8 for use on particular projects.

CHAPTER 9:
Article 9 Payments and Completion

Tamara Lindsay
Peter J. Marino
Stephen D. Marso

ARTICLE 9 PAYMENTS AND COMPLETION

As the title indicates, Article 9 contains provisions relating to project payments and completion. These provisions detail the processes related to payment applications, progress payments, final payments, substantial completion, and final completion, as well as the dispute-resolution procedures applicable to payment disputes.

Regular and orderly progress payments are absolutely critical to a construction project. Disruption in the payment stream can and often does lead to delays, claims, and litigation. As with the prior editions, the 2017 Form A201 Article 9 payment provisions generally are directed toward balancing the contractor's need to keep the stream of payments flowing to itself and its subcontractors against the interests of the owner (and often the owner's lender(s)) in getting the job built in a timely fashion, properly, and within budget.

§ 9.1 CONTRACT SUM

§ 9.1.1 The Contract Sum is stated in the Agreement and, including authorized adjustments, is the total amount payable by the Owner to the Contractor for performance of the Work under the Contract Documents.

Section 9.1 simply references the Contract Sum agreed upon by the parties in the base agreement document (A101 or the like) as the total amount, subject to authorized adjustments, that the owner is required to pay the contractor for the work. There was no substantive change to this section in the current 2017 edition of the A201. The text remains identical to the 2007 version. The section designation was changed from 9.1 to 9.1.1 to accommodate Section 9.1.2, discussed below. There has been very little case law or commentary on this section as it is a very general provision that sets the stage for the remaining payment provisions in Article 9.

§ 9.1.2 If unit prices are stated in the Contract Documents or subsequently agreed upon, and if quantities originally contemplated are materially changed so that application of such unit prices to the actual quantities causes substantial inequity to the Owner or Contractor, the applicable unit prices shall be equitably adjusted.

This new Section 9.1.2 is essentially the same provision that appeared at Section 7.3.4 in the 2007 edition of the A201. It is what is sometimes called a "Variation in Estimated Quantities," or VEQ clause, which addresses a practical situation sometimes encountered in construction agreements that utilize unit pricing; namely, if the actual number of units is confirmed to be meaningfully more or less than the parties anticipated at the time of contracting, the agreed-upon per unit price may not fairly and accurately reflect the cost of the work. This issue can arise in unit-priced contracts regardless of whether such pricing is included in the agreement at signing or added later through change order or other contract modification.

A common item to which unit pricing is often applied is earthwork on a site where rock is likely to be encountered, but the amount of rock is indeterminate at the time of contract execution.[1] Section 9.1.2 allows both parties to point to an

1. *See, e.g.*, Celco Constr. Corp. v. Town of Avon, 26 N.E.3d 202 (Mass. App. Ct. 2015).

express contract provision allowing them to argue for the adjustment of previously agreed-upon unit pricing should the circumstances on the project change from what was initially contemplated.

VEQ clauses, especially on public projects, often specify a percentage variation from the anticipated quantities before an adjustment in unit prices may be sought. A fairly typical arrangement is to allow either party to request repricing in the event of a 15 percent or greater variation in actual quantities."[2] The drafters of the 2017 edition of the A201 rejected this specific-percentage approach, however, in favor of the more subjective standard. Under the 2017 form, equitable adjustment of a unit price is triggered when quantities increase or decrease in such a "material" way as to render application of that unit price "substantially inequitable" to either the owner or contractor. If this two-part requirement is met, this section provides that the unit price "shall be equitably adjusted."

Both the terms "materially changed" and "substantial inequity" are legal terms of art, and there will no doubt be disputes over their application in particular circumstances. Obviously, there has been no case law to date interpreting this new section, though that will likely change in the near term as disputes arise over its application. While there have been no cases under this new provision, similar unit price adjustment provisions have been addressed in several cases, with differing and very fact-specific and court-specific results.[3]

§ ~~§~~9.2 SCHEDULE OF VALUES

> Where the Contract is based on a stipulated sum or Guaranteed Maximum Price, the Contractor shall submit a schedule of values to the Architect before the first Application for Payment, ~~a schedule of values~~ allocating the entire Contract Sum to the various portions of the Work ~~and~~. The schedule of values shall be prepared in ~~such~~the form, and supported by ~~such~~the data to substantiate its accuracy ~~as~~, required by the Architect ~~may require~~. This schedule, unless objected to by the Architect, shall be used as a basis for reviewing the Contractor's Applications for Payment. Any changes to the schedule of values shall be submitted to the Architect and supported by such data to substantiate its accuracy as the Architect may require, and unless objected to by the Architect, shall be used as a basis for reviewing the Contractor's subsequent Applications for Payment.

Section 9.2 applies to stipulated-sum and guaranteed-maximum-price contracts. It provides that, before the contractor's first application for payment, the contractor must submit a schedule of values that allocates the entire Contract Sum to the various portions of the work. The architect and owner use the schedule of values to help measure the progress of the project against the contractor's corresponding requests for payment. This section makes it clear that, unless the schedule of values submitted by the contractor is objected to by the architect, it will be used as the basis for the architect's review of payment applications.

---

2. 3 Philip L. Bruner, Patrick J. O'Connor, Bruner & O'Connor on Construction Law § 8.4 (2002).
3. *See, e.g.*, Pavers, Inc. v. Bd. of Regents of Univ. of Neb., 755 N.W.2d 400, 408–10 (Neb. 2008) (affirming the trial court's order that contractor be compensated at cost for excess quantities instead of at the full unit price); Bellevue Tech. Tower, LLC v. DPR Constr., Inc., No. 52695-2-I, 2005 WL 39813, at *6–7 (Wash. Ct. App. Jan. 10, 2005) (stating that an increase in the quantity of contaminated soil removed from the project site from 500 cubic yards to 24,555 cubic yards "was undisputedly a material change," but refusing to allow equitable adjustment because the owner did not request it until after the 21-day time limit for claims).

The 2017 drafters revised this section to make several minor stylistic changes. They also added a provision specifically allowing for changes to the schedule of values during the course of a project as long as the changes are submitted to the architect, supported by such data as the architect may require to substantiate accuracy, and not objected to by the architect. Allowing for changes allows the contractor to "correct" the original schedule of values under the contract to match the actual progress of the work. How an architect responds to a contractor's submission of a revised schedule of values could impact potential delay or other claims on a project, and this provision could help in avoiding disputes involving allegations of "advance" or false billing sometimes tied to alleged unbalanced bids.[4]

§ 9.3 APPLICATIONS FOR PAYMENT

> **§ 9.3.1** At least ten days before the date established for each progress payment, the Contractor shall submit to the Architect an itemized Application for Payment prepared in accordance with the schedule of values, if required under Section 9.2, for completed portions of the Work. ~~Such~~The application shall be notarized, if required, and supported by ~~such~~all data substantiating the Contractor's right to payment ~~as~~that the Owner or Architect ~~may~~ require, such as copies of requisitions, and releases and waivers of liens from Subcontractors and ~~material~~ suppliers, and shall reflect retainage if provided for in the Contract Documents.

This section sets forth the basic requirements for the submission of applications for payment and makes clear that applications may be submitted only for completed portions of the work. It also provides the owner or architect with the authority to dictate the level and type of data substantiating the contractor's right to payment, and it specifically references copies of subcontractor requisitions. This section was revised in 2017 to add a specific reference to "releases and waivers of liens" as an additional example of supporting data that may be required with all pay applications. This change seems to track common practice today of nearly all owners, lenders, and architects to require that partial lien releases and waivers be submitted with each interim application for payment, and not simply in connection with the final payment application or in exchange for final payment.

Universal Concrete Prods. Corp. v. Turner Constr. Co.,[5] is a relatively recent Fourth Circuit Court of Appeals opinion that cited the 2007 version of Section 9.3.1. *Universal Concrete* involved a subcontractor that was attempting to invalidate as ambiguous the "pay-when-paid" clause contained in its subcontract. The Court of Appeals, affirming the district court, held that the particular pay-when-paid clause at issue was unambiguous and thus enforceable under applicable Virginia law. It supported the holding by noting that the subcontract incorporated by reference the provisions of the AIA prime contract and cited Section 9.3.1's requirement that the contractor submit "requisitions from Subcontractors" with its invoices, noting that this provision "contemplate[s] the likelihood that the general contractor and subcontractor will enter into a pay-when-paid contract." This *Universal Concrete* ruling is obviously specific to the facts and the jurisdiction, though interesting nonetheless.

---

4. *See, e.g.*, J & D Maint. & Servs. v. United States, 45 Fed. Cl. 532 (1999) (assessing treble damages under the False Claims Act against a federal contractor due to advance billing where the contractor admitted to intentionally front-loading the bid).

5. 595 F.3d 527 (4th Cir. 2010).

§ 9.3.1.1 As provided in Section 7.3.9, such applications may include requests for payment on account of changes in the Work that have been properly authorized by Construction Change Directives, or by interim determinations of the Architect, but not yet included in Change Orders.

Section 9.3.1.1 essentially reiterates Section 7.3.9 confirming that a contractor may seek payment for Construction Change Directives (CCDs) and other interim determinations of the architect even though such have not been included in a fully executed change order, and even though full agreement on pricing of a change may not yet be reached. The 2017 version of this section is unchanged from the prior 2007 version. This provision helps keep contract funds flowing under circumstances that might otherwise put financial stress on the contractor or its subcontractors when some portion of a change item or change pricing is in dispute—which happens often. There is a fuller discussion of the issues arising from CCDs and payment therefor in the chapter on Article 7 of the Desk Book, as the appellate cases in this area typically analyze these issues under Section 7.3.9.

§ 9.3.1.2 Applications for Payment shall not include requests for payment for portions of the Work for which the Contractor does not intend to pay a Subcontractor or ~~material~~ supplier, unless such Work has been performed by others whom the Contractor intends to pay.

This section is one of several in the A201 indicative of a trend in industry form agreements to help ensure that payments to contractors appropriately flow down to subcontractors. It does so by prohibiting a contractor from including a request for payment for portions of work unless it intends to pay the subcontractor for such work, or intends to pay a replacement subcontractor for the work at issue. This section obviously is also protective of the owner by reducing the likelihood that an owner will be paying (or overpaying) for work that the contractor knows is or may be defective, or likely to result in a lien, bond, or other subcontractor claim that impacts the project. The section was added to the A201 in the 2007 edition and remains substantially in the same form in the 2017 edition.

§ 9.3.2 Unless otherwise provided in the Contract Documents, payments shall be made on account of materials and equipment delivered and suitably stored at the site for subsequent incorporation in the Work. If approved in advance by the Owner, payment may similarly be made for materials and equipment suitably stored off the site at a location agreed upon in writing. Payment for materials and equipment stored on or off the site shall be conditioned upon compliance by the Contractor with procedures satisfactory to the Owner to establish the Owner's title to such materials and equipment or otherwise protect the Owner's interest, and shall include the costs of applicable insurance, storage, and transportation to the site, for such materials and equipment stored off the site.

This section addresses payment for on-site and off-site stored materials and equipment. This section clearly establishes the contractor's right to be paid for materials and equipment delivered and stored at the project site for later incorporation into the work. It does not further detail or qualify the timing or conditions surrounding payment for on-site materials and equipment, and issues and claims can often arise as to stored materials. Common issues involve damage to or degradation of on-site stored materials.[6]

As to payment for project-related materials and equipment stored off-site, this provision requires that the contractor obtain advance approval of the owner, and that the off-site storage location be agreed upon in writing. Issues as to title to stored materials will be discussed in connection with the related Section 9.3.3 below.

> **§ 9.3.3** The Contractor warrants that title to all Work covered by an Application for Payment will pass to the Owner no later than the time of payment. The Contractor further warrants that upon submittal of an Application for Payment all Work for which Certificates for Payment have been previously issued and payments received from the Owner shall, to the best of the Contractor's knowledge, information, and belief, be free and clear of liens, claims, security interests, or encumbrances, in favor of the Contractor, Subcontractors, ~~material~~ suppliers, or other persons or entities ~~making a claim by reason of having~~that provided labor, materials, and equipment relating to the Work.

This section works together with prior Section 9.3.2 to establish that title passes to the owner as to all work (including on-site or off-site stored materials addressed in Section 9.3.2) at the time of payment by the owner for such work. This section also includes a warranty from the contractor to the owner that to the best of the contractor's "knowledge, information, and belief, all work included in an Application for Payment is free of liens, claims, security interests or other encumbrances" by any party in the contractor chain—including subcontractors, material and equipment suppliers, and laborers. This provision often comes into play in the context of the bankruptcy of a subcontractor or supplier, where other parties (or their lenders) downstream in the contract chain assert Uniform Commercial Code (UCC) security interests or other liens or claims typically in off-site or on-site materials or equipment. Depending upon the nature of the security interest and the jurisdiction and/or applicable law involved, claimants may attempt to repossess materials and equipment, which can create disruption and delay on a project. These situations are often further complicated by the broad bankruptcy "stay" that comes into effect under Section 523 of the U.S. Bankruptcy Code if one of the parties in the contract or supply chain files for bankruptcy.

This section, combined with the newly revised Section 9.3.1 (that now specifically encourages releases and waivers of liens from subcontractors and suppliers), is an attempt to minimize disputes and disruption surrounding issues of "title" to work (i.e., materials and equipment). This Section 9.3.3 was not revised in any material way in the 2017 edition.

A sampling of case law details the frequency with which these issues can arise.[7]

6. *Compare* Boh Bros. Constr. Co. v. Louisiana, 9 So. 3d 982 (La. Ct. App. 2009) (requiring the owner to compensate the contractor for damage to stored materials from Hurricane Katrina), *with* McPhee Elec. Ltd. v. Konover Constr. Corp., No. CV075009694S, 2009 WL 4846555, at *22 (Conn. Super. Ct. Oct. 22, 2009) (finding that under subcontract, electrical subcontractor was responsible for replacing stored materials that had been stolen). The practitioner should also see the payment terms in the Owner-Contractor Agreement, for example, Section 5.16.1 of the A101–2017, which permits billing for stored materials as part of payment application process.

7. *See, e.g.*, *In re* Roby, No. 12-10450-TEC, 2013 WL 2452178 (Bankr. N.D. Cal. June 4, 2013) (finding contractor's debt to owner nondischargeable due to fraud after contractor, in effort to comply with Section 9.3.3, falsely told owners he had used past progress payments to pay sums owed to subcontractors); *In re* Watermark Marina of Wilmington, LLC, No. 08-08103-8-JRL, 2009 WL 4405184 (Bankr. E.D.N.C. Nov. 24, 2009) (rejecting argument that contractor's compliance with Section 9.3.3 waived its lien rights as to not-yet-paid-for work); E. Elec. & Heating, Inc. v. Pike Creek Prof'l Ctr., Nos. 85L-AP-21, 85L-MY-1, 85C-MR-79, 1987 WL 9610 (Del. Super. Ct. Apr. 7, 1987) (holding that the contractor's failure to comply with Section 9.3.3 constituted a material breach of the contract); Luczak Bros., Inc. v. Generes, 451 N.E.2d 1267 (Ill. App. Ct. 1983) (rejecting the argument that Section 9.3.3 itself operates as a waiver of Subcontractors' liens).

§ 9.4 CERTIFICATES FOR PAYMENT

§ 9.4.1 The Architect will, within seven days after receipt of the Contractor's Application for Payment, either (1) issue to the Owner a Certificate for Payment in the full amount of the Application for Payment, with a copy to the Contractor~~,~~; or (2) issue to the Owner a Certificate for Payment for such amount as the Architect determines is properly due, ~~or~~and notify the Contractor and Owner ~~in writing~~ of the Architect's reasons for withholding certification in ~~whole or in~~ part as provided in Section 9.5.1; or (3) withhold certification of the entire Application for Payment, and notify the Contractor and Owner of the Architect's reason for withholding certification in whole as provided in Section 9.5.1.

This section establishes the basic obligation of the architect to respond to an application for payment within seven days of its submission. Under this section, the architect must either approve payment in part or in full, triggering the owner's obligation to make payment, or notify the contractor and owner as to the architect's reasons for withholding payment certification. The A201 does not address how long the owner has to make payments once the architect approves an application. That payment term must be inserted in the A101 (see Section 5.1.3), A103 (see Section 12.1.3), A111, or other "base" form agreement used for a project. (See Section 9.6.1 and discussion thereof.)

Section 9.4.1 is clear that payment certification may be withheld only for one or more of the nine reasons enumerated in Section 9.5.1. The bases for withholding certification are discussed in detail later in this chapter in connection with Section 9.5.1. While the only change in the text of this section since the 2007 edition is the deletion of the requirement that notice from the architect to the contractor and owner of its reasons for withholding payment certification must be "in writing," new Section 1.6 of the A201 requires that *all* notices are to be in writing.

There is limited case law specifically addressing the general requirements of Section 9.4.1. Most of the case law in this area assesses the merits of the specific bases under 9.5.1 that are cited by the architect for withholding payments in a particular case. Having said that, the earlier version of this provision was cited in one appellate court decision.[8]

§ 9.4.2 The issuance of a Certificate for Payment will constitute a representation by the Architect to the Owner, based on the Architect's evaluation of the Work and the data ~~comprising~~in the Application for Payment, that, to the best of the Architect's knowledge, information, and belief, the Work has progressed to the point indicated ~~and that~~, the quality of the Work is in accordance with the Contract Documents, and that the Contractor is entitled to payment in the amount certified. The foregoing representations are subject to an evaluation of the Work for conformance with the Contract Documents upon Substantial Completion, to results of subsequent tests and inspections, to correction of minor deviations from the Contract Documents prior to completion, and to specific qualifications expressed by the Architect. ~~The issuance of a Certificate for Payment will further constitute a representation that the Contractor is entitled to payment in the amount certified.~~ However, the issuance of a Certificate for Payment will not be a representation that the Architect has (1) made exhaustive or continuous on-site inspections to check the

8. *See* Envirocorp Well Servs., Inc. v. Camp Dresser & McKee, Inc., No. IP 99-1575-C-T/G, 2000 U.S. Dist. LEXIS 16088 (S.D. Ind. Oct. 25, 2000) (court upheld architect's right to withhold payment to contractor where architect and owner recommended withholding payment from a particular subcontractor, and contractor nonetheless approved payment to that subcontractor; court further held that the subcontractor was entitled to payment from the contractor (not the owner)).

quality or quantity of the Work; (2) reviewed construction means, methods, techniques, sequences, or procedures; (3) reviewed copies of requisitions received from Subcontractors and ~~material~~ suppliers and other data requested by the Owner to substantiate the Contractor's right to payment; or (4) made examination to ascertain how or for what purpose the Contractor has used money previously paid on account of the Contract Sum.

This section establishes the extent of the architect's representations as to its review of payment applications and related certifications. It contains limited representations as to the architect's evaluation of the work in place on the project as well as review of the supporting "data in the Application for Payment." The 2007 edition of this section, which has been carried forward without substantive change in the 2017 edition, qualifies the scope of the architect's representations as to both percentage completion and conformance of the construction to the design. The limited nature of the representations since the 2007 revisions to this section have led more owners to engage separate owner's representatives or other construction consultants to take on more robust construction oversight obligations. These provisions should be understood in conjunction with an architect's agreed-upon contract administration services, if any, included in the owner's agreement with the architect (see, for example, the AIA B101 form agreement provisions regarding contract administration services, basic services, and additional services).

This provision typically comes into play if and when latent construction defects are identified on a project.[9]

§ 9.5 DECISIONS TO WITHHOLD CERTIFICATION

Section 9.5 contains four subsections that prescribe when an architect may withhold payment certification to a contractor, provide for the applicable dispute resolution procedures, and address the means by which an owner may nevertheless pay subcontractors and suppliers.

§ 9.5.1 The Architect may withhold a Certificate for Payment in whole or in part, to the extent reasonably necessary to protect the Owner, if in the Architect's opinion the representations to the Owner required by Section 9.4.2 cannot be made. If the Architect is unable to certify payment in the amount of the Application, the Architect will notify the Contractor and Owner as provided in Section 9.4.1. If the Contractor and Architect cannot agree on a revised amount, the Architect will promptly issue a Certificate for Payment for the amount for which the Architect is able to make such representations to the Owner. The Architect may also withhold a Certificate for Payment or, because of subsequently discovered evidence, may nullify the whole or a part of a Certificate for Payment previously issued, to such extent as may be necessary in the Architect's opinion to protect the Owner from loss for which the Contractor is responsible, including loss resulting from acts and omissions described in Section 3.3.2, because of

.1 defective Work not remedied;

.2 third party claims filed or reasonable evidence indicating probable filing of such claims, unless security acceptable to the Owner is provided by the Contractor;

9. *See, e.g.*, Sweeney Co. v. Engineers-Constructors, Inc., 823 F.2d 805 (4th Cir. 1987) (finding that architect certifications associated with progress payments were "good evidence," though not conclusive proof, that subcontractor's work had been substantially completed and of generally acceptable quality); RLI Ins. Co. v. Indian River Sch. Dist., 556 F. Supp. 2d 356 (D. Del. 2008) (rejecting a claim for negligent misrepresentation associated with architect's certifying payment for defective or unfinished work).

.3 failure of the Contractor to make payments properly to Subcontractors or suppliers for labor, materials or equipment;
.4 reasonable evidence that the Work cannot be completed for the unpaid balance of the Contract Sum;
.5 damage to the Owner or a ~~separate contractor~~Separate Contractor;
.6 reasonable evidence that the Work will not be completed within the Contract Time, and that the unpaid balance would not be adequate to cover actual or liquidated damages for the anticipated delay; or
.7 repeated failure to carry out the Work in accordance with the Contract Documents.

Section 9.5.1 prescribes nine different grounds upon which an architect may withhold, in whole or in part, a payment certification to a Contractor; all of the nine grounds must be based on the contractor's work.[10] In addition to the seven grounds listed in the section, the following two additional grounds are incorporated by reference to Section 9.4.2: (1) the Work has not progressed to the point indicated in the pay application; and (2) the quality of the Work for which payment is being requested is not in accordance with the Contract Documents. The architect may also nullify, in whole or in part, a previously issued payment certification for payment, but only for any of the seven listed grounds listed in the section.

If the architect does not certify payment in the amount requested in the application, he or she is required, within seven days after receiving the application, to notify, in writing, the owner and contractor with the reasons for withholding certification. Although not explicitly required, Section 9.5.1 does anticipate that the architect and contractor will communicate with one another in an attempt to reach an agreement on a revised amount that the architect is able to certify. If agreement on a revised amount cannot be reached, the architect is required to certify the amount he or she believes is payable to the contractor.

Payment-certification decisions by architects are important not only to the owner and contractor, but also to the architect. At least one court has held that if, despite knowledge of grounds under Section 9.5.1 upon which certification should be withheld, an architect nevertheless certifies payment, the architect can be liable for damages.[11]

The 2017 amendments to Section 9.5.1 add suppliers to the group of payees to whom nonpayment by a contractor can serve as a basis for an architect to withhold payment certification. The amendments also make a change from "separate contractor" to "Separate Contractor," a defined term. (See Section 6.1.7.).

~~§ 9.5.2 When the above~~**§ 9.5.2** When either party disputes the Architect's decision regarding a Certificate for Payment under Section 9.5.1, in whole or in part, that party may submit a Claim in accordance with Article 15.

Section 9.5.2 is a new provision with the 2017 amendments. It permits the owner and contractor to invoke the claims procedures under Article 15 to resolve disputes over an architect's decision on a payment certification.

10. North Am. Roofing Servs. v. Menard, Inc., 997 N.E.2d 1087 (Ind. Ct. App. 2013) (stating that the withholding of certification because of the filing of third-party claims is appropriate only if the claims arise from the contractor's work, and holding that once the claims are dismissed then withholding of certification is no longer appropriate).

11. *See* Boren v. Thompson & Assocs., 999 P.2d 438 (Okla. 2000) (holding that architect could be liable to unpaid subcontractor for improperly certifying payments to the contractor when architect had knowledge that contractor had failed to procure a statutorily required payment bond and was not paying subcontractors).

§ 9.5.3 When the reasons for withholding certification are removed, certification will be made for amounts previously withheld.

Section 9.5.3 directs that an architect must certify any previously withheld payments once the once the grounds for withholding certification no longer exist.[12] The 2017 amendments make a nonsubstantive deletion of the word "above."

§ 9.5.~~3~~4 If the Architect withholds certification for payment under Section 9.5.1.3, the Owner may, at its sole option, issue joint checks to the Contractor and to any Subcontractor or ~~material or equipment suppliers~~supplier to whom the Contractor failed to make payment for Work properly performed or material or equipment suitably delivered. If the Owner makes payments by joint check, the Owner shall notify the Architect and the ~~Architect will~~Contractor shall reflect such payment on ~~the~~its next ~~Certificate~~Application for Payment.

Section 9.5.4 provides a joint-check mechanism by which payments can be made to subcontractors and suppliers who bear no fault for an architect's decision to withhold a payment certification. Its 2017 amendments enlarge the category of suppliers to whom an owner may issue joint checks by including those suppliers who did not technically qualify as "material and equipment" suppliers under the 2007 version.

Despite the owner having the right to issue joint checks, the contractor, as a required joint payee on the checks, will be the party who ultimately decides whether subcontractors and suppliers receive any proceeds from joint checks. The proceeds actually received by subcontractors and suppliers can be a very important issue for owners, depending on the jurisdiction. Under the "California Rule," all joint-check payees are deemed to have received as proceeds the full amount of the joint check regardless of the amount actually received. In contrast, under the "Federal Rule," a payee is deemed to have received as proceeds only the amount it actually receives.[13]

The 2017 amendments also removed the requirement that the architect reflect the joint-check payment on the next Certificate for Payment and replaced it with the requirement that the contractor reflect such joint-check payment on its next Application for Payment. The 2017 amendments also make a nonsubstantive grammatical change from the word "suppliers" to "supplier."

§ 9.6 PROGRESS PAYMENTS

Section 9.6 contains seven subsections that address issues relating to progress payments by owners to contractors; progress payments by contractors to subcontractors and suppliers; and the ability of owners, subcontractors, and suppliers to obtain information relating to payments received and made by contractors.

§ 9.6.1 After the Architect has issued a Certificate for Payment, the Owner shall make payment in the manner and within the time provided in the Contract Documents, and shall so notify the Architect.

12. *See North Am. Roofing Servs.*, 997 N.E.2d 1087; United States v. Dodd, 978 F. Supp. 2d 404 (M.D. Pa. 2013).
13. *See* Iowa Supply Co. v. Grooms & Co. Constr., Inc., 428 N.W.2d 662 (Iowa 1988) (discussing both rules).

Section 9.6.1 remains unchanged from its 2007 version. It provides that once an architect has issued a payment certification, the owner must make the progress payment to the contractor as provided in the Contract Documents and notify the architect of the payment. Article 5 of the A101 Standard Form of Agreement Between Owner and Contractor prescribes the means and deadlines by which progress payments must be made, and it typically requires contractors to submit monthly payment applications and owners to make monthly progress payments.[14]

§ 9.6.2 The Contractor shall pay each Subcontractor, no later than seven days after receipt of payment from the Owner, the amount to which the Subcontractor is entitled, reflecting percentages actually retained from payments to the Contractor on account of the Subcontractor's portion of the Work. The Contractor shall, by appropriate agreement with each Subcontractor, require each Subcontractor to make payments to Sub-subcontractors in a similar manner.

Section 9.6.2 requires contractors to make progress payments to their subcontractors no later than seven days after contractors receive payments from the owner for the subcontractors' work, and it permits contractors to retain from subcontractors the same percentage of payment the owner retained from the contractor with respect to the subcontractors' work. Although it does not specifically mention suppliers, this section does apply to suppliers per Section 9.6.5. Also, this section requires contractors to include in their subcontract documents provisions that require the subcontractors to make payments to their sub-subcontractors in the same manner.

This section, in conjunction with Section 9.6.7, can be interpreted as creating an express trust in favor of subcontractors and suppliers.[15] It is also a conditional-payment clause, with courts interpreting it as a pay-when-paid clause and not a pay-if-paid clause.[16] Most courts hold that to qualify as a pay-if-paid clause, conditional-payment language must make explicitly clear that subcontractors and suppliers bear the financial risk of nonpayment by the owner and that payment by the owner is a condition-precedent to the contractor's payment obligations.[17]

---

14. *See* Abbey Villas Dev. Corp. v. Cite Contractors, Inc., 716 N.E.2d 91 (Ind. Ct. App. 1999) (relying on custom and practices in the construction industry to interpret the following provision as requiring progress payments in the amount of 90 percent of each invoice, and rejecting owner's argument that the provision permitted all payments to be withheld until the entire project was 90 percent completed: "Payment [to] be made at 90% upon completion of the work with the 10% balance being released when work is accepted by the appropriate Agency and/or accepted by Owner").

15. Westview Invs., Ltd. v. U.S. Bank, 138 P.3d 638 (Wash. Ct. App. 2006).

16. Framingham Heavy Equip. Co. v. John T. Callahan & Sons, Inc., 807 N.E.2d 851 (Mass. App. Ct. 2004). *See also* Evans, Mechwart, Hambleton & Tilton, Inc. v. Triad Architects, Ltd., 965 N.E.2d 1007 (Ohio Ct. App. 2011) (interpreting AIA Document C141–1997 §§ 12.5 and 13.4.3 as pay-when-paid clause); International Eng'g Servs. v. Scherer Constr. & Eng'g of C. Fla., LLC, 74 So. 3d 531 (Fla. Dist. Ct. App. 2011) (interpreting as a pay-when-paid clause a provision similar to Section 9.6.2).

17. *See* Thos. J. Dyer Co. v. Bishop Int'l Engineering Co., 303 F.2d 655 (6th Cir. 1962); Sloan & Co. v. Liberty Mut. Ins. Co., 653 F.3d 175 (3d Cir. 2011) (Pennsylvania law) (interpreting provision containing "condition precedent" language as a pay-if-paid clause); Earthdata Int'l of N.C., L.L.C. v. STV Inc., 159 F. Supp. 2d 844 (E.D. Pa. 2001) (interpreting subcontract language that said payment will be made "when such payment is received" from owner as creating a pay-when-paid clause, and reasoning that if the contractor wanted a pay-if-paid clause it should have used "words such as 'condition,' 'if and only if,' or 'unless and until' that would unambiguously convey that intention").

Even if Section 9.6.2 contained the necessary language to qualify as a pay-if-paid clause, its enforceability would vary from state to state.[18]

The 2017 amendments make a nonsubstantive punctuation change by including commas around the phrase "no later than seven days after receipt of payment from the Owner" in the first sentence.

§ 9.6.3 The Architect will, on request, furnish to a Subcontractor, if practicable, information regarding percentages of completion or amounts applied for by the Contractor and action taken thereon by the Architect and Owner on account of portions of the Work done by such Subcontractor.

Section 9.6.3 remains unchanged from its 2007 version. It requires an architect, to the extent practicable, to furnish to a requesting subcontractor information about the subcontractor's work included by the contractor in its payment application as well as the action taken by the architect and owner on that payment application. Although it does not specifically mention suppliers, the section does apply to suppliers per Section 9.6.5. This provides subcontractors and suppliers with an important investigatory tool to determine if a contractor is properly paying its subcontractors and suppliers. If, for example, the contractor represented in a given payment application that the subcontractor's work was 50 percent completed and the owner paid the contractor accordingly, such information would be very valuable to a subcontractor and would enable it to take immediate steps to obtain the money it is owed if the contractor had only paid for 25 percent completion of the work.[19]

§ 9.6.4 The Owner has the right to request written evidence from the Contractor that the Contractor has properly paid Subcontractors and ~~material and equipment~~ suppliers amounts paid by the Owner to the Contractor for subcontracted Work. If the Contractor

18. In the following states, pay-if-paid clauses are unenforceable: California: Wm. R. Clarke Corp. v. Safeco Ins. Co., 938 P.2d 372 (Cal. 1997) (public policy); Capitol Steel Fabricators, Inc. v. Mega Constr. Co., 68 Cal. Rptr. 2d 672 (Ct. App. 1997) (public policy). Illinois: Brown & Kerr, Inc. v. St. Paul Fire & Marine Ins. Co., 940 F. Supp. 1245 (N.D. Ill. 1996) (mechanic's lien statute). Missouri: Mo. Ann. Stat. § 431.183 (mechanic's liens). New York: West-Fair Elec. Contractors v. Aetna Cas. & Sur. Co., 661 N.E.2d 967 (N.Y. 1995) (public policy). Nevada: Lehrer McGovern Bovis, Inc. v. Bullock Insulation, Inc., 197 P.3d 1032 (Nev. 2008). North Carolina: Am. Nat'l Elec. Corp. v. Poythress Commer. Contractors, Inc., 604 S.E.2d 315 (N.C. Ct. App. 2004). Virgin Islands: Shearman & Assoc. v. Continental Cas. Co., 901 F. Supp. 199 (D.V.I. 1995) (public policy). Wisconsin: Wis. Stat. § 779.135(3).

In the following states, pay-if-paid clauses are enforceable: Arkansas: Brown v. Maryland Cas. Co., 442 S.W.2d 187 (Ark. 1969). Connecticut: Lindade Constr., Inc. v. Cont'l Cas. Co., Nos. X10=UWYCV5008768S, X10=UWY-CV5008767S, 2009 Conn. Super. LEXIS 758 (Feb. 5, 2009). Colorado: Main Elec., Ltd. v. Printz Servs. Corp., 980 P.2d 522 (Colo. 1999). Delaware: McAnulla Elec. Constr., Inc. v. Radius Techs., LLC, No. N10C-03-076 PLA, 2010 Del. Super. LEXIS 407 (Sept. 24, 2010). Florida: Peacock Constr. Co. v. Modern Air Conditioning, Inc., 353 So. 2d 840 (Fla. 1977); WMS Constr., Inc. v. Palm Springs Mile Assocs., Ltd., 762 So. 2d 973 (Fla. Dist. Ct. App. 2000). Georgia: Associated Mechanical Corp., Inc. v. Martin K. Eby Constr. Co, Inc., 67 F. Supp. 2d 1375 (M.D. Ga. 1999); Powell Co. v. McGarey Group, LLC, 508 F. Supp. 2d 1202 (N.D. Ga. 2007). Kentucky: New Amsterdam Cas. Co. v. Allen Co., 446 S.W.2d 278 (Ky. Ct. App. 1969). Louisiana: S. States Masonry, Inc. v. J.A. Jones Constr. Co., 507 So. 2d 198 (La. 1987); Vector Elec. & Controls, Inc. v. JE Merit Constructors, Inc., 2006 La. App. Unpub. LEXIS 2 (Nov. 8, 2006). Maryland: Gilbane Bldg. Co. v. Brisk Waterproofing Co., Inc., 585 A.2d 248 (Md. Ct. Spec. App. 1991); Architectural Systems, Inc. v. Gilbane Bldg. Co., 760 F. Supp. 79 (D. Md. 1991). Massachusetts: A.J. Wolfe Co. v. Baltimore Contractors, Inc., 244 N.E.2d 717 (Mass. 1969); Edward G. Sawyer Co. v. FKI Logistex Integration, Inc., No. CA 04-1185, 2007 Mass. Super. LEXIS 625 (July 25, 2007). Michigan: Christman Co. v. Anthony S. Brown Dev. Co., 533 N.W.2d 838 (Mich. Ct. App. 1995). New Jersey: Avon Bros., Inc. v. Tom Martin Constr. Co., Inc., 2000 N.J. Super. Unpub. LEXIS 1 (Aug. 30, 2000). New Mexico: MidAmerican Constr. Mgmt., Inc. v. Mas Tech N.A., Inc., 436 F.3d 1257 (10th Cir. 2006). Ohio: Power & Pollution Serv., Inc. v. Suburban Power Piping Corp., 598 N.E.2d 69 (Ohio Ct. App. 1991); Evans, Mechwart, Hambleton & Tilton, Inc. v. Triad Architects, Ltd., 965 N.E.2d 1007 (Ohio Ct. App. 2011). Tennessee: Koch v. Constr. Tech., Inc., 924 S.W.2d 68 (Tenn. 1996). Texas: MidAmerican Constr. Mgmt., Inc. v. Mas Tech N.A., Inc., 436 F.3d 1257 (10th Cir. 2006). Virginia: Galloway Corp. v. S.B. Ballard Constr. Co., 464 S.E.2d 349 (Va. 1995). West Virginia: Wellington Power Corp. v. CNA Sur. Corp., 614 S.E.2d 680 (W. Va. 2005).

19. *See* Badger State, Inc. v. Keller Constr. Co., No. 2011AP701, 2012 Wisc. App. LEXIS 593 (July 26, 2012) (holding that Sections 9.6.2 and 9.6.3 "created a contract that intended to benefit subcontractors on that job as a class").

fails to furnish such evidence within seven days, the Owner shall have the right to contact Subcontractors and suppliers to ascertain whether they have been properly paid. Neither the Owner nor Architect shall have an obligation to pay, or to see to the payment of money to, a Subcontractor or supplier, except as may otherwise be required by law.

Section 9.6.4 is a sister provision to Section 9.6.3. It gives owners (in contrast to subcontractors and suppliers in Section 9.6.3) the right to obtain information to ensure that contractors are properly paying their subcontractors and suppliers. It permits an owner to request from contractors written evidence that they have properly paid their subcontractors and suppliers. If a contractor does not furnish the evidence within seven days of the request, the owner then has the right to contact the subcontractors and suppliers to determine if they have been properly paid.

The section also states that neither the owner nor the architect has any responsibility with respect to payments to subcontractors or suppliers, except as otherwise provided by law. This, in conjunction with Section 1.1.2, is an important provision for limiting liability of owners and architects because it provides a contractual counterargument against, for example, the argument that owners and architects owe payment-related duties to subcontractors and suppliers.[20]

The 2017 amendments to Section 9.6.4 enlarge the category of suppliers an owner may contact about contractor payments. The category now includes those suppliers that did not technically qualify as "material and equipment" suppliers under the 2007 version. The amendments also make explicit what was already implicit in Section 9.6.4: owners may contact suppliers (in addition to subcontractors) to determine if a contractor has paid them, but owners do not have any obligation to pay suppliers or to ensure that suppliers are being paid. Finally, the 2017 amendments make a nonsubstantive punctuation change by including commas around the phrase "or to see to the payment of money to" in the last sentence.

§ 9.6.5 ~~Contractor~~The Contractor's payments to ~~material and equipment~~ suppliers shall be treated in a manner similar to that provided in Sections 9.6.2, 9.6.3 and 9.6.4.

Section 9.6.5 provides that Sections 9.6.2, 9.6.3, and 9.6.4 apply to suppliers. Therefore, wherever the word "Subcontractor(s)" appears in those sections, one should insert immediately thereafter the words "and Supplier(s)."

The 2017 amendments to Section 9.6.5 enlarge the category of suppliers to which the Sections 9.6.2, 9.6.3, and 9.6.4 apply. The category now includes those suppliers that did not technically qualify as "material and equipment" suppliers under the 2007 version. The amendments also make a nonsubstantive change from "Contractor payments" to "The Contractor's payments."

§ 9.6.6 A Certificate for Payment, a progress payment, or partial or entire use or occupancy of the Project by the Owner shall not constitute acceptance of Work not in accordance with the Contract Documents.

20. *But see* Souder v. Cannon, 235 S.W.3d 841 (Tex. App. 2007) (holding that city officials could be liable to unpaid subcontractors when the officials approved payment despite knowledge that contractor was not paying subcontractors); Boren v. Thompson & Assocs., 999 P.2d 438 (Okla. 2000) (holding that architect could be liable to unpaid subcontractor for improperly certifying payments to the contractor when architect had knowledge that contractor had failed to procure a statutorily required payment bond and was not paying subcontractors).

Section 9.6.6 remains unchanged from its 2007 version. It states that an architect's issuance of a payment certification, an owner making a progress payment, or an owner's partial or full use or occupancy of the project does not constitute acceptance of nonconforming work. Because Section 12.3 permits an owner to accept nonconforming work, Section 9.6.6 is an important provision for owners because it makes clear that the listed actions do not constitute acceptance.[21]

> **§ 9.6.7** Unless the Contractor provides the Owner with a payment bond in the full penal sum of the Contract Sum, payments received by the Contractor for Work properly performed by Subcontractors ~~and/~~or provided by suppliers shall be held by the Contractor for those Subcontractors or suppliers who performed Work or furnished materials, or both, under contract with the Contractor for which payment was made by the Owner. Nothing contained herein shall require money to be placed in a separate account and not commingled with money of the Contractor, ~~shall~~ create any fiduciary liability or tort liability on the part of the Contractor for breach of trust, or ~~shall~~ entitle any person or entity to an award of punitive damages against the Contractor for breach of the requirements of this provision.

Section 9.6.7 creates contractual priority in favor of subcontractors and suppliers to any money received by a contractor for its subcontractors' and suppliers' work, unless the contractor has furnished a payment bond. It accomplishes this priority by requiring the contractor to hold such money for its subcontractors and suppliers. Such priority would become important if, for example, a contractor's lender attempted to seize the money from the contractor for payment on outstanding debt. In such a circumstance, if a subcontractor could trace the money as being money paid by the owner for the subcontractor's work, then the subcontractor could have priority over the lender to the money.

This section, in conjunction with Section 9.6.2, can be interpreted as creating an express trust in favor of subcontractors and suppliers.[22] Notwithstanding the creation of a trust, the section does not create any fiduciary, tort, or punitive-damage liability against a contractor for a breach nor does it prohibit a contractor from commingling such money with its other funds or require it to place such money in a separate account. Therefore, subcontractors and suppliers are limited to using the section against non-contractor third parties, such as against a contractor's lender in a priority dispute.[23]

The 2017 amendments make a nonsubstantive change by describing suppliers as "providing" work instead of "performing" work, and they make a nonsubstantive grammatical change by deleting the word "shall" in two locations.

> **§ 9.6.8** Provided the Owner has fulfilled its payment obligations under the Contract Documents, the Contractor shall defend and indemnify the Owner from all loss, liability, damage or expense, including reasonable attorney's fees and litigation expenses, arising

21. *See* Colin C & M Corp. v. Bacon Constr. Co., No. BRCV2005-01047A, 2007 Mass. Super. LEXIS 562 (Dec. 7, 2007).
22. Westview Invs., Ltd. v. U.S. Bank, 138 P.3d 638 (Wash. Ct. App. 2006).
23. *Id.* at 642–45 (holding that a bank who has "knowledge sufficient to require inquiry whether funds deposited by a general contractor into its bank account are trust funds cannot, as against the subcontractors, set off the funds to pay an indebtedness owed the bank by the general contractor" where "the construction contracts at issue [Sections 9.6.2 and 9.6.7] created express trusts designating progress payments made for the benefit of subcontractors to be trust funds"); *id.* at 644–45 (stating that the language in Section 9.6.7 protecting contractor against trust-related liability and permitting it to commingle funds does not negate the creation of a trust enforceable by subcontractor against a third-party bank).

> out of any lien claim or other claim for payment by any Subcontractor or supplier of any tier. Upon receipt of notice of a lien claim or other claim for payment, the Owner shall notify the Contractor. If approved by the applicable court, when required, the Contractor may substitute a surety bond for the property against which the lien or other claim for payment has been asserted.

Section 9.6.8 is a new provision with the 2017 amendments. If the owner has fulfilled its payment obligations to the contractor, it requires the contractor to indemnify the owner from any lien and payment claims of subcontractors and suppliers. It also requires the owner to notify the contractor upon the owner's receipt of lien and payment claims, and it permits the contractor to bond off lien and payment claims against the property upon approval of a court.

§ 9.7 FAILURE OF PAYMENT

> If the Architect does not issue a Certificate for Payment, through no fault of the Contractor, within seven days after receipt of the Contractor's Application for Payment, or if the Owner does not pay the Contractor within seven days after the date established in the Contract Documents, the amount certified by the Architect or awarded by binding dispute resolution, then the Contractor may, upon seven additional days' ~~written~~ notice to the Owner and Architect, stop the Work until payment of the amount owing has been received. The Contract Time shall be extended appropriately and the Contract Sum shall be increased by the amount of the Contractor's reasonable costs of ~~shut-down~~shutdown, delay and start-up, plus interest as provided for in the Contract Documents.

Section 9.7 provides the contractors with significant rights and leverage against the owners for an architect's improper failure to issue a payment certification or for an owner's improper failure to pay certified amounts or amounts awarded in binding dispute resolution. If an architect improperly fails, within seven days after receipt of the contractor's payment application, to certify payment then the contractor can stop work upon seven days' additional notice to the owner and architect.[24] Similarly, upon the same seven days' notice to the owner and architect, the contractor can stop work if the owner fails, within seven days after the payment deadline in the contract documents, to pay amounts certified by the architect or fails to pay amounts awarded in binding dispute resolution.[25] Notably, unlike the deadline to pay certified amounts, there is no express deadline in the A201 by which an owner must pay amounts awarded in binding dispute resolution.[26]

In addition to stopping work, the section requires that a contractor receive an appropriate increase in the Contract Time and an increase in the Contract Sum, including applicable interest, for the contractor's reasonable shutdown, delay, and start-up costs.

The 2017 amendments appear to eliminate the requirement that the contractor's notice of stopping work be made in writing, but new Section 1.6 requires all notices to be in writing. The amendments also make a nonsubstantive spelling change from "shut down" to "shutdown."

24. General Ins. Co. of Am. v. K. Capolino Constr. Corp., 983 F. Supp. 403 (S.D.N.Y. 1997).
25. Commonwealth Constr. Co. v. Cornerstone Fellowship Baptist Church, No. 04L-10-101 RRC, 2006 Del. Super. LEXIS 349 (Aug. 31, 2006).
26. Snyder v. Eanes Indep. Sch. Dist., 860 S.W.2d 692 (Tex. App. 1993).

§ 9.8 SUBSTANTIAL COMPLETION

§ 9.8.1 Substantial Completion is the stage in the progress of the Work when the Work or designated portion thereof is sufficiently complete in accordance with the Contract Documents so that the Owner can occupy or utilize the Work for its intended use.

The 2017 amendments to the AIA A201 do not include any substantive or nonsubstantive changes to Sections 9.8.1., 9.8.2, 9.8.3, 9.8.4, or 9.8.5.

Nonetheless, Section 9.8.1 remains a vital provision, as substantial completion not only triggers the contractor's right to receive final payment but also shifts the risk of loss from the contractor to the owner. It is often the case that the contractor's warranties begin to run from the date of substantial completion, and the ability to assess liquidated damages typically ends on this date.[27] The date of substantial completion is also extremely important for determining whether claims have been timely asserted.

Many courts will look strictly to the date of the architect's certificate of substantial completion in determining if and when substantial completion has occurred and whether a party's claims are untimely based on the expiration of a statute of limitations period that began to run from the date of substantial completion. However, the date of substantial completion can be inferred "by examining the factual situation at the construction site and determining whether the Owner has received the applicable permits."[28]

When challenging a party's claims based on the expiration of the statute of limitations, it is important that the architect's certificate of substantial completion be attached as an exhibit and properly authenticated. Otherwise, the court may rule that the date of substantial completion—the date from which the statute of limitations begins to run—has not been sufficiently established.[29]

Other courts have also considered the importance of the existence of documentary proof of the date of substantial completion for purposes of determining whether claims have been timely pursued.[30]

It is also possible for parties to contractually shorten the relevant time period for timely pursuing claims with the date of substantial completion still governing the commencement of that limitations period.[31]

27. *See, e.g.*, All Seasons Const., Inc. v. Mansfield Housing Auth., 920 So. 2d 413 (La. Ct. App. 2006) (holding that if work was completed by substantial completion date then liquidated damages could not be deducted from final payment).

28. MR Hotels, LLC v. LLW Architects, Inc., No. M2015-00840-COA-R9-CV, 2016 WL 4070050, at *6 (Tenn. Ct. App. July 28, 2016); Louisville/Jefferson Cty. Metro Gov't v. HNTB Corp., No. 3:07-CV-52-R, 2007 WL 1100743 (W.D. Ky. Apr. 11, 2007) (finding that substantial completion had occurred because the baseball field at issue was open and games continued to be played without lighting systems complications). *See also* PIH Beaverton, LLC v. Super One, Inc., 323 P.3d 961 (Or. 2014) (discussing interplay between Section 9.8.1 and Oregon Revised Statute 12.135).

29. *See* Sunset Presbyterian Church v. Brockamp & Jaeger, Inc., 325 P.3d 730 (Or. 2014); *see also* 15th Place Condo. Ass'n v. South Campus Dev. Team, LLC, 14 N.E.3d 592 (Ill. App. Ct. 2014) (holding that affidavit contesting date of substantial completion did not create genuine issue of material fact where opposing affidavit was from project architect and attached copies of correspondence discussing actual date of substantial completion).

30. *MR Hotels*, 2016 WL 4070050, at *6; Homsey Architects, Inc. v. Nine Ninety Nine, LLC, No. 4412-VCP, 2010 WL 2476298 (Del. Ch. June 14, 2010); Fed. Ins. Co. v. Konstant Architecture Planning, Inc., 902 N.E.2d 1213 (Ill. App. Ct. 2009); Brookridge Apartments, Ltd. v. Universal Contractors, Inc., 844 S.W.2d 637 (Tenn. Ct. App. 1992) (discussing interplay of date of substantial completion and TENN. CODE ANN. §§ 28-3-201, -202); Trinity Church v. Lawson-Bell, 925 A.2d 720 (N.J. Super. Ct. App. Div. 2007) (rejecting plaintiff's claims that architect had fraudulently issued certificate of substantial completion based on plaintiff's erroneous interpretation of substantial completion to mean substantial compliance with the contract or an absence of defects); Northridge Homes, Inc. v. John W. French & Assoc., Inc., No. 97-2281-H, 1999 WL 1260285 (Mass. Super. Ct. Nov. 15, 1999); Gillespie Cmty. Unit Sch. Dis. No. 7, Macoupin Cty. v. Wight & Co., No. 4-11-1001, 2012 WL 7037447 (Ill. App. Ct. Sept. 11, 2012) (granting summary judgment in favor of architect that claims were time-barred).

31. *See* KBE Bldg. Corp. v. Const. Servs. of NC, Inc., No. 0763, Sept. Term, 2013, 2015 WL 5813848 (Md. Ct. Spec. App. Oct. 5, 2015).

Many jurisdictions also provide that a statute of repose commences at the date of substantial completion.[32] However, if the date of issuance and date of completion are left blank on a certificate of substantial completion, the court may find that the incomplete certificate is irrelevant for a determination of whether a claim has been timely asserted.[33]

§ 9.8.2 When the Contractor considers that the Work, or a portion thereof which the Owner agrees to accept separately, is substantially complete, the Contractor shall prepare and submit to the Architect a comprehensive list of items to be completed or corrected prior to final payment. Failure to include an item on such list does not alter the responsibility of the Contractor to complete all Work in accordance with the Contract Documents.

Section 9.8.2 "include steps that the contractor, architect, and owner of the property must take before the architect issues a Certificate of Substantial Completion."[34] Essentially, Section 9.8.2 directs the contractor to create an itemized "punch list"—"A punch list is an itemized list of finish work, corrections, repairs, and services to be performed in order to complete a construction contract."[35] This is the first step in the process of a project being deemed substantially complete by the architect.

In determining which steps a contractor should undertake in order to avoid the assessment of liquidated damages, at least one court previously held that to the extent a contractor believed that portions of its work were substantially complete, the contractor would be wise to take the steps outlined in Section 9.8.2 to start the process of having the project deemed substantially complete.[36]

The Southern District of New York has also held that correspondence inviting the contractor to compile a punch list, in accordance with Section 9.8.2, was evidence that substantial completion had been achieved prior to that same contractor being terminated and the surety taking over.[37]

§ 9.8.3 Upon receipt of the Contractor's list, the Architect will make an inspection to determine whether the Work or designated portion thereof is substantially complete. If the Architect's inspection discloses any item, whether or not included on the Contractor's list, which is not sufficiently complete in accordance with the Contract Documents so that the Owner can occupy or utilize the Work or designated portion thereof for its intended use, the Contractor shall, before issuance of the Certificate of Substantial Completion, complete or correct such item upon notification by the Architect. In such case, the Contractor shall then submit a request for another inspection by the Architect to determine Substantial Completion.

32. *See* Russo Farms, Inc. v. Vineland Bd. of Educ., 675 A.2d 1077 (N.J. 1996) (citing Ark. Code Ann. § 16-56-112; Cal. Civ. Proc. Code § 337.1; Conn. Gen. Stat. § 52-584a; Haw. Rev. Stat. § 657-8; Ind. Code § 34-4-20-2; Ky. Rev. Stat. Ann. § 413.135; Minn. Stat. Ann. § 541.051; Mont. Code Ann. § 27-2-208; N.H. Rev. Stat. Ann. § 508:4-b; N.M. Stat. Ann. § 37-1-27; S.D. Codified Laws Ann. § 15-2A-3; Tenn. Code Ann. § 28-3-202; Wis. Stat. Ann. § 893.89; Wyo. Stat. § 1-3-111).

33. *See* Town of Kearny v. Brandt, 67 A.3d 601 (N.J. 2013).

34. *Sunset*, 325 P.3d at 733.

35. J.A. Sullivan Corp. v. Com., 494 N.E.2d 374, 375 (Mass. 1986).

36. Comstock Potomac Yard, L.C. v. Balfour Beatty Const., LLC, 694 F. Supp. 2d 468 (E.D. Va. 2010) (judgment vacated). *See also* Younglove Const., LLC v. PSD Dev., LLC, No. 3:08CV1447, 2010 WL 4570194 (N.D. Ohio Nov. 3, 2010) (contractor failed to give notice of substantial completion and, therefore, was not entitled to retainage).

37. Travelers Cas. & Sur. Co. v. White Plains Public Sch., No. 3:07CV8144, 2007 WL 935612 (S.D.N.Y. Jan. 26, 2007).

As discussed earlier, although the contractor initiates the process of a project being deemed substantially complete through its generation of a punch list, it is the architect who actually determines the date of substantial completion and whether the project is "sufficiently complete" to allow the owner to "occupy and utilize" the work performed by the contractor. If substantial completion has been attained, the architect is legally obligated to issue the certificate of substantial completion.[38] Failure of the architect to fulfill its duty to inspect the work and specify additional corrections can also constitute a waiver of the standard condition precedent to final payment—the condition that the architect approve the work and issue a certificate of substantial completion.[39]

If, however, the architect finds an item during its inspection that is either on the punch list or not on the punch list but that would otherwise prevent the owner from occupying or using the property for the property's intended use, the architect is authorized to direct the contractor to complete/correct the item and may withhold the issuance of the certificate of substantial completion until such corrective or completion measures are completed. To the extent the architect directs the contractor to complete or correct an item prior to its issuance of a certificate of substantial completion, the contractor, after having performed such work, is required to request another inspection by the architect.

> **§ 9.8.4** When the Work or designated portion thereof is substantially complete, the Architect will prepare a Certificate of Substantial Completion that shall establish the date of Substantial Completion~~, shall~~; establish responsibilities of the Owner and Contractor for security, maintenance, heat, utilities, damage to the Work and insurance~~,~~; and ~~shall~~ fix the time within which the Contractor shall finish all items on the list accompanying the Certificate. Warranties required by the Contract Documents shall commence on the date of Substantial Completion of the Work or designated portion thereof unless otherwise provided in the Certificate of Substantial Completion.

Assuming the architect has received the contractor's punch list and determined that substantial completion has been achieved, the next step in this process is the architect's issuance of the certificate of substantial completion. It is important to note that the issuance of the certificate of substantial completion does not absolve the contractor from its obligations to address issues on the punch list. Once the final punch list and certificate of substantial completion are presented to and accepted by the owner and contractor, payment for retainage from the owner can be requested.[40]

When the conditions set forth in Sections 9.8.1. through 9.8.4 are fulfilled, the architect is obliged to issue the certificate of substantial completion. Section 9.8.4 of the AIA A201 "directly links the date of substantial completion to the issuance of the architect's Certificate of Substantial Completion."[41]

---

38. *See* Haugen v. Raupach, 260 P.2d 340 (Wash. 1953) ("If the architect is satisfied that there has been substantial completion of the contract it then becomes his duty to issue the certificate of completion, and if he does not do so, his conduct is regarded as arbitrary and capricious.").

39. *See, e.g.*, Hartford Elec. Applicators of Thermalux, Inc. v. Alden, 363 A.2d 135 (Conn. 1975).

40. *See infra* Section 9.8.5.

41. KBE Bldg. Corp. v. Const. Servs. of NC, Inc., No. 0763, Sept. Term, 2013, 2015 WL 5813848, at *19 (Md. Ct. Spec. App. Oct. 5, 2015). *But see* Whiting-Turner Contracting Co. v. Capstone Dev. Corp., No. WDQ-12-3730, 2013 WL 5423953 (D. Md. Sept. 26, 2013) (interpreting additional contractual provisions as requiring owner approval before project deemed substantially complete).

In *Travelers Casualty & Surety Co. v. White Plains Public Schools*,[42] the court noted that an examination of the circumstances surrounding the architect's withholding of the certificate of substantial completion not only may have violated the architect's code of ethics and professional conduct but also likely constituted constructive fraud. Three years later, in *Board of Education of Woodford County v. D.W. Wilburn, Inc.*,[43] the Kentucky Court of Appeals opined on the impact of an architect's failure to deliver a final punch list, in conjunction with a number of other failures on the part of the project owner, in a trial court's decision to award the contractor virtually all sums it claimed in relation to a project that was delayed months beyond the original substantial completion date. Therein, the court held the owner had breached the contract in a number of ways, including refusing to sign valid change orders or the certificate of substantial completion. Although not the entire basis of the court's decision, the court in *Wilburn* likely referenced the failure of the architect to deliver a final punch list alongside a timeline for completion as further evidence of the owner's misconduct in not paying the contractor.

Some commentators have asked, "What happens if the architect fails to designate responsibility for security or maintenance, etc."? Better yet, what if the architect places all of the responsibilities enumerated in Section 9.8.4 on the contractor? The short answer is, of course, that the parties can proceed to dispute resolution in accordance with Section 15 of the A201, but ideally these issues would be negotiated and resolved prior to contract execution.[44]

> **§ 9.8.5** The Certificate of Substantial Completion shall be submitted to the Owner and Contractor for their written acceptance of responsibilities assigned to them in ~~such~~the Certificate. Upon such acceptance, and consent of surety~~;~~ if any, the Owner shall make payment of retainage applying to ~~such~~the Work or designated portion thereof. Such payment shall be adjusted for Work that is incomplete or not in accordance with the requirements of the Contract Documents.

The 2017 amendments to the A201 do not include changes to Section 9.8.5. If an owner improperly disburses funds without confirming with the surety, a surety may be justified in refusing to honor its bond.[45]

Additionally, it is not only the issuance of the certificate of substantial completion that governs applicable deadlines or start dates; the owner and contractor must also approve the start or end of their respective responsibilities.[46]

§ 9.9 PARTIAL OCCUPANCY OR USE

> **§ 9.9.1** The Owner may occupy or use any completed or partially completed portion of the Work at any stage when such portion is designated by separate agreement with the Contractor, provided such occupancy or use is consented to by the insurer ~~as required under~~

42. No. 3:07CV8144, 2007 WL 935612 (S.D.N.Y. Jan. 26, 2007).
43. Nos. 2008-CA-2422, 2009-CA-50, 2010 WL 2010760 (Ky. Ct. App. May 21, 2010).
44. Werner Sabo, Legal Guide AIA Documents § 4.54 (Substantial Completion: ¶ 9.8), at 3 n.280 (5th ed. Supp. 2017-1).
45. *See* Stonington Water Street Assoc., LLC v. Hodess Bldg. Co., Inc., 792 F. Supp. 2d 253 (D. Conn. 2011).
46. Dynalectic Co. v. Whittenberg Const. Co., No. 5:06CV208-R, 2008 WL 2415875 (W.D. Ky. June 12, 2008) (analyzing Section 9.3.1 and 9.8.5 in determining that it was apparent that final payment was conditioned upon completion as well as approval by owner and architect).

~~Section 11.3.1.5~~ and authorized by public authorities having jurisdiction over the Project. Such partial occupancy or use may commence whether or not the portion is substantially complete, provided the Owner and Contractor have accepted in writing the responsibilities assigned to each of them for payments, retainage, if any, security, maintenance, heat, utilities, damage to the Work and insurance, and have agreed in writing concerning the period for correction of the Work and commencement of warranties required by the Contract Documents. When the Contractor considers a portion substantially complete, the Contractor shall prepare and submit a list to the Architect as provided under Section 9.8.2. Consent of the Contractor to partial occupancy or use shall not be unreasonably withheld. The stage of the progress of the Work shall be determined by written agreement between the Owner and Contractor or, if no agreement is reached, by decision of the Architect.

In the event that the owner wishes to occupy all or a portion of the project before the project is finally complete, the owner may do so, but only if the parties agree and establish the various parameters set forth in Section 9.9.1. The practice in the industry has been for these responsibilities to be allocated informally. But this practice allows the contractor to successfully argue that the owner's partial occupancy should constitute acceptance of the entire project. This, in turn, compromises the parties' ability to enforce the contract documents.

Moreover, if the owner elects to partially occupy the project, the architect will likely be required to perform more than two inspections (inspection at substantial and final completion are standard). It is advisable that architects include additional inspections for partial occupancy in their list for additional services for which they expect to be compensated.

The changes in Section 9.9.1 show that insurer consent is no longer required under Section 11.3.1.5 but that the same consent to partial occupancy may be required pursuant to the insurance and bonds exhibit to the contract. It is also important to note that the contractor is prohibited from unreasonably withholding consent to partial occupancy.

§ 9.9.2 Immediately prior to such partial occupancy or use, the Owner, Contractor, and Architect shall jointly inspect the area to be occupied or portion of the Work to be used in order to determine and record the condition of the Work.

The 2017 amendments to the A201 do not contain any changes for Section 9.9.2. The requirements of Section 9.9.2 are fairly straightforward; if the owner desires partial occupancy, a joint inspection is mandatory so that the progress of the work is appropriately documented. The joint inspection requirement is meant to avoid disputes, not to ensure quality control.[47] An easy example of the type of dispute the joint inspection requirement is meant to protect against is a dispute as to whether the work was installed improperly in the first instance or whether any nonconforming work was caused by damage by the owner after partial occupancy or use.[48]

§ 9.9.3 Unless otherwise agreed upon, partial occupancy or use of a portion or portions of the Work shall not constitute acceptance of Work not complying with the requirements of the Contract Documents.

47. *See* 2 Bruner & O'Connor, *supra* note 2, § 5:188 (Dec. 2016 update).
48. *Id.*

There are also no changes to Section 9.9.3 in the 2017 amendments. Whereas Section 9.9.3 primarily concerns risk mitigation, Section 9.9.3 relates more to risk allocation. In conjunction with Section 9.9.6, Section 9.9.3 was devised in order to ensure that occupancy or use of partial work is not deemed to be acceptance of nonconforming work by the owner.[49] At least one case has interpreted partial occupancy by the owner to constitute acceptance of various changes in the work where owner's "protests appear to have been merely expressions of some dissatisfaction" and owner did not demand that any work be redone or replaced.[50]

Section 9.9.3 is also related to the Section 9.10.4.2, which further clarifies that the issuance of final payment by the owner does not waive the owner's claims for nonconforming work.

§ 9.10 FINAL COMPLETION AND FINAL PAYMENT

§ 9.10.1 Upon receipt of the Contractor's ~~written~~ notice that the Work is ready for final inspection and acceptance and upon receipt of a final Application for Payment, the Architect will promptly make such inspection ~~and, when~~. When the Architect finds the Work acceptable under the Contract Documents and the Contract fully performed, the Architect will promptly issue a final Certificate for Payment stating that to the best of the Architect's knowledge, information and belief, and on the basis of the Architect's on-site visits and inspections, the Work has been completed in accordance with ~~terms and conditions of~~ the Contract Documents and that the entire balance found to be due the Contractor and noted in the final Certificate is due and payable. The Architect's final Certificate for Payment will constitute a further representation that conditions listed in Section 9.10.2 as precedent to the Contractor's being entitled to final payment have been fulfilled.

The only change in the 2007 amendments to this section was the addition of the words "the Contractor's" in the first sentence, which clarified that the notice concerning the work being ready for final inspection was expected to be generated by the contractor. Failure by the contractor to submit a request for final payment to the architect (and construction manager) has been held to mean that the owner is entitled to withhold payment.[51] The 2017 amendments contain no additional changes.

As discussed in relation to Section 9.9.1, inspections by the architect at substantial and final completion are standard. If no additional inspections are needed for partial occupancy, the inspection mandated in Section 9.10.1 would constitute the architect's second and final inspection. An interesting wrinkle is that Section 9.10.1 requires the architect to evaluate and determine that the work is conforming, yet this determination of the architect does not mean that any nonconforming work that may exist is accepted. Section 4.2.6 permits the architect to reject nonconforming work, but there is no corresponding provision in the A201 that allows an architect to accept nonconforming work. Quite the contrary, the determination by the architect that the work is conforming for purposes of a final inspection under Section 9.10.1 means only that the owner is now required to issue final payment.[52] Again, Section 9.10.4.2 allows the owner to issue this payment without being deemed to have accepted nonconforming work.

49. *Id. See id.* § 5:189.
50. Shimek v. Vogel, 105 N.W.2d 677, 680 (N.D. 1960).
51. *See* Boro Const. v. Ridley Sch., 992 A.2d 208 (Pa. Super. Ct. 2008).
52. *See infra* Section 9.10.3.

Some courts (correctly) interpret this section as meaning that the architect's certificate for final payment is typically a condition precedent to a contractor being entitled to receive payment.[53] In *Kilianek v. Kim*,[54] the court overturned an arbitration award that had been issued in favor of a contractor because the architect's certificate for final payment had not been issued. "[T]he obligation of the owner to make final payment is not binding as it was the subject matter of a condition precedent which was never satisfied."[55]

While the contractor's receipt of the architect's certificate for final payment should be a condition precedent to receiving payment, issuance of final payment by the owner without having this same certificate in hand can have serious consequence. For instance, in *Friendswood Independent School District v. National Surety Corp.*,[56] the court held that the surety was relieved of its obligations under the relevant performance bond because the owner for the project had issued final payment prior to obtaining the architect's certificate of final payment. Somewhat relatedly, to the extent the owner makes changes or additions without consulting with the architect, the owner can be held to have waived the requirement of an architect's certificate for final payment.[57]

As *Ferrante Immobiliare, LLC v. Pace*,[58] alludes to, issuance of the architect's certificate for final payment can also be important for calculating whether a cause of action is barred by the applicable statute of limitations. In *Mainline Electric Corp. v. East Quogue Union Free School District*,[59] the court held that a contractor's claim for amounts due was not statutorily barred, despite the fact that much of the work had been completed for years, because final payment was not due until after issuance of the architect's certificate for final payment, and the certificate had not yet been issued.

> **§ 9.10.2** Neither final payment nor any remaining retained percentage shall become due until the Contractor submits to the Architect (1) an affidavit that payrolls, bills for materials and equipment, and other indebtedness connected with the Work for which the Owner or the Owner's property might be responsible or encumbered (less amounts withheld by Owner) have been paid or otherwise satisfied, (2) a certificate evidencing that insurance required by the Contract Documents to remain in force after final payment is currently in effect ~~and will not be canceled or allowed to expire until at least 30 days' prior written notice has been given to the Owner~~, (3) a written statement that the Contractor knows of no ~~substantial~~ reason that the insurance will not be renewable to cover the period required by the Contract Documents, (4) consent of surety, if any, to final payment ~~and (5)~~, (5) documentation of any special warranties, such as manufacturers' warranties or specific Subcontractor warranties, and (6) if required by the Owner, other data establishing payment or satisfaction of obligations, such as receipts~~,~~ and releases and waivers of liens,

53. *See* Laurel Race Course, Inc. v. Regal Const. Co., Inc., 333 A.2d 319 (Md. 1975) ("[W]here payments under a contract are due only when the certificate of an architect or engineer is issued, production of the certificate becomes a condition precedent to liability of the owner for materials and labor in the absence of fraud or bad faith."); Russell H. Lankton Const. Co. v. LaHood, 493 N.E.2d 714 (Ill. App. Ct. 1986) (arbitration award was contingent upon issuance of architect's final certification).

54. 548 N.E.2d 598 (Ill. App. Ct. 1989).

55. *Id.* at 601.

56. 423 S.W.2d 95 (Tex. Civ. App. 1967).

57. *See* Steffek v. Wichers, 507 P.2d 274 (Kan. 1973); *see also* Formigli Corp. v. Fox, 348 F. Supp. 629 (E.D. Pa. 1972) (owner's repeated issuance of payment without consent of architect operated to waive requirement of architect's certificate for final payment); Ferrante Immobiliare, LLC v. Pace, 891 N.Y.S.2d 27 (App. Div. 2009) (holding that architect's issuance of certificate for final payment was waived, for purposes of calculating date that statute of limitations began to run, because owner and contractor previously entered into settlement agreement without involvement from architect).

58. 891 N.Y.S.2d 27.

59. 849 N.Y.S.2d 92 (App. Div. 2007).

claims, security interests, or encumbrances arising out of the Contract, to the extent and in such form as may be designated by the Owner. If a Subcontractor refuses to furnish a release or waiver required by the Owner, the Contractor may furnish a bond satisfactory to the Owner to indemnify the Owner against such lien. ~~If such lien~~, claim, security interest, or encumbrance. If a lien, claim, security interest, or encumbrance remains unsatisfied after payments are made, the Contractor shall refund to the Owner all money that the Owner may be compelled to pay in discharging ~~such lien~~the lien, claim, security interest, or encumbrance, including all costs and reasonable attorneys' fees.

Section 9.10.2 is intended to protect the owner from "mechanics' liens and insured losses by obligating the contractor to provide certain assurances."[60] The only change to Section 9.10.2 is the addition of item (5)—"documentation of any special warranties, such as manufacturers' warranties or specific subcontractor warranties"—between consent of the surety and other data that may be required by the owner. This addition appears to be in response to owners' increased practice of withholding final payment or retention until delivery of applicable warranties. This is especially common on more complex projects. The 2017 amendments, therefore, simply codify what was already being demanded by owners in practice.

Weighing in on what the words "consent of the surety, if any" mean in the context of Section 9.10.2, at least one court has held that "when a surety was involved, the surety's consent to final payment [is] required."[61]

§ 9.10.3 If, after Substantial Completion of the Work, final completion thereof is materially delayed through no fault of the Contractor or by issuance of Change Orders affecting final completion, and the Architect so confirms, the Owner shall, upon application by the Contractor and certification by the Architect, and without terminating the Contract, make payment of the balance due for that portion of the Work fully completed, corrected, and accepted. If the remaining balance for Work not fully completed or corrected is less than retainage stipulated in the Contract Documents, and if bonds have been furnished, the written consent of the surety to payment of the balance due for that portion of the Work fully completed and accepted shall be submitted by the Contractor to the Architect prior to certification of such payment. Such payment shall be made under terms and conditions governing final payment, except that it shall not constitute a waiver of ~~claims~~Claims.

The changes in Section 9.10.3 clarify that the payment owed by the owner to the contractor is for work performed between the time of substantial completion and the time of final completion, not just for work that has been completed and accepted but also for any work that has been corrected and accepted after substantial completion. In this respect, the 2017 amendments attempt to address those instances where remedial work has been performed and accepted subsequent to substantial completion but prior to final completion. There is a debate within the industry as to what constitutes a "material" delay.

60. 2 Bruner & O'Connor, *supra* note 2, § 5:191 (Dec. 2016 Update).

61. Great Amer. Ins. Co. v. Norwin Sch. Dist., 544 F.3d 229, 249 (3d Cir. 2008). See *International Engineering Services, Inc. v. Const. & Engineering of Central Florida, LLC*, 74 So. 3d 531 (Fla. Dist. Ct. App. 2011) for a discussion of ambiguity created when Section 9.10.2 is read in conjunction with a pay-when-paid provision in a contract.

§ 9.10.4 The making of final payment shall constitute a waiver of Claims by the Owner except those arising from

.1 liens, Claims, security interests, or encumbrances arising out of the Contract and unsettled;

.2 failure of the Work to comply with the requirements of the Contract Documents; ~~or~~

.3 terms of special warranties required by the Contract Documents~~.~~; or

~~§~~ **.4** audits performed by the Owner, if permitted by the Contract Documents, after final payment.

Commentators have pointed out that Section 9.10.4 is meant to protect owners from the common law consequences associated with issuance of final payment. In particular, "[w]ithout a contract provision clearly dealing with the question of acceptance, disputes may arise over whether the owner sufficiently reserved the right to make a claim once it had paid the contractor."[62]

Section 9.10 has been updated to include subsection .4 as an additional circumstance under which claims by the owner are not waived upon issuance of final payment. The addition of subsection .4 correlates with a rise in the industry of owner audits on construction projects, a practice that many believe to be an essential function of project controls. Since a number of audits are not initiated (let alone completed) prior to issuance of final payment, Section 9.10.4 offers owners the ability to issue final payment but still preserve claims that may arise upon completion of an audit.

To the extent the form A201 is modified to state that issuance of final payment constitutes a waiver of all claims related to nonconforming work, it has been held that such a broad waiver was enforceable and prohibited claims by the owner related to allegedly nonconforming work.[63]

Section 9.10.4 can also be important for contractors performing work in states that have adopted the so-called "accepted work doctrine," which generally provides that contractors are no longer liable to individuals injured on the property once the contractor's work has been completed and accepted by the owner.[64]

§ 9.10.5 Acceptance of final payment by the Contractor, a Subcontractor, or ~~materials~~a supplier, shall constitute a waiver of claims by that payee except those previously made in writing and identified by that payee as unsettled at the time of final Application for Payment.

The 2017 amendments to Section 9.10.5 delete the adjective "material," which had previously qualified those types of suppliers that this provision governed. With this deletion, other suppliers, such as labor or equipment suppliers, could fall under the terms of this section. This deletion is also consistent with deletions elsewhere in Chapter 9—Sections 9.3.1, 9.3.1.2, 9.3.3, 9.5.4, 9.6.4, 9.6.5, and 9.6.7.

62. 2 Bruner & O'Connor, *supra* note 2, § 5:193 (Dec. 2016 update). *See also* Younglove Const., LLC v. PSD Dev., LLC, No. 3:08CV1447, 2010 WL 4568716 (N.D. Ohio Nov. 3, 2010) (interpreting language similar to Section 9.10.4 for proposition that use of property as intended did not operate as waiver of claims against contractor for nonconforming work); City of Kimberling City v. Leo Journagan Const. Co., Inc., 337 S.W.3d 48 (Mo. Ct. App. 2011).

63. *See* Mountain View/Evergreen Imp. & Serv. Dist. v. Casper Concrete Co., 912 P.2d 529 (Wyo. 1996).

64. *See, e.g.*, Lamb v. D.S. Duggins Welding, Inc., 729 S.E.2d 724 (N.C. Ct. App. 2012) (holding on matter of first impression that accepted work doctrine barred claims by contractor's employee against subcontractor and sub-subcontractor); Neiman v. Leo A. Daly Co., 148 Cal. Rptr. 3d 818 (Ct. App. 2012). *But see* Davis v. Baugh Indus. Contractors, Inc., 150 P.3d 545 (Wash. 2007).

Whereas final payment by the owner does not operate to affect the owner's right to pursue certain claims, acceptance of final payment by the contractor can automatically operate as a waiver. For instance, in *Jackman Construction, Inc. v. Town of Baggs, Wyoming*,[65] the court held that acceptance of final payment by contractor in an amount that included a deduction for liquidated damages constituted a waiver by the contractor for claims of additional monies owed. This is somewhat analogous to the concept of "accord and satisfaction." Depending on the jurisdiction, some courts have held that acceptance of final payment constitutes waiver by the contractor of unsettled claims.

An interesting question remains as to how far in advance a contractor must identify in writing the amounts that remain disputed before accepting final payment. It is conceivable that this provision could be interpreted to mean that as long as the contractor informs the owner of the disputed amount prior to acceptance (regardless of whether the disputed amount was set forth on the payment application), the contractor has adequately preserved its claims. This type of interpretation appears to violate the spirit of the rules, however, and would not be the most prudent course of action.

CONCLUSION

As discussed throughout this chapter, many of the changes to Article 9 have been in response to emerging trends in the construction industry; the 2017 amendments now include provisions related to VEQ clauses and audit provisions. Some timing issues remain unresolved with the updates. For instance, the A201 still does not address how long the owner has to make payment once the architect approves an application or issues related to timing for payment of on-site materials and equipment. The time for payment should be addressed in the owner-contractor agreement. The article is, nonetheless, consistent with regard to expanding the categories of suppliers that fall under its provisions beyond just "material" suppliers. The overarching theme to the amendments in Article 9 appears to be avoiding and resolving disputes as easily as possible.

65. 278 P.3d 247 (Wyo. 2012).

CHAPTER 10:
Article 10 Protection of Persons and Property

Cathy Lilford Altman
Kate Glaze

ARTICLE 10 PROTECTION OF PERSONS AND PROPERTY

Article 10 covers three topics that address protection of people and property—(1) safety procedures and precautions, (2) hazardous materials, and (3) emergencies. The responsibilities, indemnities, agreements, and representations in this article are often implicated when a person is injured on the project site, the work itself is damaged, or other property is damaged. Clarity and delineation of roles and responsibilities can help avoid lengthy conflict in the event of personal injury, property damage, or resulting schedule delays on the project.

§ 10.1 SAFETY PRECAUTIONS AND PROGRAMS

> The Contractor shall be responsible for initiating, maintaining, and supervising all safety precautions and programs in connection with the performance of the Contract.

This section allocates responsibility for safety precautions and programs to the contractor. The A201–2017 made no revisions to this language from the 2007 version. For new construction, allocation of full responsibility for safety to the contractor makes sense given the contractor's control over the persons and the site. For renovations or additions to occupied buildings, contractors may push for different allocation of certain responsibilities.

Parties are generally not liable for the actions of their independent contractors.[1] Tension exists for owners who, wishing to retain enough control or oversight to ensure proper safety precautions and programs, still wish to retain the independent contractor relationship.[2] For example, courts have held that an owner's right to stop work that the owner knows to be unsafe is not the same as retaining authority over the manner, method, or details of an independent contractor's work; therefore the "retained control" exception of a statute generally protecting a party from liability for its independent contractor's actions is not invoked.[3] Similarly, a construction manager (with a direct contract with the owner) who was required to review the general contractor's safety procedures and to stop work when it becomes aware of unsafe conditions was found not to owe a duty to the general contractor's subcontractor.[4]

Parties often make the following revisions to this provision:

- Add language requiring that the contractor comply with the owner's safety programs or requirements, or requiring that the contractor present its own written program to owner for approval.
- Add language limiting the contractor's responsibility for certain procedures or programs if the owner requires their use even after the contractor has advised against them as dangerous or insufficient.
- Add an affirmative statement that the owner is not responsible for safety precautions.
- Add language including site security within the scope of contractor's responsibilities under this provision. Depending on the site, security can be a major concern and, if so, additional language should be added throughout

1. Fifth Club, Inc. v. Ramirez, 196 S.W.3d 788, 791 (Tex. 2006).
2. Koch Refining Co. v. Chapa, 11 S.W.3d 153 (Tex. 1999).
3. Yazujian v. Jacobs Project Mgmt. Co., C.A. No. 12-1450, 2013 U.S. Dist. LEXIS 159447 (E.D. Pa. Nov. 6, 2013) (interpreting a contract between a general contractor and independent contractor and applying specific Pennsylvania statute).
4. Lee v. GDH, LLC, 25 N.E.3d 761 (Ind. Ct. App. 2015) (contractor's contract included language similar to Section 10.1; also finding that the construction manager did not assume a duty through its actions of managing the project).

this article to include security measures within contractor's control and expense.

§ 10.2 SAFETY OF PERSONS AND PROPERTY

> § **10.2.1** The Contractor shall take reasonable precautions for safety of, and shall provide reasonable protection to prevent damage, injury, or loss to
>
> **.1** employees on the Work and other persons who may be affected thereby;
>
> **.2** the Work and materials and equipment to be incorporated therein, whether in storage on or off the site, under care, custody, or control of the Contractor, a Subcontractor, or ~~the Contractor's Subcontractors or~~a Sub-~~subcontractors~~sub-contractor; and
>
> **.3** other property at the site or adjacent thereto, such as trees, shrubs, lawns, walks, pavements, roadways, structures, and utilities not designated for removal, relocation, or replacement in the course of construction.

This section identifies the people and property the contractor must protect during construction. The revisions to the 2017 version of the A201 are nonsubstantive.

As written, this section requires the contractor to take reasonable precautions and provide reasonable protection for employees, other people who might be affected by the work, the work itself, and materials and equipment to be incorporated in the work (including both those on- and off-site so long as they are in contractor's or its subcontractors' or suppliers' control), and other property at the site or adjacent to it including landscaping, structure, and utilities.

Owners might expand the description of people/property to be protected to include all personnel or all people on site and to include the entire project. Owners might replace "reasonable" with "necessary" to heighten the standard to which the contractor is held. Contractors, conversely, might seek to exclude trespassers from the definition of people it must keep safe, or limit required safety precautions to commercially reasonable standards. Owners' project-specific requirements such as fencing, covered walkways, and signage should be addressed in this section. Finally, while it is likely already included within the provision, it may be wise to add specific language addressing protection of finished work to this provision.

> § **10.2.2** The Contractor shall comply with, and give notices required by applicable laws, statutes, ordinances, codes, rules and regulations, and lawful orders of public authorities, bearing on safety of persons or property or their protection from damage, injury, or loss.

Section 10.2.2 requires the contractor to comply with applicable laws bearing on safety and protection of people and property from damage, injury, and loss. There are no substantive revisions to this section in the 2017 A201.

This section is similar to Section 3.7.2, which requires that the contractor comply with all laws applicable to the performance of the work. It differs from Section 3.2.3, which states that the contractor does not need to confirm that the contract documents comply with all laws but does need to notify the owner if it has actual knowledge of noncompliance. Section 10.2.2, like 3.7.2, is an absolute requirement that the contractor comply with safety laws and does not limit the contractor's obligations to only those laws about which it actually has knowledge.

The parties might add language to make it clear that the contractor shall be responsible for any fines, penalties, or charges assessed by any regulatory body because of violations of safety or health regulations related to the work as well as any schedule impacts due to such violations.

> § **10.2.3** The Contractor shall implement, erect, and maintain, as required by existing conditions and performance of the Contract, reasonable safeguards for safety and protection, including posting danger signs and other warnings against hazards; promulgating safety regulations; and notifying the owners and users of adjacent sites and utilities of the safeguards.

Section 10.2.3 elaborates on Section 10.2.1 and requires that the contractor take certain actions as reasonable precautions for safety. The A201–2017 adds clarification language that does not substantively change the provision. Adding the word "implement" seems appropriate to clearly include the safety programs required under Section 10.1. Adding "of the safeguards" at the end clarifies what information the contractor needs to provide owners of adjacent sites and avoids implying that the contractor needs to give such owners additional information about the project.

Owners might add an affirmative statement that the contractor is responsible for all of the costs associated with such precautions and responsible for promptly repairing any damage to the adjacent sites. Contractors might be concerned about such expenses and try to tie this provision to Section 3.7.4 (Concealed or Unknown Conditions) to make clear that it is entitled to additional compensation if additional safety precautions for adjacent owners are necessary because of a concealed condition. Owners concerned about neighboring structures might add specific language about shoring up, bracing, underpinning, and protecting foundations and nearby structures during the work.

Owners could add an acknowledgment about the nature of the property into this section; for example, notice that the property is a school, is in a dense urban environment, is located near a railroad track, or has another known risk. All parties should consider adding language here to prohibit the contractor's employees, subcontractors, and anyone else from carrying weapons onsite or bringing alcohol on-site, keeping in mind state-specific limits on such restrictions.[5]

> § **10.2.4** When use or storage of explosives or other hazardous materials or equipment, or unusual methods are necessary for execution of the Work, the Contractor shall exercise utmost care and carry on such activities under supervision of properly qualified personnel.

Section 10.2.4 does not expressly require that the contractor give the owner notice when it is doing certain dangerous activities. There is, however, model language available to address the issue in the A503 Supplementary Conditions. There are no changes to this provision in the 2017 A201. Note that this provision applies to "hazardous equipment" and "unusual methods," both of which are broad terms not generally used elsewhere in the A201. Also, parties frequently add notice language to this provision requiring that the contractor notify the owner before beginning the dangerous activity.

5. *See, e.g.*, TEX. PENAL CODE § 30.07.

§ **10.2.5** The Contractor shall promptly remedy damage and loss (other than damage or loss insured under property insurance required by the Contract Documents) to property referred to in Sections 10.2.1.2 and 10.2.1.3 caused in whole or in part by the Contractor, a Subcontractor, a Sub-subcontractor, or anyone directly or indirectly employed by any of them, or by anyone for whose acts they may be liable and for which the Contractor is responsible under Sections 10.2.1.2 and 10.2.1.3~~, except damage or loss~~. The Contractor may make a Claim for the cost to remedy the damage or loss to the extent such damage or loss is attributable to acts or omissions of the Owner or Architect or anyone directly or indirectly employed by either of them, or by anyone for whose acts either of them may be liable, and not attributable to the fault or negligence of the Contractor. The foregoing obligations of the Contractor are in addition to the Contractor's obligations under Section 3.18.

Section 10.2.5 requires the contractor to repair or otherwise remedy any damage or loss to the property described in Sections 10.2.1.2 and 10.2.1.3 caused by the contractor's (or its subcontractor's) actions except when (1) the damage or loss is insured under property insurance required in the contract, or (2) damage or loss is attributable to the acts or omissions of the owner, architect, or their employees and not attributable to the contractor's negligence. This section also makes it clear that these obligations are in addition to the contractor's general indemnity obligations in Section 3.18. The 2017 revised language makes it explicit that the contractor must repair all damage for which it is at least partially responsible, but may then be entitled to make a claim for the repair costs attributable to damage partially caused by the owner, architect, or their employees.

Some parties delete the language excluding damages or losses insured by property insurance, though contractors would resist such a change. A court recently harmonized the insurance language in this section with the waiver of subrogation in the A201–2007 Section 11.3.7 (numbered as Section 11.4.7 in the case), finding that together they place responsibility for damage to non-work on the contractor while the owner (through its insurance) is responsible for damage to the work.[6]

§ **10.2.6** The Contractor shall designate a responsible member of the Contractor's organization at the site whose duty shall be the prevention of accidents. This person shall be the Contractor's superintendent unless otherwise designated by the Contractor in writing to the Owner and Architect.

Section 10.2.6 requires the contractor to designate a person at the site whose duty is the prevention of accidents and provides that such person will be the superintendent unless the contractor designates someone else in writing. The 2017 A201 makes no revisions to this section.

§ **10.2.7** The Contractor shall not permit any part of the construction or site to be loaded so as to cause damage or create an unsafe condition.

6. Copper Mt., Inc. v. Industrial Sys., 208 P.3d 692 (Colo. 2009).

This section states that the contractor shall not permit any part of the site to be loaded so as to cause damage or create an unsafe condition, and it has not been revised in the A201 2017. Some parties expand this provision to include conditions at risk of becoming unsafe.

Loading refers to burdens imposed on unfinished buildings as a result of the construction process, such as the weight of the workers, equipment, and building materials, and they can range from very short term (minutes) to long term (days or weeks). Contractors must consider appropriate bracing, the amount of the load, the duration of the load, and the placement of the load. The contract could be revised with additional loading details if the parties are aware of particular risk or concern related to the project.

§ 10.2.8 INJURY OR DAMAGE TO PERSON OR PROPERTY

If either party suffers injury or damage to person or property because of an act or omission of the other party, or of others for whose acts such party is legally responsible, ~~written~~ notice of ~~such~~the injury or damage, whether or not insured, shall be given to the other party within a reasonable time not exceeding 21 days after discovery. The notice shall provide sufficient detail to enable the other party to investigate the matter.

Section 10.2.8 provides that any party who suffers an injury or loss because of an act or omission of another party (or of one for whom the other party is responsible) must give written notice of the injury or damage in detail within a reasonable time not exceeding 21 days after discovery. The 2017 A201 revises "such" to "the" with no substantive impact. Note that notice must be given within a reasonable time that cannot exceed 21 days, but that a court could find to be less than 21 days given the circumstances. If the parties intend for timely notice hereunder to be required in order to make a claim against the other party, they should add clarification language to this provision. Of course, some state statutes prohibit the enforcement of contract provisions that effectively limit a party's access to courts based on notice provisions or shortened limitations.[7]

Many contracts add a section here requiring that the contractor give notice to the owner within a short time period ranging from "immediately" to "within 48 hours" about any accident related to the work that results in any kind of injury to a person or property. These provisions can include specific people to notify, steps to be taken, and information to include in the event of an accident.

§ 10.3 HAZARDOUS MATERIALS AND SUBSTANCES

§ 10.3.1 The Contractor is responsible for compliance with any requirements included in the Contract Documents regarding hazardous materials or substances. If the Contractor encounters a hazardous material or substance not addressed in the Contract Documents and if reasonable precautions will be inadequate to prevent foreseeable bodily injury or death to persons resulting from a material or substance, including but not limited to asbestos or polychlorinated biphenyl (PCB), encountered on the site by the Contractor, the Contractor shall, upon recognizing the condition, immediately stop Work in the affected area and ~~report the condition to~~notify the Owner and Architect ~~in writing~~of the condition.

7. Douglas S. Oles, *Lack of Notice as a Defense to Construction Claims*, 32 CONSTR. LAW. 6 (Winter 2012). *See, e.g.*, MISS. CODE. ANN. § 15-1-5; TEX. CIV. PRAC. & REM. CODE § 16.070(a).

Section 10.3 addresses hazardous materials—both regulating their use and allocating risk and responsibility if they are discovered. As additional background, it is worth remembering that a new build project will not involve asbestos-containing materials but might involve subsurface contamination, while demolition/renovation could involve such hazardous materials as asbestos-containing materials or lead-based paints. Section 10.3 does not distinguish between these job types, and the distinction will be important when analyzing this section for a particular project.

The foundational rationale in Section 10.3 is workplace safety, rather than regulatory compliance. Given this, an additional hazard that is not specifically addressed and might be considered is "vapor intrusion" (commonly, the passage of hydrocarbon or organic compounds such as radon vapors from the subsurface into enclosed spaces). Vapor intrusion is not typically considered dangerous to the temporary construction team, but it instead affects the long-term users or inhabitants. Given the potential future bodily injuries related to vapor intrusion, it may arguably be contemplated as a Section 10.3 "hazardous material." Vapor intrusion is regulated differently from state to state, and no preemptive federal program creates uniformity.

Lastly, while section 10.3 addresses the discovery or importation of "hazardous materials or substances," it does *not* address a common, nearly universal environmental issue at greenfield construction sites: compliance with EPA's storm water pollution prevention program at certain construction sites where "clearing, grading, and excavation" occur.[8] This federal Water Quality Act of 1987 program[9] has extremely broad application at covered construction sites, jurisdiction over which is frequently delegated to state authorities. Under the program, the owner and contractor (as site operator) share responsibilities for permitting and compliance associated with control of storm water discharges as the result of the regulation's obligatory permitting of discharges, with no clear regulatory delineation of which party is solely responsible.[10] This shared responsibility for program liabilities makes a clear contract allocation and indemnity especially important in greenfield construction projects of sufficient size, where the program would most commonly apply.

Section 10.3.1 requires the contractor to comply with all provisions in the contract related to hazardous materials and to stop work if hazardous materials are encountered and cannot be safely handled. The A201–2017 adds the word "substances," although the purpose of doing so is not entirely clear. Note that the terms "hazardous materials" and "hazardous substances" are not defined as many practitioners would reasonably expect. The term "hazardous substance" is a defined term under CERCLA (Comprehensive Environmental Response, Compensation, and Liability Act, or Superfund) and many baby Superfund statutes at the state level.[11] Conversely, the reach of "hazardous materials" is more uncertain—arguably any material, however benign in the abstract, will be perceived to be "hazardous" if it causes an injury. Therefore, the practitioner may wish to minimize the potential for an interpretation debate by adding his or her own definitions of these terms.

Finally, NESHAP (National Emissions Standards for Hazardous Air Pollutants) regulations[12] frequently require preconstruction surveys for asbestos-containing materials before a building permit can be issued. Thus, asbestos will be a known risk before a building permit is issued, and the contract should address the allocation (in both time and cost) of this work specifically, as suggested by the language in this section requiring that the contractor comply with "any requirements included

8. Implemented through the definition at 40 C.F.R. § 122.26(14)(x).
9. Codified as amendments to the Clean Water Act, 33 U.S.C. §§ 1251–1387.
10. *See* 40 C.F.R. § 122.26(a).
11. 40 C.F.R § 302.4. *See, e.g.*, Cal. Health & Safety Code § 25501.
12. 40 C.F.R. § 61.140.

in the Contract Documents regarding hazardous materials or substances." The same principle would apply to contaminated raw land.

> **§ 10.3.2** Upon receipt of the Contractor's ~~written~~ notice, the Owner shall obtain the services of a licensed laboratory to verify the presence or absence of the material or substance reported by the Contractor and, in the event such material or substance is found to be present, to cause it to be rendered harmless. Unless otherwise required by the Contract Documents, the Owner shall furnish in writing to the Contractor and Architect the names and qualifications of persons or entities who are to perform tests verifying the presence or absence of ~~such~~the material or substance or who are to perform the task of removal or safe containment of ~~such~~the material or substance. The Contractor and the Architect will promptly reply to the Owner in writing stating whether or not either has reasonable objection to the persons or entities proposed by the Owner. If either the Contractor or Architect has an objection to a person or entity proposed by the Owner, the Owner shall propose another to whom the Contractor and the Architect have no reasonable objection. When the material or substance has been rendered harmless, Work in the affected area shall resume upon written agreement of the Owner and Contractor. By Change Order, the Contract Time shall be extended appropriately and the Contract Sum shall be increased ~~in~~by the amount of the Contractor's reasonable additional costs of ~~shut-down~~shutdown, delay, and start-up.

This section imposes duties on the owner to take certain actions if hazardous materials are discovered at the project site. There are minor revisions to the language in the 2017 version that do not have a substantive impact on the provision.

Specifically, Section 10.3.2 imposes the unilateral and unconditional duty on the owner to "verify the presence or absence of the [hazardous material]" and "to perform the task of removal or safe containment." Owners often revise this provision to retain the right to discontinue or suspend the project in case contamination is discovered, a critically important issue because the discovered condition is per se unknowable and unquantifiable. Few, if any, owners or contractors would expect to be chained to the project after the discovery of hazardous materials that might substantially disrupt project feasibility. Granting the owner the right to choose between suspension of work and abandonment of the project should tie into other recovery provisions for the contractor, such as termination for convenience in Section 14.4.

The owner's remedial obligation is stated in what many practitioners will find to be limited terms: "cause [hazardous materials] to be rendered harmless" and "the task of removal or safe containment." Practitioners active in the environmental field will recognize that risk-based approaches prevail in the field today, and those can result in active monitoring, exposure controls, institutional controls, and like remedies that do not result in removal of hazardous materials, per se, from contaminated property. To the extent that Section 10.3.2 will govern the parties' responsibility with respect to the particular remedy applied on the work site, it might be advisable to permit a more nuanced approach to legal compliance and safety assurance, assuming of course that the measures applied comply with applicable code, laws, and regulations.

Additionally, the contractor may want to retain the right to reasonably approve such remediation to the extent that it is expected to return its workers to the site.

§ **10.3.3** To the fullest extent permitted by law, the Owner shall indemnify and hold harmless the Contractor, Subcontractors, Architect, Architect's consultants, and agents and employees of any of them from and against claims, damages, losses, and expenses, including but not limited to attorneys' fees, arising out of or resulting from performance of the Work in the affected area if in fact the material or substance presents the risk of bodily injury or death as described in Section 10.3.1 and has not been rendered harmless, provided that such claim, damage, loss, or expense is attributable to bodily injury, sickness, disease or death, or to injury to or destruction of tangible property (other than the Work itself), except to the extent that such damage, loss, or expense is due to the fault or negligence of the party seeking indemnity.

In Section 10.3.3 the owner is obligated to indemnify the contractor, subcontractors, architect, architect's consultants, and agents and employees of any of them under certain conditions (which include the necessary condition that the environmental condition in question "has not been rendered harmless"). There are no substantive revisions in the A201–2017 version of Section 10.3.3. The owner's and contractor's indemnity provisions (Sections 10.3.3 and 10.3.5, respectively) are not drafted to be exactly reciprocal because the owner's section deals with detected pre-existing conditions while the contractor's section deals with hazardous materials brought to the site by the contractor.

Another unusual aspect of the indemnity structure is that while most environmental regulations are founded on no-fault or strict liability principles (e.g., CERCLA), indemnity carve-outs in the A201 are based on "the fault or negligence" of the party seeking indemnity. The result is logical, but perhaps unexpected, given the prevalence of no-fault principles in the field, and drafters may want to address this difference in the indemnity language.

§ **10.3.4** The Owner shall not be responsible under this Section 10.3 for hazardous materials or substances the Contractor brings to the site unless such materials or substances are required by the Contract Documents. The Owner shall be responsible for hazardous materials or substances required by the Contract Documents, except to the extent of the Contractor's fault or negligence in the use and handling of such materials or substances.

Section 10.3.4 allocates the risk for hazardous materials brought on site and not required by the contract documents. If hazardous materials are required by the contract documents, the risk is allocated to the owner except to the extent of the contractor's fault or negligence in handling them. The A201–2017 adds the word "hazardous" in two places, making its references consistent with the rest of Section 10.3.

§ **10.3.5** The Contractor shall ~~indemnify~~reimburse the Owner for the cost and expense the Owner incurs (1) for remediation of ~~a material~~hazardous materials or ~~substance~~substances the Contractor brings to the site and negligently handles, or (2) where the Contractor fails to perform its obligations under Section 10.3.1, except to the extent that the cost and expense are due to the Owner's fault or negligence.

In this section, the contractor agrees to reimburse the owner for two categories of losses. First, the contractor is liable for remediation expenses if it brings hazardous materials on site and negligently handles them. If this provision is narrowly construed, then "remediation" is a limited term that arguably omits both regulatory response activity, such as field investigation, and reporting. Given the prevalence of regulatory response in the aftermath of a release of hazardous substances, this limitation could be significant.

Second, the contractor is liable for its failure to perform its obligations under section 10.3.1 to comply with all requirements in the contract documents regarding hazardous materials and to halt work if a dangerous condition arises, except to the extent the losses are due to the owner's own fault or negligence. This too is narrow in the absence of another provision specifically addressing a concern for a particular project.

The A201–2017 replaces "indemnify" with "reimburse" and adds and pluralizes "hazardous materials or substances" without making a substantive revision to the provision.

> **§ 10.3.6** If, without negligence on the part of the Contractor, the Contractor is held liable by a government agency for the cost of remediation of a hazardous material or substance solely by reason of performing Work as required by the Contract Documents, the Owner shall ~~indemnify~~reimburse the Contractor for all cost and expense thereby incurred.

The A201 allocates the risk of a compliance order by a governmental agency on the owner, absent the contractor's simple negligence. The A201–2017 replaces "indemnify" with "reimburse." Some thought should be given to clarifying the contractor's and owner's contractual positions under unilateral administrative orders, administrative consent decrees, and other trappings of administrative law as remediation alone might not capture these expenses (similar to section 10.3.4's limitation).

§ 10.4 EMERGENCIES

> In an emergency affecting safety of persons or property, the Contractor shall act, at the Contractor's discretion, to prevent threatened damage, injury, or loss. Additional compensation or extension of time claimed by the Contractor on account of an emergency shall be determined as provided in Article 15 and Article 7.

Section 10.4 completes Article 10 by requiring the contractor to act in its discretion to prevent threatened damage, injury, or loss in an emergency, and provides that any claims the contractor has to additional time or money because of the emergency will be determined as provided in Articles 15 and 7. There are no revisions to this section in the A201–2017. As written, the contractor's discretion should be tied to the standard of care, so parties should modify this provision if they intend to hold the contractor to a higher or lower standard of care during emergencies. Additionally, some drafters add a notice provision requiring that the contractor give the owner prompt notice of any emergency.

CONCLUSION

Article 10 of the A201 addresses the parties' duties and relationship when construction projects go wrong in the worst ways—accidents, injuries, and property damages. Although Article 10 is rarely the subject of intense dispute during negotiations, practitioners cannot overlook its importance. Parties to projects that involve higher-than-average physical risks for injury, hazardous waste, and accidents should certainly take a fresh look at these provisions to ensure they have properly allocated and planned for these risks.

CHAPTER 11:
Article 11 Insurance and Bonds

Anatoly M. Darov
William E. Kelley, Jr.
Sam Laurin

ARTICLE 11 INSURANCE AND BONDS

The most significant change to the 2017 version of the A201 is the movement of most of the A201's substantive insurance and bond provisions out of the A201 and into a new Exhibit A to the parties' agreement (for example, the A101). Exhibit A specifically identifies the insurance and bond policy requirements and amounts in a check box format and makes it easier for project participants to place additional optional coverages. The stand-alone Exhibit A should encourage the project participants to more thoroughly examine and consider the proposed insurance program for the project and more appropriately tailor insurance limits and optional policies. Indeed, many parties have long been adopting custom-made insurance exhibits in lieu of the original Article 11 language to accomplish these very goals.

Unlike the 2007 version, which simply listed all required policies in a single-spaced subparagraph format in former Section 11.1.1, Exhibit A dedicates a separate section for each required policy, which makes the insurance section easier to navigate. The setup of Exhibit A also puts more emphasis on identifying specific limits, unlike original Section 11.1.2, which merely required "limits of liability in the Contract Documents or required by law."

The specifics of Exhibit A will be addressed in detail later in this chapter. The first portion of this chapter will focus on the changes and additions to the insurance requirements that remain in the body of A201 under "Article 11—Insurance and Bonds."

The portion of Article 11 that remains in the A201 is mostly focused on the following three areas: (1) a description of insurance that the contractor is required to purchase and maintain, (2) a description of the insurance that the owner is required to purchase and maintain, and (3) the details of the parties' waiver of subrogation.

Because the details of the contractor's insurance requirements are now in Exhibit A, the general requirements for the insurance that the contractor is required to provide has gone from nearly two pages in the 2007 version to just four paragraphs. To illustrate the magnitude of the changes to Article 11, Section 11.1 is reproduced here in both its revised and comparative forms.

(As Revised)

§ 11.1 CONTRACTOR'S INSURANCE AND BONDS

§ 11.1.1 The Contractor shall purchase and maintain insurance of the types and limits of liability, containing the endorsements, and subject to the terms and conditions, as described in the Agreement or elsewhere in the Contract Documents. The Contractor shall purchase and maintain the required insurance from an insurance company or insurance companies lawfully authorized to issue insurance in the jurisdiction where the Project is located. The Owner, Architect, and Architect's consultants shall be named as additional insureds under the Contractor's commercial general liability policy or as otherwise described in the Contract Documents.

§ 11.1.2 The Contractor shall provide surety bonds of the types, for such penal sums, and subject to such terms and conditions as required by the Contract Documents. The Contractor shall purchase and maintain the required bonds from a company or companies lawfully authorized to issue surety bonds in the jurisdiction where the Project is located.

§ 11.1.3 Upon the request of any person or entity appearing to be a potential beneficiary of bonds covering payment of obligations arising under the Contract, the Contractor shall promptly furnish a copy of the bonds or shall authorize a copy to be furnished.

§ 11.1.4 Notice of Cancellation or Expiration of Contractor's Required Insurance. Within three (3) business days of the date the Contractor becomes aware of an impending or actual cancellation or expiration of any insurance required by the Contract Documents, the Contractor shall provide notice to the Owner of such impending or actual cancellation or expiration. Upon receipt of notice from the Contractor, the Owner shall, unless the lapse in coverage arises from an act or omission of the Owner, have the right to stop the Work until the lapse in coverage has been cured by the procurement of replacement coverage by the Contractor. The furnishing of notice by the Contractor shall not relieve the Contractor of any contractual obligation to provide any required coverage.

(Comparative)

§ 11.1 CONTRACTOR'S ~~LIABILITY~~ INSURANCE AND BONDS

§ 11.1.1 The Contractor shall purchase ~~from~~ and maintain ~~in a company or companies lawfully authorized to do business in the jurisdiction in which the Project is located such~~ insurance ~~as will protect the Contractor from claims set forth below which may arise out~~ of ~~or result from the Contractor's operations and completed operations under the Contract~~the types ~~and for which the Contractor may be legally liable, whether such operations be by the Contractor or by a Subcontractor or by anyone directly or indirectly employed by any of them, or by anyone for whose acts any of them may be liable:~~

~~.1 Claims under workers' compensation, disability benefit~~limits of liability, containing the endorsements, and ~~other similar employee benefit acts that are applicable to the Work to be performed;~~

~~.2 Claims for damages because of bodily injury, occupational sickness or disease, or death of the Contractor's employees;~~

~~.3 Claims for damages because of bodily injury, sickness or disease, or death of any person other than the Contractor's employees;~~

~~.4 Claims for damages insured by usual personal injury liability coverage;~~

~~.5 Claims for damages, other than to the Work itself, because of injury to or destruction of tangible property, including loss of use resulting therefrom;~~

~~.6 Claims for damages because of bodily injury, death of a person or property damage arising out of ownership, maintenance or use of a motor vehicle;~~

~~.7 Claims for bodily injury or property damage arising out of completed operations;~~ subject to the terms and

~~.8 Claims involving contractual liability insurance applicable to the Contractor's obligations under Section 3.18.~~

~~**§ 11.1.2** The insurance required by Section 11.1.1 shall be written for not less than limits of liability specified~~ conditions, as described in the Agreement or elsewhere in the Contract Documents ~~or required by law, whichever coverage is greater. Coverages, whether written on an occurrence or claims-made basis, shall be maintained without interruption from the date of commencement of the Work until the date of final payment and termination of any coverage required to be maintained after final payment, and, with respect to the Contractor's completed operations coverage, until the expiration of the period for correction of Work or for such other period for maintenance of completed operations coverage as specified in the Contract Documents.~~

~~**§ 11.1.3** Certificates of insurance acceptable to the Owner shall be filed with the Owner prior to commencement of the Work and thereafter upon renewal or replacement of each required policy of insurance. These certificates and the insurance policies required by this Section 11.1 shall contain a provision that coverages afforded under the policies will not be canceled or allowed to expire until at least 30 days' prior written notice has been given to the Owner. An additional certificate evidencing continuation of liability coverage, including coverage for completed operations, shall be submitted with the final Application for Payment as required by Section 9.10.2 and thereafter upon renewal or replacement of such coverage until the expiration of the time required by Section 11.1.2. Information concerning reduction of coverage on account of revised limits or claims paid under the General Aggregate, or both, shall be furnished by the Contractor with reasonable promptness.~~

~~**§ 11.1.4** The Contractor shall cause the commercial liability coverage required by the Contract Documents to include (1) the Owner, the Architect and the~~. The Contractor shall purchase and maintain the required insurance from an insurance company or insurance companies lawfully authorized to issue insurance in the jurisdiction where the Project is located. The Owner, Architect, and Architect's consultants shall be named as additional insureds ~~for claims caused in whole or in part by the Contractor's negligent acts or omissions during the Contractor's operations; and (2) the Owner as an additional insured for claims caused in whole or in part by the Contractor's negligent acts or omissions during the Contractor's completed operations.~~

~~§ 11.2 OWNER'S LIABILITY INSURANCE~~

~~The Owner shall be responsible for purchasing and maintaining the Owner's usual liability insurance.~~

~~§ 11.3 PROPERTY INSURANCE~~

~~**§ 11.3.1** Unless~~ under the Contractor's commercial general liability policy or as otherwise ~~provided, the Owner shall purchase and maintain, in a company or companies lawfully authorized to do business in the jurisdiction in which the Project is located, property insurance written on a builder's risk "all-risk" or equivalent policy form in the amount of the initial Contract Sum, plus value of subsequent Contract Modifications and cost of materials supplied or installed by others, comprising total value for the entire Project at the site on a replacement cost basis without optional deductibles. Such property insurance shall be maintained, unless otherwise provided in the Contract Documents or otherwise agreed in writing by all persons and entities who are beneficiaries of such insurance, until final payment has been made as provided in Section 9.10 or until no person or entity other than the Owner has an insurable interest in the property required by this Section 11.3 to be covered, whichever is later. This insurance shall include interests of the Owner, the Contractor, Subcontractors and Sub-subcontractors in the Project~~described in the Contract Documents.

~~**§ 11.3.1.1** Property insurance shall be on an "all-risk" or equivalent policy form and shall include, without limitation, insurance against the perils of fire (with extended coverage) and physical loss or damage including, without duplication of coverage, theft, vandalism, malicious mischief, collapse, earthquake, flood, windstorm, falsework, testing and startup, temporary buildings and debris removal including demolition occasioned by enforcement of any applicable legal requirements, and shall cover~~

~~reasonable compensation for Architect's and Contractor's services and expenses required as a result of such insured loss.~~

~~§ 11.3.1.2 If the Owner does not intend to purchase such property~~ **§ 11.1.2** The Contractor shall provide surety bonds of the types, for such penal sums, and subject to such terms and conditions as required by the Contract Documents. The Contractor shall purchase and maintain the required bonds from a company or companies lawfully authorized to issue surety bonds in the jurisdiction where the Project is located.

[Note. Section 11.1.2 formerly appeared at Section 11.4.]

§ 11.1.3 Upon the request of any person or entity appearing to be a potential beneficiary of bonds covering payment of obligations arising under the Contract, the Contractor shall promptly furnish a copy of the bonds or shall authorize a copy to be furnished.

The revised Section 11.1.1 does not limit the insurance required for a project only to what is specified in Exhibit A. Additional insurance requirements may be contained anywhere in the Contract Documents. The owner's option to require the contractor to provide performance and payment bonds remains in Section 11.1.2. Section 11.1.3 is a new provision that allows any potential beneficiary of the bond to request a copy of the bond. As a result, any potential bond claimant is likely a third-party beneficiary of Section 11.1.3.

Like the contractor's insurance requirements in the A201, the owner's insurance requirements have gone from nearly two pages to a few paragraphs.

§ 11.1.4 Notice of Cancellation or Expiration of Contractor's Required Insurance. Within three (3) business days of the date the Contractor becomes aware of an impending or actual cancellation or expiration of any insurance required by the Contract Documents, the Contractor shall provide notice to the Owner of such impending or actual cancellation or expiration. Upon receipt of notice from the Contractor, the Owner shall, unless the lapse in coverage arises from an act or omission of the Owner, have the right to stop the Work until the lapse in coverage has been cured by the procurement of replacement coverage by the Contractor. The furnishing of notice by the Contractor shall not relieve the Contractor of any contractual obligation to provide any required coverage.

Section 11.1.4 imposes certain obligations on a contractor if the contractor becomes aware that any required insurance under the Contract Documents will be cancelled. It also corrects a potential inconsistency between the A201 and standard certificates of insurance.

A certificate of insurance is a "document acknowledging that an insurance policy has been written, and setting forth in general terms what the policy covers."[1] "[A] certificate of insurance is not a separate and distinct independent contract but rather evidence of the underlying agreement that references the coverage of

1. Black's Law Dictionary (7th ed. 1999).

the policy in general terms."[2] Nonetheless, "[i]t is well settled that an insurer may be equitably estopped from denying coverage where the party for whose benefit the insurance was procured reasonably relied upon the provisions of an insurance certificate to that party's detriment."[3] Conversely, though, some jurisdictions have rejected this estoppel argument and held that where a certificate of insurance expressly states that it does not alter the coverage of the underlying policy, the certificate of insurance will not be deemed to change the underlying insurance policy.[4]

So what happens when the A201 requirements for certificates of insurance conflict with the standard or available certificates of insurance used by the insurance industry? The changes in Section 11.1.4 are intended to address one such situation. In this regard, prior Section 11.1.3 of the A201–2007 provided that the contractor would file certificates of insurance with the owner, and that the certificates of insurance "shall contain a provision that coverages afforded under the policies will not be canceled or allowed to expire until at least 30 days' prior written notice has been given to the Owner." This language afforded potential protection to the owner by virtue of the required notice to be provided prior to cancellation or expiration of an insurance policy maintained by the contractor. The language not only made notice a part of the contractual obligations, but in fact made it a requirement for the certificate of insurance that evidenced the required insurance coverages. When the A201–2007 was drafted, the standard Association for Cooperative Operations Research and Development (ACORD) 25 Certificate of Liability Insurance form contained the following language:

> SHOULD ANY OF THE ABOVE DESCRIBED POLICIES BE CANCELLED BEFORE THE EXPIRATION DATE THEREOF, THE ISSUING INSURER WILL ENDEAVOR TO MAIL ____ DAYS WRITTEN NOTICE TO THE CERTIFICATE HOLDER NAMED TO THE LEFT, BUT FAILURE TO DO SO SHALL IMPOSE NO OBLIGATION OR LIABILITY OF ANY KIND UPON THE INSURER, ITS AGENTS OR REPRESENTATIVES.[5]

Thus, the language in the ACORD 25 Certificate of Liability Insurance form matched the requirement of Section 11.1.3 of the A201–2007 regarding the language to be placed on that form as to pre-cancellation notice to be provided to the certificate holder.

One of the challenges with this language on the certificate of insurance form was that it purported to require notice from the insurer to the additional insured—even if the insurance policy did not require or allow for such notice. In other words, it had the potential effect of modifying the terms of the underlying coverage—even though a certificate of insurance is merely evidence of coverage, and not a separate and distinct contract for insurance. This, in essence, created conflict between the certificate of insurance language and the insurance policy language.

---

2. General Acc. Ins. Co. of America v. American Nat'l Fireproofing, Inc., 716 A.2d 751 (R.I. 1998).

3. Marlin v. Wetzel County Bd. of Educ., 212 S.E.2d 462, 472 (W. Va. 2002) (quoting Lenox Realty v. Excelsior Ins. Co., 679 N.Y.S.2d 749, 750 (App. Div. 1998)).

4. *See* Mulvey Const., Inc. v. Bituminous Cas. Corp., 571 F. App'x 150, 158 (4th Cir. 2014); Mountain Supply v. Reliance Ins. Co., 933 F.2d 882, 889 (10th Cir. 1991); T.H.E. Ins. Co. v. City of Alton, 227 F.3d 802, 806 (7th Cir. 2000); TIG Ins. Co. v. Sedgwick James of Washington, 184 F. Supp. 2d 591, 597–98 (S.D. Tex. 2001); G.E. Tignall & Co., Inc. v. Reliance Nat'l Ins. Co., 102 F. Supp. 2d 300, 304 (D. Md. 2000).

5. https://www.acord.org/standards/forms/Documents/ACORDCertificatesFAQ_201004.pdf.

In 2009, the ACORD Certificate of Liability Insurance form underwent a significant and substantive change. The prior language regarding notice of cancellation was deleted and replaced with the following:

> SHOULD ANY OF THE ABOVE DESCRIBED POLICIES BE CANCELLED BEFORE THE EXPIRATION DATE THEREOF, NOTICE WILL BE DELIVERED IN ACCORDANCE WITH THE POLICY PROVISIONS.[6]

The result was that the widely used ACORD Certificate of Liability Insurance form was suddenly in conflict with the requirement of Section 11.1.3 of the A201–2007. Adept parties using the A201–2007 had to modify the language to bring it into conformance with the change on the ACORD form. However, parties that left Section 11.1.3 of the A201–2007 unmodified created potential conflicts between the contract requirements and the language actually used on the certificate of insurance as to pre-cancellation or coverage modification notice.

The new Section 11.1.4 in the A201–2017 document no longer requires that certain language be included on the certificate of insurance. Instead, the burden is shifted directly to the contractor to notify the owner of pending cancellations or expirations of insurance policies and coverages required under the contract. Further, where a contractor fails to procure insurance coverage or allows required insurance coverage to lapse, the owner can suspend the work until the required insurance is procured by the contractor.

The result for practitioners is that resolution of pre-cancellation notice issues will require review of the applicable insurance policy. As noted in *Pinnacol Assurance v. Hoff*,[7] where a certificate of insurance contained language stating "SHOULD ANY OF THE ABOVE DESCRIBED POLICIES BE CANCELLED BEFORE THE EXPIRATION THEREOF, NOTICE WILL BE DELIVERED IN ACCORDANCE WITH THE POLICY PROVISIONS," the parties "consigned the entire question of notice, including to whom it must be given, to the provisions of the policy being cancelled."

The A201–2017 provides much more detail for the scenario where the owner is required to purchase certain insurance and chooses not to do so.

§ 11.2 OWNER'S INSURANCE

§ 11.2.1 The Owner shall purchase and maintain insurance of the types and limits of liability, containing the endorsements, and subject to the terms and conditions, as described in the Agreement or elsewhere in the Contract Documents. The Owner shall purchase and maintain the required insurance ~~required by the Contract and~~from an insurance company or insurance companies lawfully authorized to issue insurance in the jurisdiction where the Project is located.

§ 11.2.2 Failure to Purchase Required Property Insurance. If the Owner fails to purchase and maintain the required property insurance, with all of the coverages and

6. *Id.*
7. 375 P.3d 1214, 1222 (Colo. 2016).

in the ~~amount~~amounts described ~~above~~in the Agreement or elsewhere in the Contract Documents, the Owner shall ~~so~~ inform the Contractor in writing prior to commencement of the Work. ~~The~~Upon receipt of notice from the Owner, the Contractor may ~~then effect~~delay commencement of the Work and may obtain insurance that will protect the interests of the Contractor, Subcontractors, and Sub-Subcontractors in the Work. When the failure to provide coverage has been cured or resolved, the Contract Sum and Contract Time shall be equitably adjusted. In the event the Owner fails to procure coverage, the Owner waives all rights against the Contractor, Subcontractors, and Sub-subcontractors ~~in the Work, and by appropriate Change Order the~~to the extent the loss to the Owner would have been covered by the insurance to have been procured by the Owner. The cost ~~thereof~~ of the insurance shall be charged to the Owner~~. If~~ by a Change Order. If the Owner does not provide written notice, and the Contractor is damaged by the failure or neglect of the Owner to purchase or maintain the required insurance ~~as described above, without so notifying the Contractor in writing, then~~, the Owner shall ~~bear~~reimburse the Contractor for all reasonable costs ~~property~~and damages attributable thereto.

[Note. Section 11.2.2 formerly appeared at Section 11.3.1.2.]

Section 11.2.1 focuses on the contractor's right to purchase insurance that Contract Documents require the owner to purchase if the owner declines to purchase the insurance. Section 11.2.1 adds the specific requirement that the owner must give notice that it will not purchase the required insurance before the commencement of the work. This is an appropriate change because it provides the specific deadline to give the notification. Section 11.2.2 describes the contractor's rights under the Contract Documents if the owner does not purchase the required insurance, which include delaying the commencement of the work and requesting an equitable adjustment in the Contract Time and Contract Sum until the required insurance is obtained. More importantly, this section provides that if the owner fails to procure the required coverage, it waives all claims to the extent the claims would have been covered by the insurance. It also provides that the owner is responsible to the contractor for all damages incurred by the contractor if the owner fails to give the required notice.[8]

§ 11.2.3 Notice of Cancellation or Expiration of Owner's Required Property Insurance. Within three (3) business days of the date the Owner becomes aware of an impending or actual cancellation or expiration of any property insurance required by the Contract Documents, the Owner shall provide notice to the Contractor of such impending or actual cancellation or expiration. Unless the lapse in coverage arises from an act or omission of the Contractor: (1) the Contractor, upon receipt of notice from the Owner, shall have the right to stop the Work until the lapse in coverage has been cured by the procurement of replacement coverage by either the Owner or the Contractor; (2) the Contract Time and Contract Sum shall be equitably adjusted; and (3) the Owner waives all rights against the Contractor, Subcontractors, and Sub-subcontractors to the extent any loss to the Owner would have been covered by the insurance had it not expired or been cancelled. If the Contractor purchases replacement coverage, the cost of the insurance shall be charged to the Owner by an appropriate Change Order. The furnishing of notice by the Owner shall not relieve the Owner of any contractual obligation to provide required insurance.

8. The waiver of claims and liability to the contractor are important because an owner is usually going to purchase the insurance required by contract or request that the contractor provide it well before the work commences. If the required owner insurance is not in place when the contract is signed, then it is likely that the owner will never communicate that fact to the contractor.

Section 11.2.3 provides the contractor with new rights if any owner-furnished insurance policy—except the owner's liability policy—is cancelled or expires. Under this provision, the owner has three business days from the date it becomes aware of an impending or actual cancellation or expiration to notify the contractor of that cancellation or expiration. Upon receipt of such notice, the contractor may stop work until the lapse is cured provided the lapse is not caused from an act or omission of the contractor. The provision is silent on whether the lapse must be an "actual" lapse or an "impending" lapse before contractor's right to suspend work arises. If properly exercised by the contractor, it is entitled to a commensurate extension of the Contract Time equal to the length of suspension of work as well as an increase of the Contract Sum for reasonable costs of shutdown, delay, and start-up. The provision also waives the owner's rights against contractor and first- and second-tier subcontractors to the extent any loss would have been covered by lapsed insurance coverages. The contractor may also purchase replacement coverage at the owner's cost.

This provision is entirely new with no corollary in the A201–2007. If used without modification, it provides the contractor with an important remedy when the owner fails to maintain its required policies. Beginning in 2009, the ACORD 24/25 form of certificate of insurance[9] no longer includes a notice of cancellation. The current version of the ACORD 24 and 25 certificate refers instead to the notice of cancellation provisions of the insurance policy. In practice, this provision will require the owner to provide notice of expiring or cancelled policies even though such expiration or cancellation may not result in a lapse of coverage.

§ 11.3 WAIVERS OF SUBROGATION

§ 11.3.1 The Owner and Contractor waive all rights against (1) each other and any of their subcontractors, sub-subcontractors, agents, and employees, each of the other; (2) the Architect and Architect's consultants; and (3) Separate Contractors, if any, and any of their subcontractors, sub-subcontractors, agents, and employees, for damages caused by fire, or other causes of loss, to the extent those losses are covered by property insurance required by the Agreement or other property insurance applicable to the Project, except such rights as they have to proceeds of such insurance. The Owner or Contractor, as appropriate, shall require similar written waivers in favor of the individuals and entities identified above from the Architect, Architect's consultants, Separate Contractors, subcontractors, and sub-subcontractors. The policies of insurance purchased and maintained by each person or entity agreeing to waive claims pursuant to this section 11.3.1 shall not prohibit this waiver of subrogation. This waiver of subrogation shall be effective as to a person or entity (1) even though that person or entity would otherwise have a duty of indemnification, contractual or otherwise, (2) even though that person or entity did not pay the insurance premium directly or indirectly, or (3) whether or not the person or entity had an insurable interest in the damaged property.

§ 11.3.52 If during the Project construction period the Owner insures properties, real or personal or both, at or adjacent to the site by property insurance under policies separate from those insuring the Project, or if after final payment property insurance is to be provided on the completed Project through a policy or policies other than those insuring the Project during the construction period, to the extent permissible by such policies, the

9. ACORD publishes eight forms of insurance certificates. ACORD 24 is the property insurance form. ACORD 25 is the liability insurance form.

Owner ~~shall waive~~<u>waives</u> all rights in accordance with the terms of Section 11.3.~~7~~<u>1</u> for damages caused by fire or other causes of loss covered by this separate property insurance~~. All separate policies shall provide this waiver of subrogation by endorsement or otherwise~~.

The waiver of subrogation provision is present in all current AIA contracts for construction as well as many prior versions. The reason for the waiver is as follows:

> A waiver of subrogation clause is a risk-shifting provision premised upon the recognition that it is economically inefficient for parties to a contract to insure against the same risk. A subrogation waiver encourages parties to anticipate risks and to procure insurance covering those risks and also facilitates and preserves economic relations and activity. Because a property Owner can generally acquire insurance to protect the property against fire and other perils, in the context of a construction contract, the waiver of subrogation clause shifts the ultimate risk of loss resulting from such perils to the Owner to the extent the damages are covered by insurance. The intent is to avoid disruption during construction and provide certainty and eliminate litigation by having the contracting parties look only to the Owner's insurance for protection in the event of loss resulting from fire or other perils. In other words, a waiver of subrogation clause substitutes the protection of insurance for the uncertain and expensive protection of liability litigation.[10]

Given the use of the AIA waiver of subrogation and judicial acceptance of its enforceability, construction lawyers should assume that if the AIA waiver language is present and the work is damaged during the construction, the waiver of subrogation will be enforceable.

The changes in the waiver of subrogation language in Section 11.3.1, along with a change in Exhibit A, make clear that the waiver of subrogation applies to (1) insured losses to property other than the work as defined in the A201 and (2) losses covered by property insurance after the work is completed and accepted. The waiver is now expanded to any insurance required by the Contract Documents and is applicable "to property at or adjacent to the project site." This change addresses a split in decisions throughout the country regarding whether the waiver of subrogation applied to any insurance obtained for the project or just the insurance obtained for the work (i.e., builder's risk). The current majority view is that the waiver applies to any insurance in place that covers a loss: the "any insurance" approach. The minority view is that the waiver applies only to damages to the work.[11] This distinction was based on the language in Section 11.3.7 of the A201–2007 version, which provided that "[t]he Owner and Contractor waive all rights . . . for damages caused by fire or other perils to the extent covered by property insurance pursuant to this Section 11.3 or other property insurance applicable to the work." Exhibit A now provides in Section 2.3.3.3 that "[i]f the Work involves remodeling an

---

10. TX C.C. Inc. v. Wilson/Barnes General Contractors Inc., 233 S.W.3d 562, 567–68 (Tex. App. 2007) (citations omitted) (citing Ind. Ins. Co. v. Erhlich, 880 F. Supp. 513, 518 (W.D. Mich. 1994); Acadia Ins. v. Buck Constr., 756 A.2d 515, 519, 520 (Me. 2000); Temple EasTex, Inc. v. Old Orchard Creek Partners, Ltd., 848 S.W.2d 724, 730 (Tex. App. 1992); and Walker Eng'g, Inc. v. Bracebridge Corp., 102 S.W.3d 837, 841 (Tex. App. 2003)).

11. The decision by the Indiana Supreme Court in *Board of Commissioners of County of Jefferson v. Teton Corp.*, 30 N.E.3d 711, 715–18 (Ind. 2015), provides an overview of the reasoning behind the "any insurance" approach and the "Work versus non-Work" approach. The Indiana Supreme Court adopted the "any insurance" approach in its decision. *Id.* at 718. The split in the decisions as to whether to follow the "any insurance" or "Work versus non-Work" reasoning was recognized in the recent decision of *Turner Construction Co. v. BFPE International, Inc.*, Civ. No. JKB-15-368, 2016 U.S. Dist. LEXIS 39161 (D. Md. March 25, 2016), which also decided to follow the "any insurance" approach.

existing structure or constructing an addition to an existing structure, the Owner shall purchase and maintain, until the expiration of the period for correction of work as set forth in Section 12.2.2 of the General Conditions, 'all-risk' property insurance, on a replacement cost basis. . . ." The minority view on "work versus non-work" is probably no longer viable if insurance covers a loss to non-work property and the parties used the revised language in the contract.

The 2017 changes also likely resolve another conflict in court decisions: whether the waiver of subrogation clause applies to damages that are covered by policies purchased by the owner after the completion of construction when the standard AIA subrogation language is used. The current majority rule is that the waiver of subrogation does apply to post-completion damage based largely on the language from Section 11.3.5 of the 2007 version that, "if after final payment insurance is to be provided on the completed project during the construction period, the Owner shall waive all rights in accordance with the terms of Subparagraph 11.3.7 [waiver of subrogation]."[12]

Most cases hold that the foregoing language meant that if the owner purchased property insurance post-completion, the waiver of subrogation applied.[13] The minority view holds that the waiver is not applicable unless there was a specific contract requirement that the owner procure property insurance after completion.[14] This view is based on an interpretation that the language "if after final payment insurance is to be provided" did not implicate the waiver because it did not require the owner to purchase the insurance. Because the A201–2007 version did not require the owner to purchase property insurance, courts that followed the minority view held that the waiver of subrogation did not apply. The language in A.2.3.1.3 in Exhibit A appears to address this issue because it provides that:

> § A.2.3.1.3 Unless the parties agree otherwise, upon Substantial Completion, the owner shall continue the insurance required by Section A.2.3.1 or, if necessary, replace the insurance policy required under Section A.2.3.1 with property insurance written for the total value of the project that shall remain in effect until expiration of the period for correction of the Work set forth in Section 12.2.2 of the General Conditions.

If this language is used and there is a loss post-completion, then the waiver of subrogation should apply because the owner is now required to procure property insurance for the total value of the project post-completion unless the parties otherwise agree. As a practical matter it is extremely unlikely that an owner would consciously decide not to have post-completion property insurance (which would be for the full insurable value), so the waiver of subrogation should apply.[15]

Whether the indemnification provision in Section 3.18 of the A201 affects the waiver of subrogation also has been addressed by courts. In *Trump-Equitable Fifth Ave. Co. H.R.H. Construction Corp.*,[16] for example, the insurer seeking to subrogate on behalf of the owner against the general contractor argued that the general contractor's agreement to indemnify the owner overrode the waiver of subrogation. The court

12. *E.g.*, *TX. C.C. Inc.*, 233 S.W.3d 562.
13. *Id.*
14. Probably the leading case applying the minority view was *Lumbermens Mutual Casualty Co. v. Grinnell Corp.*, 477 F. Supp. 2d 327 (D. Ma. 2007).
15. In addition to the previously cited cases, this conclusion is buttressed by the decision by the Maryland Court of Appeals in *John L. Mattingly Construction Co. Inc. v. Hartford Underwriters Insurance Co.*, 999 A.2d 1066 (Md. 2010). In *Mattingly*, the court found that a modified AIA contract was ambiguous as to whether the waiver of subrogation applied to post completion losses. One factor in the Court's decision that the contract was ambiguous was that the language "is to be provided" was not present in the contract at issue. *Id.* at 1075–76.
16. 485 N.Y.S.2d 65 (App. Div. 1985).

summarily rejected this claim by correctly noting the difference between first-party coverage (for example, builders risk coverage) and third-party coverage (commercial general liability to cover third-party claims).[17] In *Trump-Equitable* there was no dispute that the general contractor purchased commercial general liability coverage and that the owner had purchased the builder's risk that covered and paid the claim. The court held that there was no third-party claim and the indemnification provision was not implicated because the owner had been fully compensated for its loss.[18]

The waiver of subrogation clause also applies to another A201 provision that construction lawyers often address: the waiver of consequential damages. The differences and similarities between a waiver of subrogation and waiver of consequential damages are often addressed when it is alleged that the party causing the property damage was grossly negligent. In that scenario, the insurer tries to avoid the applicability of the waiver of subrogation provision by arguing that the waiver of subrogation clause is akin to the waiver of consequential damages clause. Insurers make this argument because a waiver of consequential damages provisions is often considered an exculpatory clause that does not necessarily bar a claim for gross negligence.

The Nebraska Supreme Court rejected this argument in *Lexington Ins. Co. v. Entrex Communication Services*.[19] While that court noted similarities between an exculpatory clause and a waiver of subrogation, it also found significant differences. Exculpatory clauses preclude all recovery for damages while a waiver of subrogation applies only to losses covered by insurance, reducing the risk that an injured party will be left uncompensated. The court also noted the important policy reasons behind a waiver of subrogation and that these policy reasons do not apply to a broader clause.[20]

The changes to the waiver of subrogation language do not necessarily address the issue of whether the waiver applies to a loss paid by a standard commercial general liability policy. This issue arose in *Tellespen Builder, L.P. v. Kendall/Heaton Insurance*.[21] In *Tellespen*, the Texas Court of Appeals held that a waiver of subrogation that applied to any damage "covered by property or equipment insurance" was ambiguous as to whether it applied to a subrogation claim made by a general contractor's commercial general liability insurer against a subcontractor.

In *Tellespen*, the owner made a claim against the general contractor for defective work that caused structural and water damage. The general contractor performed the repairs and in turn was reimbursed by its general liability insurer. The general contractor's insurer then brought a claim against the design professional and subcontractors claiming that they caused the damages.[22] The design professional and subcontractors argued that the waiver in the applicable contract applied. The waiver of subrogation stated that it applied to any "loss or damage to the extent covered by property or equipment insurance." The subcontractors and design professionals therefore argued that because the general liability policy covered "property damage," they were entitled to the protection provided by the waiver.[23]

The general contractor in turn argued that property insurance has a specific meaning: a first-party policy that protects the insured's property, not a policy that protects an insured from third-party claims. The general contractor also argued that

17. *Id.* at 68.
18. *Id.*
19. 749 N.W.2d 124 (Neb. 2008).
20. *Id.* at 129. The *Trump-Equitable Fifth Ave. Co.* decision, discussed above, reached the same conclusion. *Trump-Equitable Fifth Ave. Co.*, 485 N.Y.S.2d 65.
21. 325 S.W.3d 692 (Tex. App. 2010).
22. *Id.* at 694–95.
23. *Id.* at 696.

if the waiver of subrogation provision were applied then it would render two general indemnity clauses meaningless if the waiver applied to third-party claims.[24] The Texas Court of Appeals held that the waiver of subrogation was ambiguous based on the contract language alone. The court concluded that both parties' arguments had merit and that the general contractor was not required to purchase property insurance.[25] Given the growing trend of states to find coverage for property damage under a commercial general liability (CGL) policy, the applicability of the waiver of subrogation to a CGL policy may become more prevalent.[26]

§ 11.4 LOSS OF USE, BUSINESS INTERRUPTION, AND DELAY IN COMPLETION INSURANCE

> The Owner, at the Owner's option, may purchase and maintain insurance that will protect the Owner against loss of use of the Owner's property, or the inability to conduct normal operations, due to fire or other causes of loss. The Owner waives all rights of action against the Contractor and Architect for loss of use of the Owner's property, due to fire or other hazards however caused.

The new Section 11.4 largely mirrors the previous Section 11.3.3. The only substantive difference between 11.4 and the previous 11.3.3 is that in 11.3.3, the waiver covered consequential damages while 11.4 does not apply the waiver to consequential damages. This change was probably made because parties sometime strike the waiver of consequential damages in Article 15 of the A201. Presumably the deletion of "consequential damages" in Section 11.4 was to avoid a situation where the parties waive consequential damages in Article 15, but the waiver of subrogation still covered consequential damages.

§ ~~§ 11.3.8~~11.5 ADJUSTMENT AND SETTLEMENT OF INSURED LOSS

> **§ 11.5.1** A loss insured under the ~~Owner's~~ property insurance required by the Agreement shall be adjusted by the Owner as fiduciary and made payable to the Owner as fiduciary for the insureds, as their interests may appear, subject to requirements of any applicable mortgagee clause and of Section 11.~~3.10~~5.2. The Owner shall pay the Architect and Contractor ~~shall pay Subcontractors~~ their just shares of insurance proceeds received by the ~~Contractor~~Owner, and by appropriate agreements~~, written where legally required for validity,~~ the Architect and Contractor shall ~~require Subcontractors to~~ make payments to their ~~Sub-subcontractors~~consultants and Subcontractors in similar manner.
>
> **§ 11.5.2** Prior to settlement of an insured loss, the Owner shall notify the Contractor of the terms of the proposed settlement as well as the proposed allocation of the insurance proceeds. The Contractor shall have 14 days from receipt of notice to object to the proposed settlement or allocation of the proceeds. If the Contractor does not object,

24. *Id.* at 697.
25. *Id.* at 699.
26. See the recent decisions by the New Jersey Supreme Court in *In Cypress Point Condominium Association, Inc. v. Adria Towers, L.L.C.*, No. C-220 Sept. Term 2015, 076348, 2016 WL 4131662, at *8 (N.J. Aug. 4, 2016), and by the Iowa Supreme Court in *National Security Corp. v Westlake Investments, Inc.*, 880 N.W.2d 724 (Iowa 2016), which continue the general trend of finding that a CGL policy does cover construction defects in certain instances.

the Owner shall settle the loss and the Contractor shall be bound by the settlement and allocation. Upon receipt, the Owner shall deposit the insurance proceeds in a separate account and make the appropriate distributions. Thereafter, if no other agreement is made or the Owner does not terminate the Contract for convenience, the Owner and Contractor shall execute a Change Order for reconstruction of the damaged or destroyed Work in the amount allocated for that purpose. If the Contractor timely objects to either the terms of the proposed settlement or the allocation of the proceeds, the Owner may proceed to settle the insured loss, and any dispute between the Owner and Contractor arising out of the settlement or allocation of the proceeds shall be resolved pursuant to Article 15. Pending resolution of any dispute, the Owner may issue a Construction Change Directive for the reconstruction of the damaged or destroyed Work.

The one substantive change in Section 11.5.1 is the language "required by the Agreement," which affects the waiver of subrogation as already discussed. This change raises the issue of whether the owner actually has the ability to allocate proceeds from a payment of a loss, if all project participants and not just the owner are considered insureds,[27] which is what Exhibit A currently requires. If all of the parties are insureds under the policy, would an insurer ever agree to pay the amount of the loss solely to the owner? In addition to this issue, the changes provide more detail on how to address a contractor's objection to an owner's allocation of proceeds. The requirement that the owner provide a bond in performance of its duties in allocating proceeds has been deleted.

ARTICLE A.1 GENERAL

The Owner and Contractor shall purchase and maintain insurance, and provide bonds, as set forth in this Exhibit. As used in this Exhibit, the term General Conditions refers to AIA Document A201™–2017, General Conditions of the Contract for Construction.

This provision is a general statement indicating that Exhibit A specifies the obligations of the owner and the contractor relative to bonds and insurance. Purchasing and maintaining insurance does not imply any other obligations or conditions of coverage, which obligations and conditions must be set forth in Exhibit A.[28]

ARTICLE A.2 OWNER'S INSURANCE

§ A.2.1 GENERAL

Prior to commencement of the Work, the Owner shall secure the insurance, and provide evidence of the coverage, required under this Article A.2 and, upon the Contractor's request,

27. There is legitimate debate whether all of the project participants should be named insureds under a builder's risk policy. Bell, Dunn, Costner, *Confronting Conventional Wisdom on Builders Risk: From Named-Insured Status to Concurrent Causation,* 31 ABA Forum on the Construction Industry (Fall 2011); Coombs, *Should Builders Risk Policies Cover Nonowner Entities as Insureds? You Bet, and Here is Why,* IRMI Expert Commentary (Feb. 2012); Bell. Dunn, Costner, *Responding to a Contrary Position: Minimizing the Number of Named Insureds on the Builders Risk Policy,* ABA Forum on the Construction Industry (September 2012).

28. West Bend Mutual Ins. Co. v. Athens Construction Co., 29 N.E.3d 636 (Ill. App. Ct. 2015) (obligation to purchase and maintain insurance without specifying additional insured requirements did not require subcontractor to list contractor has additional insured and met requirements of the contract).

provide a copy of the property insurance policy or policies required by Section A.2.3. The copy of the policy or policies provided shall contain all applicable conditions, definitions, exclusions, and endorsements.

This provision obligates the owner to secure and provide evidence of insurance required of the owner pursuant to Article A.2 of the exhibit. Such coverage may include both liability insurance (under Article A.2.2) and property insurance (under A.2.3). This article requires the owner to provide a complete copy of the property insurance policy or policies, inclusive of all applicable coverage conditions, definitions, exclusions, and endorsements, to the contractor upon request. The text of this section is similar to portions of Section 11.2 and 11.3.1 through Section 11.3.6 in the A201–2007. Under Section 11.1.3 of the A201–2007, the owner was obligated to file a copy of each property insurance policy with the contractor, whether or not requested by the contractor, prior to the occurrence of any exposure to loss.

Evidence of insurance in the form of a certificate of insurance does not constitute an insurance contract between the carrier and the certificate holder or the insured indicated on the certificate.[29] Further, coverages indicated on a certificate of insurance will not prevail over those actually existing under the policy terms.[30] See Section 11.1.4 in this chapter for further discussion of certificates of insurance.

Several cases have focused on the failure to procure required insurance coverage. In *Coluccio Construction Co. v. King County*,[31] a public entity was held liable for damages sustained by the general contractor when it failed to procure property insurance for a public works project. The court concluded that by failing to supply the required insurance, the owner assumed all risk of loss for the amounts that would have been covered by insurance. Liability in such cases also may fall to the insurance broker.[32] In *SMI Owen Steel Co. v. Marsh USA, Inc.*,[33] an insurance broker responsible for administering a contractor-controlled insurance program supplied insurance certificates and other program documentation to a subcontractor that incorrectly indicated that the subcontractor was insured under a professional liability policy. The court found the broker negligent for failing to procure promised insurance and awarded damages to the subcontractor.

§ A.2.2 LIABILITY INSURANCE

The Owner shall be responsible for purchasing and maintaining the Owner's usual general liability insurance.

This language is unchanged from the 2007 version. Parties engaged in projects undertaken where the owner maintains ongoing operations on or about the project site that are unrelated to the contractor's work should be careful to ensure that the

29. United States Pipe & Foundry Co. v. United States Fidelity & Guaranty Co., 505 F.2d 88 (5th Cir. 1974) (certificate of insurance did not constitute a valid contract).
30. Lezak & Levy Wholesale Meats, Inc. v. Illinois Employees Ins. Co., 460 N.E.2d 475 (Ill. App. Ct. 1984).
31. 150 P.3d 1147 (Wis. Ct. App. 2007).
32. *See also* Colorado Casualty Ins. v. Safety Control Co., 288 P.3d 764 (Ariz. Ct. App. 2012).
33. 520 F.3d 432, 443–44 (5th Cir. 2008).

owner carries appropriate liability coverages and limits for that other work. Consideration also should be given to whether the owner should be required to extend additional insured status to the contractor, as several courts have considered whether an obligation to obtain insurance includes an agreement to add another party as additional insured and have generally found that it does not.[34]

§ A.2.3 REQUIRED PROPERTY INSURANCE

§ A.2.3.1 Unless this obligation is placed on the Contractor pursuant to Section A.3.3.2.1, the Owner shall purchase and maintain, from an insurance company or insurance companies lawfully authorized to issue insurance in the jurisdiction where the Project is located, property insurance written on a builder's risk "all-risks" completed value or equivalent policy form and sufficient to cover the total value of the entire Project on a replacement cost basis. The Owner's property insurance coverage shall be no less than the amount of the initial Contract Sum, plus the value of subsequent Modifications and labor performed and materials or equipment supplied by others. The property insurance shall be maintained until Substantial Completion and thereafter as provided in Section A.2.3.1.3, unless otherwise provided in the Contract Documents or otherwise agreed in writing by the parties to this Agreement. This insurance shall include the interests of the Owner, Contractor, Subcontractors, and Sub-subcontractors in the Project as insureds. This insurance shall include the interests of mortgagees as loss payees.

This provision is similar to the 2007 version in that it requires the owner to purchase and maintain property insurance on a builder's risk "all risks" or equivalent form unless otherwise provided in Section A.3.3.2.1. Builder's risk insurance is first-party coverage that protects an insured from loss to property caused by a peril covered by the policy. The coverage amount is required to be the original Contract Sum plus all subsequent modifications to the contract and any work performed by others plus the value of any furnishings, fixtures, materials, and equipment located at the project site. The coverage amount must be adequate to cover the entire project at the site on a replacement value basis. Unlike the A201–2007 version, the new provision requires the coverage to be maintained until substantial completion or as otherwise specified in the Contract Documents, or agreed to by the parties. Property insurance provided under this section must include all contractors and subcontractors as insureds but does not require these parties to be listed as loss payees.

An "all risks" policy on a completed value or equivalent form is required. Although some builder's risk policies are still written on a "named peril" or "covered cost" basis, the "all risks" approach is required by the AIA documents and is the most common approach. Property policies may be written on a "completed value form" or "reporting form" basis. Completed value forms are typically used when insurance is purchased for a specific project, whereas reporting forms are typically used when the owner or contractor are involved in numerous projects on an ongoing basis.[35]

Property insurance required under this section must include the interests of loss payees and mortgagees. The failure of a builder's risk policy to name certain parties that have an insurable interest in the property is problematic in that it may

34. Brillante v. R.W. Granger & Sons, Inc., 772 N.E.2d 74, 76 (Mass. App. Ct. 2002).

35. *See* 2 Philip L. Bruner, Patrick J. O'Connor, Bruner & O'Connor on Construction Law § 5:220 (2002).

result in a suit by the builder's risk carrier seeking subrogation from a construction participant who would otherwise be immune from such an action as an insured under the policy.[36]

§ A.2.3.1.1 Causes of Loss. The insurance required by this Section A.2.3.1 shall provide coverage for direct physical loss or damage, and shall not exclude the risks of fire, explosion, theft, vandalism, malicious mischief, collapse, earthquake, flood, or windstorm. The insurance shall also provide coverage for ensuing loss or resulting damage from error, omission, or deficiency in construction methods, design, specifications, workmanship, or materials. Sub-limits, if any, are as follows:

(Indicate below the cause of loss and any applicable sub-limit.)

Cause	Sub-Limit

Unlike the 2007 version of this clause, which listed various perils to be covered under the "all-risks" coverage, the 2017 version lists perils that cannot be excluded, all of which arise from fortuitous events. This is consistent with court interpretations of builder's risk insurance policies as requiring a fortuitous event for coverage to apply.[37]

§ A.2.3.1.2 Specific Required Coverage. The insurance required by this Section A.2.3.1 shall provide coverage for loss or damage to falsework and other temporary structures, and to building systems from testing and startup. The insurance shall also cover debris removal, including demolition occasioned by enforcement of any applicable legal requirements; and reasonable compensation for the Architect's and Contractor's services and expenses required as a result of such insured loss, including claim preparation expenses. Sub-limits, if any, are as follows:

(Indicate below type of coverage and any applicable sub-limit for specific required coverages.)

Cause	Sub-Limit

This section specifies certain required coverage for temporary construction, testing, and start-up of building systems, debris removal (including demolition required as a result of enforcement of applicable legal requirements), consultant expenses, and claim preparation expenses required as a result of such insured loss. Policy sub limits may be listed in this section to the extent they exist under the policy.

36. Metric Constructors, Inc. v. Industrial Risk Insurers, 401 S.E.2d 126 (N.C. Ct. App. 1991), *aff'd*, 410 S.E.2d 392 (N.C. 1991).

37. Coluccio Const. Co., Inc. v. King Cnty., 150 P.3d 1147, 1156 (Wash. Ct. App. 2007) ("As a condition precedent to coverage under an all risk builders' risk insurance policy, the loss-causing instrumentality must have been brought about as the result of a 'fortuitous event.'").

> **§ A.2.3.1.3** Unless the parties agree otherwise, upon Substantial Completion, the Owner shall continue the insurance required by Section A.2.3.1 or, if necessary, replace the insurance policy required under Section A.2.3.1 with property insurance written for the total value of the Project that shall remain in effect until expiration of the period for correction of the Work set forth in Section 12.2.2 of the General Conditions.

This section requires the owner to maintain the property insurance or replace such insurance with a policy written for the total value of the project through the expiration of the period for correction of the work required under Section 12.2.2.

> **§ A.2.3.1.4 Deductibles and Self-Insured Retentions.** If the insurance required by this Section A.2.3 is subject to deductibles or self-insured retentions, the Owner shall be responsible for all loss not covered because of such deductibles or retentions.

This section provides that the owner is responsible for all losses not covered on account of deductibles or self-insured retentions (SIR). The reference to SIRs is new to the 2017 edition. Both deductibles and SIRs are used to make the insured financially responsible for certain losses that would otherwise be covered by the policy. Although both approaches shift risk to the insured, deductibles and SIRs operate differently.[38] Deductibles are included in the policy limits and are rarely identified on certificates of insurance, whereas SIRs serve as the first layer of coverage before an insurer is obligated for costs to the policy limit and are always identified on certificates.[39] With regard to payment of claims, deductible amounts are generally paid by the insurance company to the claimant with reimbursement by the insured. On the other hand, the insurer has no liability to the claimant for amounts in the SIR.[40] In addition, an insurer's duty to defend does not arise in a SIR arrangement until the SIR limits have been exhausted, unlike deductibles that do not disturb the insurer's defense obligation.[41]

Because SIRs are dependent on the balance sheet of the insured, there may be no coverage for the SIR layer in the event of a bankruptcy. Without an explicit provision making insurers liable for coverage above the SIR, some courts have held that, in a bankruptcy, no coverage exists under the policy.[42]

In another case, a court considered whether an insurer could share in recoveries from contribution actions against third parties. The court found that an insurer had no right to a proportional share of funds recovered under subrogation and that the insured was entitled to retain such funds up to its SIR limit.[43]

With regarding to responsibility for deductibles and SIRs, the owner should consider modifying the AIA's standard language to allow recovery of all or part of the deductible or retention from the contractor where the loss was occasioned by the act or omission of the contractor or a party for which the contractor should be liable.[44]

---

38. 4 Bruner & O'Connor, *supra* note 35, § 11:22, at 48 n.1 (2016).
39. *Id.*
40. *Id.*
41. *Id. See also* Padilla Const. Co., Inc. v. Transportation Ins. Co., 58 Cal. Rptr. 3d 807 (Ct. App. 2007); Montgomery Ward & Co. Inc. v. Imperial Cas. & Indemnity Co., 97 Cal. Rptr. 2d 44 (Ct. App. 2000).
42. 4 Bruner & O'Connor, *supra* note 35, § 11:22 (2016) (citing Pak-Mor Mfg. Co. v. Royal Surplus Lines Ins. Co., No. SA-05-CA-135-RF, 2005 WL 3487723 (W.D. Tex. Nov. 3, 2005); and Gulf Underwriters Inc. Co. v. McClain Industries Inc., No. 273768, 2008 WL 3021134 (Mich. Ct. App. Aug. 5, 2008)).
43. Bordeaux, Inc. v. American Safety Ins. Co., 186 P.3d 1188 (Wash. Ct. App. 2008).
44. Hearst-Argyle Properties, Inc. v. Entrex Communication Services, Inc., 778 N.W.2d 465 (Neb. 2010).

§ A.2.3.2 Occupancy or Use Prior to Substantial Completion. The Owner's occupancy or use of any completed or partially completed portion of the Work prior to Substantial Completion shall not commence until the insurance company or companies providing the insurance under Section A.2.3.1 have consented in writing to the continuance of coverage. The Owner and the Contractor shall take no action with respect to partial occupancy or use that would cause cancellation, lapse, or reduction of insurance, unless they agree otherwise in writing.

This section provides that the owner must not occupy or use any portion of the work until the property insurance carriers have consented in writing to the continuance of coverage. The owner and contractor must cooperate to obtain such consent and must not undertake occupancy or use that would impair coverage. Obtaining the insurer's consent to occupancy will preclude any termination of coverage.[45]

§ A.2.3.3 INSURANCE FOR EXISTING STRUCTURES

If the Work involves remodeling an existing structure or constructing an addition to an existing structure, the Owner shall purchase and maintain, until the expiration of the period for correction of Work as set forth in Section 12.2.2 of the General Conditions, "all-risks" property insurance, on a replacement cost basis, protecting the existing structure against direct physical loss or damage from the causes of loss identified in Section A.2.3.1, notwithstanding the undertaking of the Work. The Owner shall be responsible for all co-insurance penalties.

This new provision in the 2017 edition applies to projects that involve renovations or additions to existing structures. The owner's property coverage on projects involving existing structures is now required to protect the existing structure in addition to the new work on a replacement cost basis. The necessity for replacement cost coverage is evident if a covered peril causes loss or damage to parts of the existing structure not included as part of the work. In such cases, the owner must be insured to the amount necessary to replace the entire damaged structure, including the work under the contract and the pre-existing elements. Parties furnishing property coverage must be careful to confirm coverage is afforded to existing structures on a replacement cost basis. In a loss situation, failure to obtain such coverage may result in a significant gap in funds to restore the damaged structures.

This section also makes the owner liable for any co-insurance penalties. Co-insurance provisions require the covered property to be insured to a specified percentage of its actual value, typically between 80 to 100 percent. At the time of loss, the insured's recovery will be capped at the amount of the loss times a ratio of the limit of insurance purchased to the amount of insurance required (the value of property times the coinsurance percentage), less any deductible.[46] Such penalties are imposed by carriers to create a disincentive for insured to under-insure the property.

45. Bruner & O'Connor, *supra* note 35, § 5:225 (citing McCarty v. Maryland Cas. Co., 429 F. Supp. 112 (W.D. Ark, 1976); and Hendrix v. New Amsterdam Cas. Co., 390 F.2d 299 (10th Cir. 1968)).

46. https://www.irmi.com/articles/expert-commentary/property-insurance-coinsurance.

§ A.2.4 OPTIONAL EXTENDED PROPERTY INSURANCE

The Owner shall purchase and maintain the insurance selected and described below.

(Select the types of insurance the Owner is required to purchase and maintain by placing an X in the box(es) next to the description(s) of selected insurance. For each type of insurance selected, indicate applicable limits of coverage or other conditions in the fill point below the selected item.)

☐ **§ A.2.4.1 Loss of Use, Business Interruption, and Delay in Completion Insurance**, to reimburse the Owner for loss of use of the Owner's property, or the inability to conduct normal operations due to a covered cause of loss.

☐ **§ A.2.4.2 Ordinance or Law Insurance**, for the reasonable and necessary costs to satisfy the minimum requirements of the enforcement of any law or ordinance regulating the demolition, construction, repair, replacement or use of the Project.

☐ **§ A.2.4.3 Expediting Cost Insurance**, for the reasonable and necessary costs for the temporary repair of damage to insured property, and to expedite the permanent repair or replacement of the damaged property.

☐ **§ A.2.4.4 Extra Expense Insurance**, to provide reimbursement of the reasonable and necessary excess costs incurred during the period of restoration or repair of the damaged property that are over and above the total costs that would normally have been incurred during the same period of time had no loss or damage occurred.

☐ **§ A.2.4.5 Civil Authority Insurance**, for losses or costs arising from an order of a civil authority prohibiting access to the Project, provided such order is the direct result of physical damage covered under the required property insurance.

☐ **§ A.2.4.6 Ingress/Egress Insurance**, for loss due to the necessary interruption of the insured's business due to physical prevention of ingress to, or egress from, the Project as a direct result of physical damage.

☐ **§ A.2.4.7 Soft Costs Insurance**, to reimburse the Owner for costs due to the delay of completion of the Work, arising out of physical loss or damage covered by the required property insurance: including construction loan fees; leasing and marketing expenses; additional fees, including those of architects, engineers, consultants, attorneys and accountants, needed for the completion of the construction, repairs, or reconstruction; and carrying costs such as property taxes, building permits, additional interest on loans, realty taxes, and insurance premiums over and above normal expenses.

This section is new in the 2017 edition. It provides a check-the-box approach to several additional insurance coverages that the owner may purchase in addition to the coverage afforded for direct physical loss covered by the basic property policy. Business interruption, loss of use, and delay in completion coverages are time-element coverages that generally will reimburse the insured for covered losses sustained during the "period of restoration," which has been defined as the period of time from the date of loss to the date when the property should be repaired, rebuilt, or replaced with reasonable speed and similar quality.[47]

47. 4 Bruner & O'Connor, *supra* note 35, § 11:488 (2016) (citing Lava Trading, Inc. v. Hartford Fire Ins. Co., 365 F. Supp. 2d 434 (S.D.N.Y. 2005)).

Extra expense insurance is another form of business interruption insurance designed for businesses that incur extra expenses to maintain business operations in the event of loss.

§ A.2.5 OTHER OPTIONAL INSURANCE

The Owner shall purchase and maintain the insurance selected below.

(Select the types of insurance the Owner is required to purchase and maintain by placing an X in the box(es) next to the description(s) of selected insurance.)

☐ **§ A.2.5.1 Cyber Security Insurance** for loss to the Owner due to data security and privacy breach, including costs of investigating a potential or actual breach of confidential or private information.

(Indicate applicable limits of coverage or other conditions in the fill point below.)

☐ **§ A.2.5.2 Other Insurance**

(List below any other insurance coverage to be provided by the Owner and any applicable limits.)

Cause	Sub-Limit

This section provides a check-the-box for cyber security insurance and any other insurance the owner may be required or elects to purchase. Cyber security insurance is a relatively new insurance product designed to address cyber security risk that is typically excluded from CGL policies. The breadth of coverage available in the marketplace is evolving rapidly, and specialized products responsive to the particular needs of construction are emerging. Generally, cyber security coverage provides protection against cyber-related losses, including costs arising from data destruction and theft, extortion demands, hacking, denial of service attacks, crisis management activity related to data breaches, and legal claims for defamation, fraud, and privacy violations.[48] If cyber security coverage is required, the language of Section A.2.5.1 should be closely tailored to reflect the then-current scope of coverage. Further, while most cyber security policies focus on business interruption and personal identity coverage, careful consideration should be given to cyber security risks arising from cyber attacks against critical infrastructure.

A generally accepted definition of critical infrastructure is those "systems and assets, whether physical or virtual, so vital to the United States that the incapacity or destruction of such systems and assets would have a debilitating impact on security, national economic security, national public health or safety, or any combination of those matters."[49] Instances of cyber attack against infrastructure are rare but include water systems, power grids, dams, and railway systems—all of which are constant consumers of engineering and construction services and attractive targets to cyber attack.

Another emerging area of insurance not listed in Article A.2.5 or A.2.6 relates to the use of unmanned aerial vehicles (UAVs), or drones, on construction sites.

48. *See* https://www.dhs.gov/cybersecurity-insurance.
49. https://www.whitehouse.gov/the-press-ofce/2013/02/12/executive-order-improving-critical-infrastructure-cybersecurity.

Use of drones is skyrocketing in the engineering and construction industry as the regulatory framework in the United States becomes more settled. Drones are commonly used for various forms of data collection, aerial imagery, and visual inspection of buildings and infrastructure. Insurance coverage issues relating to the use of drones abound. Options include stand-alone aircraft hull coverage—which provides the most comprehensive insurance and covers property damage, personal injury, and damage to the drone during transport. UAV endorsements to CGL policies also are available but will generally not cover damage to the aircraft, although such risk is generally limited to a few thousand dollars. If use of drones by the owner is required on a project, the details of the coverage should be specified in this section. If the contractor intends to operate a drone, the owner should specify reciprocal insurance requirements in Article A.3.2.12 of Exhibit A.

§ A.3.1 GENERAL

§ A.3.1.1 Certificates of Insurance. The Contractor shall provide certificates of insurance acceptable to the Owner evidencing compliance with the requirements in this Article A.3 at the following times: (1) prior to commencement of the Work; (2) upon renewal or replacement of each required policy of insurance; and (3) upon the Owner's written request. An additional certificate evidencing continuation of commercial liability coverage, including coverage for completed operations, shall be submitted with the final Application for Payment and thereafter upon renewal or replacement of such coverage until the expiration of the periods required by Section A.3.2.1 and Section A.3.3.1. The certificates will show the Owner as an additional insured on the Contractor's Commercial General Liability and excess or umbrella liability policy or policies.

This section is similar to Section 11.1.3 of the A201–2007 and requires the contractor to provide certificates of insurance acceptable to the owner before starting the work, upon renewal or replacement of the policy, and, new in the 2017 document, upon the owner's written request. This section eliminates the requirement that certificates contain a provision indicating that coverage will not be canceled or allowed to expire without 30 days' prior written notice having been given to the owner. As discussed in Section 2.1.2, changes to ACORD certificates to reflect insurance industry practices have made such notice requirements impractical. Certificates must now name the owner as additional insured on CGL and excess or umbrella policies.

§ A.3.1.2 Deductibles and Self-Insured Retentions. The Contractor shall disclose to the Owner any deductible or self-insured retentions applicable to any insurance required to be provided by the Contractor.

This section requires the contractor to disclose to the owner any deductible or self-insured retention applicable to the required policies and is similar to Article A.2.3.1.4 requiring the owner to make the same disclosures relative to the owner's property insurance.

§ A.3.1.3 Additional Insured Obligations. The Contractor shall cause the commercial liability coverage required by the Contract Documents to include (1) the Owner, the Architect, and the Architect's Consultants as additional insureds for claims caused in whole or in part by the Contractor's negligent acts or omissions during the Contractor's operations; and (2) the Owner as an additional insured for claims caused in whole or in part by the Contractor's negligent acts or omissions for which loss occurs during completed operations. The additional insured coverage shall be primary and non-contributory to any of the Owner's insurance policies and shall apply to both ongoing and completed operations. The additional insured coverage shall be no less than that provided by Insurance Services Office, Inc. (ISO) forms CG 20 10 07 04, CG 20 37 07 04, and, with respect to the Architect and the Architect's Consultants, CG 20 32 07 04.

This provision is similar to Section 11.1.4 of the A201–2007. The new language specifies that additional insured coverage be primary and noncontributory to the owner's insurance policies and that it apply on both ongoing and completed operations. The new language specifies the ISO coverage forms with respect to the owner and architect and its consultants, although the specified forms are not the ISO's latest versions.

Whether additional insured coverage exists is determined by the language of the policy and the relevant Contract Documents if the contract is unambiguous.[50] An intent to extend additional insured status will not be read into a contract.[51]

The reference to specific ISO forms of additional insured endorsements is significant. The scope of additional insured endorsements has evolved over the last 30 years to generally narrow the scope of coverage offered to additional insureds. The ongoing and completed operations forms referenced in Article A.3.1.3 provide the following endorsements:

> CG 20 10 07 04 (Ongoing Operations): Who is an insured is amended to include the identified persons or organizations, "but only with respect to liability for 'bodily injury,' 'property damage' or 'personal advertising injury' caused, in whole or in part, by: (1) Your acts or omissions; or (2) The acts or omissions of those acting on your behalf; in the performance of your ongoing operations for the additional insured(s) at the locations designated."
>
> CG 20 37 07 04 (Completed Operations): Who is an insured is amended to include the identified persons or organizations, "but only with respect to liability for 'bodily injury' or 'property damage' caused, in whole or in part, by 'your work' at the location designated and described in the schedule of [the endorsement] performed for that additional insured and included in the products-completed operations hazard."

Notably, the CG 20 10 07 04 and the CG 20 37 07 04 do not afford coverage to additional insureds for their sole negligence and require that the named insured have some share of liability. In 2013, ISO issued further revisions to these forms further narrowing additional insured coverage "to the extent permitted by law" and limiting such coverage to be no broader than required under the applicable contract. The insurer's liability is further limited to the amount of insurance required by the contract or available under the applicable policy, whichever is less.[52] A growing minority of jurisdictions now limit contractually required insurance coverage, i.e., additional insured status. As of June 2016, nine states have some form of limitation on contractual

50. N. Kruger, Inc. v. CNA Ins. Co., 662 N.Y.S.2d 529 (App. Div. 1997).
51. National Abatement Corp. v. National Union Fire Ins. Co. of Pittsburgh, PA, 824 N.Y.S.2d 230 (App. Div. 2006).
52. For a further discussion and analysis of additional insured endorsements, see http://www.americanbar.org/publications/under_construction/2013/august_2013/iso_additional_insured_endorsements.html.

additional insured clauses and similar legislation has recently been debated in other state legislatures around the country.[53] Although the additional insured endorsements are not expressly limited by applicable law, as are the 2013 versions, applicable law in these states will limit the extension of additional insured status.

§ A.3.2 CONTRACTOR'S REQUIRED INSURANCE COVERAGE

§ A.3.2.1 The Contractor shall purchase and maintain the following types and limits of insurance from an insurance company or insurance companies lawfully authorized to issue insurance in the jurisdiction where the Project is located. The Contractor shall maintain the required insurance until the expiration of the period for correction of Work as set forth in Section 12.2.2 of the General Conditions, unless a different duration is stated below:

(If the Contractor is required to maintain insurance for a duration other than the expiration of the period for correction of Work, state the duration.)

This language is generally a reflection of the preamble language that was previously set forth in Section 11.1.1 of the A201–2007. Notice, however, the language on the duration of the insurance. Section 11.1.2 of the A201–2007 provided that the required insurance coverages would be maintained "without interruption from the date of commencement of the Work until the date of the final payment and termination of any coverage required to be maintained after final payment, and with respect to the Contractor's completed operations coverage, until the expiration of the period for correction of Work or for such other period for maintenance of completed operations coverage as specified in the Contract Documents." This duration language is modified in the 2017 version to eliminate reference to final payment (which previously applied to coverages except for completed operations coverage), and instead to require that all required liability insurance (to be differentiated from property insurance, which is addressed in other sections) be maintained "until the expiration of the period for correction of Work as set forth in Section 12.2.2 of the General Conditions."

This change will likely have little practical effect on the way contractors procure and maintain insurance, where those policies of insurance (1) are not project specific, and (2) are regularly renewed on an annual basis. In other words, the contractor is likely to maintain most of those insurance coverages for ongoing business operations, regardless of the final payment date or expiration date of the period for correction of work. However, this language does eliminate ambiguities that might arise as to the different trigger dates set forth in Section 11.1.2 of the A201–2007 (i.e., the final payment date, and the expiration date for the period for correction of work).

§ A.3.2.2 COMMERCIAL GENERAL LIABILITY

§ A.3.2.2.1 Commercial General Liability insurance for the Project written on an occurrence form with policy limits of not less than ($) each occurrence, ($) general aggregate, and ($) aggregate for products-completed operations hazard, providing coverage for claims including

53. https://www.mwl-law.com/wp-content/uploads/2013/03/Anti-Indemnity-Statutes-In-All-50-States-00131938.pdf.

.1 damages because of bodily injury, sickness or disease, including occupational sickness or disease, and death of any person;
.2 personal injury and advertising injury;
.3 damages because of physical damage to, or destruction of, tangible property, including the loss of use of such property;
.4 bodily injury or property damage arising out of completed operations; and
.5 the Contractor's indemnity obligations under Section 3.18 of the General Conditions.

Sections 11.1.1.2, 11.1.1.3, 11.1.1.4, 11.1.1.5, and 11.1.1.7 of the A201–2007 generally required these same types of coverages, but using slightly different language. For example, Section 11.1 of the A201–2007 did not mention "Commercial General Liability" insurance by name, but rather described the types of claims that are required by coverage. Conversely, Section A.3.2.2.1 makes specific reference to procuring CGL insurance, and then further expands upon the specific scope of coverage required under the CGL insurance policy.

Another semantic difference is found in Section A.3.2.2.1.1, which makes reference to "death of any person," while the former Sections 11.1.1.2 and 11.1.1.3 made separate references to "death of the Contractor's employees" and "death of any person other than the Contractor's employees," the former language now being addressed in other provisions relating to the worker's compensation requirements. Also, the prior reference to "damages insured by usual personal injury liability coverage" found in Section 11.1.1.4 of the A201–2007 has been replaced by "personal and advertising injury" in Section A.3.2.2.1.2.

Perhaps the biggest change is the deletion of the reference to "[c]laims involving contractual liability insurance applicable to the Contractor's obligations under Section 3.18" in Section 11.1.1.8 of the A201–2007, and the addition of the new Section A.3.2.2.1.5, which now requires that the CGL insurance provide coverage for "the Contractor's indemnity obligations under Section 3.18 of the General Conditions." The difference is subtle, but may have material consequences. In this regard, the former language made specific reference to the applicable or available insurance for the contractor's contractual liability—a reference that implies that the extent of the coverage available under the policy for contractual liability is the focus. However, the new language appears to place an affirmative obligation on the contractor to procure coverage for the full extent of the contractor's contractual indemnity obligations. This could give rise to a coverage dispute, to the extent that provisions of the indemnity obligation (which is often modified by the parties) includes contractual obligations that are not covered by a contractual liability endorsement or insured contract language in standard CGL insurance policies.

For example, language in an indemnity clause that makes the contractor liable for an owner's attorney fees, expert costs, and litigation expenses may be excluded from coverage by a CGL insurance policy, under the exclusion to coverage for liability assumed by an insured under contract or agreement. In that event, the contractor could be in breach of the contractual obligation to procure insurance that covered the full extent of the contractor's indemnity obligation. This places the insurer and insured in a potentially adverse setting.

An example of the potential conflict between insurance policy language and indemnity language that includes attorney fees is found in *Golden Eagle Insurance Company v. Insurance Company of the West*.[54] There, the court addressed the

54. 121 Cal. Rptr. 2d 682 (Ct. App. 2002).

question of whether the contractual liability portions of a CGL insurance policy covered an insured's liability for defense costs, and whether such costs were "damages" within the meaning of the insurance policy. In so doing, the court noted that the policy excluded damages assumed by the insured in a contract or agreement, but that the exclusion did not apply where the contract at issue was an "insured contract." The court held that the construction subcontract agreement, which contained an indemnity provision providing for attorney fees, was an "insured contract," and further that an insured with contractual liability coverage would reasonably expect that the indemnitee's attorney fees and costs are sums the insured becomes "legally obligated to pay as *damages* because of" covered tort claims.[55] As a result, the court held that the attorney fees provided in the indemnity clause at issue were covered under the CGL policy where the underlying claim involved covered property damage.

The opposite conclusion was reached by the court in *Progressive Casualty Insurance Co. v. Brown's Crew Car of Wyoming, Inc.*[56] There, the court also addressed an indemnity and hold harmless provision that included recovery of attorney fees and the question of coverage for defense costs under a business auto insurance policy. In finding no liability for attorney fees, the court reasoned:

> At issue in this case is Brown's assumption of an obligation to pay UP's defense costs. A contractual obligation to pay defense costs, however, does not fall into the definition of an "insured contract" under the Policy. The Policy's definition of "insured contract" is limited to an agreement by the insured to assume another's "tort liability," not defense costs. If Brown's assumption of UP's tort liability were at issue, the Court may very well find that the CTS/Brown Contract was an "insured contract" under the Policy's terms. However, this case involves Brown's assumption of UP's attorneys fees [sic]. Consequently, Progressive was not required to defend UP in the Keating and Cook lawsuits under the theory that the CTS/Brown Contract was an "insured contract."[57]

In view of these cases, parties should be careful in procuring the required insurance, drafting the applicable indemnity clause to ensure that the coverage required lines up with the contractual obligation itself.

§ A.3.2.2.2 The Contractor's Commercial General Liability policy under this Section A.3.2.2 shall not contain an exclusion or restriction of coverage for the following:

.1 Claims by one insured against another insured, if the exclusion or restriction is based solely on the fact that the claimant is an insured, and there would otherwise be coverage for the claim.

.2 Claims for property damage to the Contractor's Work arising out of the products-completed operations hazard where the damaged Work or the Work out of which the damage arises was performed by a Subcontractor.

.3 Claims for bodily injury other than to employees of the insured.

.4 Claims for indemnity under Section 3.18 of the General Conditions arising out of injury to employees of the insured.

.5 Claims or loss excluded under a prior work endorsement or other similar exclusionary language.

---

55. *Id.* at 691.
56. 27 F. Supp. 2d 1288 (D. Wyo. 1998).
57. *Id.* at 1293.

.6 Claims or loss due to physical damage under a prior injury endorsement or similar exclusionary language.
.7 Claims related to residential, multi-family, or other habitational projects, if the Work is to be performed on such a project.
.8 Claims related to roofing, if the Work involves roofing.
.9 Claims related to exterior insulation finish systems (EIFS), synthetic stucco or similar exterior coatings or surfaces, if the Work involves such coatings or surfaces.
.10 Claims related to earth subsidence or movement, where the Work involves such hazards.
.11 Claims related to explosion, collapse and underground hazards, where the Work involves such hazards.

This language is new and relates to additional requirements and clarifications of the scope of CGL insurance coverage required to be procured by the contractor. In many cases, these provisions relate to common exclusions in CGL policies that have been put into place by the insurance industry in response to conflicting case law on the scope of intended insurance coverage. Some examples follow:

- **Section A.3.2.2.2.1: Claims by one insured against another insured, if the exclusion or restriction is based solely on the fact that the claimant is an insured, and there would otherwise be coverage for the claim.** A cross-liability exclusion eliminates coverage for claims or suits brought by one against another insured covered under an insurance policy. *Underwriters at Lloyds London v. STD Enterprises, Inc.*[58] Illustrative of the issue addressed by this type of exclusion, and the need for this contract provision, is the court's discussion of an indemnity lawsuit between a property owner/additional insured against a contractor/primary insured, and the resulting coverage action in *Kinsale Ins. Co. v. Georgia-Pacific, L.L.C.*:[59]

> The relevant exclusion states: "This insurance does not apply to claims or 'suits' for 'bodily injury,' 'property damage' or 'personal and advertising injury' brought by one insured against any other insured." Georgia–Pacific argues that the exclusion is inapplicable because Advanced's claim or suit is not one for property damage *brought by* one insured against another. The original property-damage claim was brought by H&E, who was a stranger to the policy. It claimed that Advanced was liable for destruction of H&E's equipment as a result of the fire at the plywood plant. Advanced did not, in turn, seek damages from Georgia–Pacific due to a property loss; it sought indemnity based on general tort principles for the property damage that occurred to another party. Advanced had no property damage, but it seeks protection from a potential duty to pay for someone else's property damages.
>
> Though Advanced does not claim property damage, the same suit contains H&E's claim for such damages. Is that enough? The district court thought it was. The court partly relied on the policy definition of a "suit," which is "a civil proceeding in which damages because of 'bodily injury,' 'property damage,' or 'personal and

58. 395 F. Supp. 2d 1142 (M.D. Fla. 2005).
59. 795 F.3d 452 (5th Cir. 2015).

> advertising injury' to which this insurance applies are alleged." We agree the litigation was a "suit" under the policy, as it was a civil proceeding in which property damage was alleged. The exclusion itself, though, requires that the claim or suit for property damage be brought by one of the insureds against another insured. The district court applied the exclusion because the court concluded that Advanced's third-party demand was both a suit and a civil proceeding that "essentially seeks reimbursement for a damages claim." The indemnity claim certainly does seek reimbursement, but that does not make the claim one for property damage brought by one insured against another.
>
> . . . .
>
> We conclude that the insured versus insured exclusion does not apply. Advanced "brought" a "claim" against another insured, Georgia–Pacific, but it was not one for "property damage." The plain meaning of the exclusion makes it inapplicable to an indemnity claim.[60]

- **Section A.3.2.2.2.2: Claims for property damage to the Contractor's Work arising out of products-completed operations hazard where the damaged Work or the Work out of which the damage arises was performed by a Subcontractor.** See *U.S. Fire Insurance Co. v. J.S.U.B., Inc.*[61] (finding that exclusion for "your work" included in the products-completed operations hazard did not apply where the work out of which the damage arose was performed on insured's behalf by a subcontractor); *Cherrington v. Erie Insurance Property and Casualty Co.*[62] (finding that exclusion for "your work" included in the products-completed operations hazard does not preclude coverage for the work of insured's subcontractors). See also *Ohio Northern University v. Charles Construction Services*[63] (enforcing coverage under the CGL policy's products-completed operations provision, which had a "your work" exclusion but also a subcontractor's work exception to that exclusion). Compare *Westfield Insurance Co. v. Custom Agri Systems, Inc.*[64] (holding that defective construction is not an "occurrence" for purposes of a CGL policy because "occurrences," as defined by the policy, require fortuity).
- **Section A.3.2.2.2.3: Claims for bodily injury other than to employees of the insured; Section A.3.2.2.2.4: Claims for indemnity under Section 3.18 of the General Conditions arising out of injury to employees of the insured.** See *Simco Enterprises, Ltd. v. James River Insurance Co.*[65] (finding that exclusion for bodily injury to insured's own employees also bars coverage for third-party indemnity claim brought against insured arising from bodily injury to insured's employee); *Merchants Mutual Insurance Co. v. Laighton Homes, LLC*[66] (finding that exclusion for bodily injury to employee arising out of and in the course of employment by the insured, "clearly and unambiguously" applies to bar coverage for general contractor's claims for indemnification of bodily injury damages that arose out of and in the course of the employee's employment by the insured subcontractor).

---

60. *Id.* at 454–55.
61. 979 So. 2d 871 (Fla. 2007).
62. 745 S.E.2d 508 (W. Va. 2013).
63. No. 5-16-01, 2017 Ohio App. LEXIS 528 (Jan. 23, 2017).
64. 979 N.E.2d 269 (Ohio 2012).
65. 566 F. Supp. 555 (E.D. Tex. 2008).
66. 899 A.2d 271 (N.H. 2006).

- **Section A.3.2.2.2.5: Claims or loss excluded under a prior work endorsement or other similar exclusionary language; Section A.3.2.2.2.6: Claims or loss due to physical damage under a prior injury endorsement or similar exclusionary language.** "From the perspective of the insurance industry, the intent of these endorsements is to eliminate coverage for injury or damage that arises from occurrences that precede policy inception."[67] See *Evanston Insurance Co. v. American Safety Indemnity Co.*[68] (in lawsuit involving alleged defects in multiple buildings—some of which occurred prior to effective date of CGL policy—court rules that prior work exclusion may be bar to some claims, but does not bar coverage for all claims, and thus insurer has duty to defend.).
- **Section A.3.2.2.2.7: Claims related to residential, multi-family, or other habitational projects, if the Work is to be performed on such a project.** See *Ment Bros. Iron Works Co., Inc. v. Interstate Fire & Casualty Co.*[69] (holding that "apartment exception" to the residential exclusion under CGL insurance policy applies to save coverage, where condominium developer solely owned all units at one time in development, thus qualifying project as "apartment" at relevant time of occurrence); *CWC Builders, Inc. v. United Specialty Insurance Co.*[70] (holding that residential exclusion applies to all portions of CGL insurance policy, and that there is no coverage for claims relating to flood damage stemming from demolition activities during construction on residential buildings).
- **Section A.3.2.2.2.8: Claims related to roofing, if the Work involves roofing.** See *Encompass Ins. Co. v. Gammon Roofing, L.L.C.*[71] (finding that roofing exclusion applies to deny coverage under CGL insurance policy where claim involved damages caused by the use of a torch during roofing work); *Anton, Ltd. v. Colony Insurance Co.*[72] (ruling that *Encompass* decision is not limited to exclude only fire damage related to roofing work, and, although the exclusion does not extend to exclude coverage for all roofing operations, the roofing exclusion applies here to deny coverage for the specific kind of roofing work the insured was engaged in); *Companion Property & Casualty Insurance Co. v. Opheim*[73] (finding that roofing exclusion applies to deny coverage under CGL insurance policy where damages at issue were caused when house flooded after a subcontractor partially removed/demolished the roof).
- **Section A.3.2.2.2.9: Exterior insulation finish systems (EIFS).** See *National American Insurance Co. v. Okemah Management Co., LLC.*[74] (in declaratory judgment action arising from construction defect claims for water leaks related to incorrect installation of EIFS system, court finds that there is no duty to defend and no coverage due to application of the EIFS Exclusion in CGL insurance policy, even where insured did not install the entire EIFS system); *National American Insurance Co. v. Gerlicher Co., LLC.*[75] (in declaratory judgment action arising from construction defect claims for water intrusion through an EIFS system, court finds that EIFS exclusion in

67. 4 Bruner & O'Connor, *supra* note 35, § 11:312 (2016).
68. 768 F. Supp. 2d 1044 (N.D. Cal. 2011).
69. 702 F.3d 118 (2d Cir. 2012).
70. 134 F. Supp. 3d 589 (D. Mass. 2015).
71. 996 So. 2d 16 (La. Ct. App. 2008).
72. 77 So. 3d 417 (La. Ct. App. 2011).
73. 92 F. Supp. 3d 539 (N.D. Tex. 2015).
74. 189 P.3d 1223 (Okla. Ct. App. 2008).
75. 260 P.3d 1279 (Okla. Ct. App. 2011).

CGL insurance policy is "clear, unambiguous" and effectively denies coverage for the loss claims asserted by building owner).

- **Section A.3.2.2.2.10: Claims related to earth subsidence or movement.** See *Hankins v. Maryland Casualty Company/Zurich American Insurance Co.*[76] (holding that "earth movement" exclusion in CGL insurance policy served to bar coverage for claim of excessive differential movement of foundation system that was the result of both upward and downward soil movements; holds that limiting applicability of the exclusion to "nature-caused" or "natural force" earth movement would be "nonsensical"). But see *Nautilus Insurance Co. v. Vuk Builders, Inc.*[77] (holding that "subsidence of land" exclusion in CGL insurance policy was ambiguous as the examples given for subsidence were "types" of earth movement rather than "causes" of earth movement; as a result, court finds that exclusion only applies to earth movement due to natural causes); *Broom v. Wilson Paving & Excavating, Inc.*[78] (finding that earth movement exclusion in CGL insurance policy is ambiguous because it could apply to both natural occurring and man-made events; as a result, court holds that policy only excludes earth movement caused by natural events).
- **Section A.3.2.2.2.11: Claims related to explosion, collapse and underground hazards, where the Work involves such hazards.** See *Maine Drilling & Blasting, Inc. v. Insurance Co. of North America*[79] (holding that although CGL insurance policy had exclusions under "explosion hazard" and "underground property damage hazard" provisions, the insured also had an "Explosives Limitation Endorsement" that added coverage back in for blasting risk, albeit at a higher deductible than underlying coverage).

> **§ A.3.2.3** Automobile Liability covering vehicles owned, and non-owned vehicles used, by the Contractor, with policy limits of not less than ($) per accident, for bodily injury, death of any person, and property damage arising out of the ownership, maintenance and use of those motor vehicles along with any other statutorily required automobile coverage.

This section makes specific reference to automobile liability insurance coverage, whereas Section 11.1.1.6 of the A201–2007 had a more general reference to this type of coverage by requiring the contractor to procure insurance for "[c]laims for damages because of bodily injury, death of a person or property damage arising out of ownership, maintenance or use of a motor vehicle." The new Section A.3.2.3 also includes a fill point for the parties to establish the required limits of the Automobile Liability insurance.

> **§ A.3.2.4** The Contractor may achieve the required limits and coverage for Commercial General Liability and Automobile Liability through a combination of primary and excess or umbrella liability insurance, provided such primary and excess or umbrella insurance policies result in the same or greater coverage as the coverages required under Section A.3.2.2 and A.3.2.3, and in no event shall any excess or umbrella liability insurance provide narrower coverage than the primary policy. The excess policy shall not require the exhaustion of the underlying limits only through the actual payment by the underlying insurers.

76. 101 So. 3d 645 (Miss. 2012).
77. 406 F. Supp. 2d 899 (N.D. Ill. 2005).
78. 356 P.3d 617 (Okla. 2015).
79. 665 A.2d 671 (Me. 1995).

This language is new, and it addresses the interrelation between primary and excess or umbrella insurance policies, including when coverage under the excess policy begins. In one representative case, the court held that an excess insurer was not required to contribute to the cost of defending the insured until insured's primary insurers had exhausted the underlying policy limits, where the policy provided that the excess insurer would not be called upon to investigate or defend any suit "except for exhaustion of underlying limits by payment of covered claims."[80]

§ A.3.2.5 Workers' Compensation at statutory limits.

The prior language relating to workers' compensation was set forth in Sections 11.1.1.1 and 11.1.1.2 of the A201–2007. That version, however, did not refer to "statutory limits," as the 2017 version does.

§ A.3.2.6 Employers' Liability with policy limits not less than ($) each accident, ($) each employee, and ($) policy limit.

This language is new. In general terms, an employers' liability insurance policy covers damages related to workplace injuries, illnesses, and deaths when those losses are not already covered under a workers' compensation policy of insurance.

§ A.3.2.7 Jones Act, and the Longshore & Harbor Workers' Compensation Act, as required, if the Work involves hazards arising from work on or near navigable waterways, including vessels and docks.

The Jones Act (also referred to as the Merchant Marine Act of 1920) is a federal law that provides that if "[a] seaman is injured in the course of employment, or if the seaman dies from the injury, the personal representative of the seaman may elect to bring a civil action at law, with the right of trial by jury, against the employer."[81] The Longshore & Harbor Workers' Compensation Act (LHWCA or Longshore Act)[82] is a workers' compensation program administered by the Office of Workers' Compensation Programs of the U.S. Department of Labor,[83] which provides for benefits to employees for disability or death, "but only if the disability or death results from an injury upon the navigable waters of the United States (including any adjoining pier, wharf, dry dock, terminal, building way, marine railway, or other adjoining area customarily used by an employer in loading, unloading, repairing, dismantling, or building a vessel)."[84] This provision applies to work on or near navigable waterways, including docks, and the special risks that may be inherent in those areas, which may not be covered under a standard CGL or workers' compensation policy.

80. Schneider Nat'l Transport v. Ford Motor Co., 280 F.3d 532 (5th Cir. 2002).
81. 46 U.S.C. § 30104 (2012).
82. 33 U.S.C. § 901 (2012).
83. *See* https://www.dol.gov/owcp/dlhwc/lhwca.htm.
84. 33 U.S.C. § 903(a) (2012).

> **§ A.3.2.8** If the Contractor is required to furnish professional services as part of the Work, the Contractor shall procure Professional Liability insurance covering performance of the professional services, with policy limits of not less than ($) per claim and ($) in the aggregate.

This language is new, and it addresses a potential coverage gap that exists if a contractor performs a "professional service" that is excluded under a CGL insurance policy. Courts have long held that a general liability policy is not a substitute for a professional "errors and omission" or malpractice policy.[85] Thus, courts have readily enforced the "professional services" exclusion in CGL insurance policies.

Of course, the challenge becomes defining what constitutes a "professional service" for the purposes of the exclusion. "A professional act or service is one arising out of a vocation, calling, occupation, or employment involving specialized knowledge, labor or skill, and the skill involved is predominantly mental or intellectual, rather than physical or manual."[86] A professional service has also been defined as "any business activity conducted by the insured, which involves specialized knowledge, labor or skill, and is predominantly mental or intellectual as opposed to physical or manual in nature."[87] "Acts which could be done by an unskilled or untrained employee are not subject to a professional services exclusion. Professional services involve discretion acquired by special training and the exercise of special judgment."[88] Several courts have found that the professional services exclusion does not apply to claims for failure to warn of a known danger and misrepresentation of a condition as safe, when it is known to be dangerous.[89]

In the context of a contractor's services, defining what services can qualify as a "professional service" may be difficult. Construction management, scheduling, coordination, pre-construction services, value engineering, design assist, delegated design, and observation activities all have aspects that are more "mental or intellectual as opposed to physical or manual in nature."[90] That does not automatically make these services "professional services," but it certainly raises questions as to whether some aspect of these activities would be excluded from coverage under a standard CGL insurance policy under the "professional services" exclusion. Thus the need for a professional liability insurance policy in certain situations.

Further confusing the issue is the duration for which the professional liability insurance has to be maintained. In this regard, professional liability insurance policies are typically "claims made" policies, meaning that the date a claim is made is the trigger for the applicable policy (as opposed to the occurrence-based nature of CGL policies, where the date of the occurrence is the trigger for the applicable policy). This is significant in relation to Section A.3.2.1 as to the required duration for the contractor to maintain insurance. In this regard, Section A.3.2.1 provides that the "Contractor shall maintain the required insurance until the expiration of the period for correction of Work as set forth in Section 12.2.2 of the General Conditions, unless a different duration is stated below."

85. Erie Ins. Group v. Alliance Environmental, Inc., 921 F. Supp. 537 (S.D. Ind. 1996).
86. Terre Haute First Nat'l Bank v. Pacific Employers Ins. Co., 634 N.E.2d 1336 (Ind. Ct. App. 1993).
87. Hurst-Rosche Engineers, Inc. v. Commercial Union Ins. Co., 51 F.3d 1336 (7th Cir. 1994).
88. Thermo Terratech v. GDC Enviro-Solutions, Inc., 265 F.3d 329 (5th Cir. 2001).
89. St. Hudson Engineers, Inc. v. Pennsylvania National Mut. Cas. Co., 909 A.2d 1156 (N.J. Super. Ct. App. Div. 2006); Gregoire v. AFB Const. Inc., et al., 478 So. 2d 538 (La. Ct. App. 1985).
90. *Hurst-Rosche Engineers, Inc.,* 51 F.3d 1336.

If the contractor is no longer on the project and no longer performing work after the expiration of the period for correction of work, there is no discernible detriment to coverage, as any occurrence that arose during the time of work would be covered under the applicable CGL insurance policy—even if a claim is not made until after expiration of the time period for maintaining the insurance (for example, job site accidents causing bodily injury, property damage, construction defects, etc.). However, for professional liability claims, the applicable policy is the policy in place at the time the claim is made. Thus, if a contractor is required to maintain professional liability insurance only "until the expiration of the period for correction of Work as set forth in Section 12.2.2 of the General Conditions," then a contractor could allow its professional liability insurance to lapse after that expiration period. In that case, the owner would potentially be without an insured remedy, while the contractor would still be in compliance with the insurance requirements.

As is common in owner/architect agreements, the parties may want to consider modifying Section A.3.2.1 as to any required professional liability insurance policy to require that such policy be maintained for an agreed-upon period of time *after* expiration of the period for correction of work.

§ A.3.2.9 If the Work involves the transport, dissemination, use, or release of pollutants, the Contractor shall procure Pollution Liability insurance, with policy limits of not less than ($) per claim and ($) in the aggregate.

This language is new. Coverage for claims of pollutants has a long history, both in terms of case law across the country and the insurance industry's response in the form of pollutant exclusions to standard CGL insurance policies. This section, however, specifically addresses the need for insurance coverage for pollution liability where the contractor is involved with the transport, dissemination, use, or release of pollutants. This type of special coverage is appropriate. "Injuries caused by hazardous materials may not be covered under the standard commercial liability policy. This is due to the incorporation of a broad pollution exclusion in such policies. Thus, while this indemnity tracts the 'bodily injury/property damage' dichotomy of the CGL policy, the risk is not likely to be transferred to a third party insurer through such a policy. Instead, a pollution liability policy may be necessary to transfer the risk."[91]

§ A.3.2.10 Coverage under Sections A.3.2.8 and A.3.2.9 may be procured through a Combined Professional Liability and Pollution Liability insurance policy, with combined policy limits of not less than ($) per claim and ($) in the aggregate.

This language is new, and it deals with the combination of professional liability and pollution liability insurance policies that may provide the required coverage.

§ A.3.2.11 Insurance for maritime liability risks associated with the operation of a vessel, if the Work requires such activities, with policy limits of not less than ($) per claim and ($) in the aggregate.

91. Bruner & O'Connor, *supra* note 35, § 5:205.

This provision applies where work specifically involves the charter of aircraft or watercraft, and the special risks inherent therein, which may not be covered under a standard CGL policy.

§ A.3.2.12 Insurance for the use or operation of manned or unmanned aircraft, if the Work requires such activities, with policy limits of not less than ($) per claim and ($) in the aggregate.

This provision applies where work specifically involves the use of a manned or unmanned aircraft, and the special risks inherent with that equipment, which may not be covered under a standard CGL policy. Unmanned aircraft are often referred to as Unmanned Aircraft Systems (UASs), Unmanned Aerial Vehicles (UAVs), or drones.[92] The Federal Aviation Administration (FAA) has recently promulgated regulations on the operation of small unmanned aircraft systems in certain circumstances, including certification of the remote pilots of these unmanned aircrafts.[93] Within that regulation, the FAA noted that:

> [E]ven though small UAS operations have the potential to pose a lower level of public risk in certain types of operations, the unmanned nature of the small UAS operations raises two unique safety concerns that are not present in manned-aircraft operations. The first safety concern is whether the person operating the small unmanned aircraft, who is physically separated from that aircraft during flight, would have the ability to see manned aircraft in the air in time to prevent a mid-air collision with the manned aircraft. . . . The second safety concern with small UAS operations is the possibility that, during flight, the person piloting the small unmanned aircraft may lose control of the aircraft due to a failure of the control link between the aircraft and the remote pilot's control station.[94]

Just as the state and federal laws and regulations relating to drone use are still developing, so, too, is the area of insurance coverage for these unmanned aircrafts. It is reported that some insurance carriers may already provide coverage in existing policies, while a small group of insurance carriers have begun providing liability coverage specifically for private drones for use in commercial operations.[95] However, it remains to be seen how the insurance industry will respond to coverage for still developing risk issues such as invasion of privacy related to use of drones.[96] Owners and contractors that contemplate use of these unmanned aircrafts for their projects will need to pay special attention to not only the developing regulations and laws, but also the potential risks and available insurance coverage for these types of activities.

This new provision attempts to address this emerging area by specifically designating insurance coverage required for these types of operations. However, the parties will need to determine what types of insurance coverage are available for these operations at the time of the contract. Upon review of the available insurance coverage for unmanned aircraft operations, the parties may need to modify this provision, as necessary, to clarify the coverage or address any material limitations or exclusions in such coverage.

92. Daniel North, *Note, Private Drones: Regulations and Insurance,* 27 Loy. Consumer L. Rev. 334, 334 (2014–2015).
93. Operation and Certification of Small Unmanned Aircraft Systems, 81 Fed. Reg. 42,063 (June 28, 2016).
94. Analysis of Public Risk Posed by Small UAS Operations, id. Section II.B., at 42,068.
95. North, *supra* note 92, at 353–54.
96. *Id.* at 354.

§ A.3.3 CONTRACTOR'S OTHER INSURANCE COVERAGE

> **§ A.3.3.1** Insurance selected and described in this Section A.3.3 shall be purchased from an insurance company or insurance companies lawfully authorized to issue insurance in the jurisdiction where the Project is located. The Contractor shall maintain the required insurance until the expiration of the period for correction of Work as set forth in Section 12.2.2 of the General Conditions, unless a different duration is stated below:
>
> ***(If the Contractor is required to maintain any of the types of insurance selected below for a duration other than the expiration of the period for correction of Work, state the duration.)***

This is a new section that addresses other insurance coverage beyond those insurance policies required in preceding text.

> **§ A.3.3.2** The Contractor shall purchase and maintain the following types and limits of insurance in accordance with Section A.3.3.1.
>
> ***(Select the types of insurance the Contractor is required to purchase and maintain by placing an X in the box(es) next to the description(s) of selected insurance. Where policy limits are provided, include the policy limit in the appropriate fill point.)***

The owner and contractor can select these additional insurance coverages through a "check the box" fill point on the contract form. When selected, the insurance is required to be procured by the contractor. When the parties do not select the insurance coverage, there is no affirmative obligation for the contractor to procure these insurance coverages.

> **§ A.3.3.2.1** Property insurance of the same type and scope satisfying the requirements identified in Section A.2.3, which, if selected in this section A.3.3.2.1, relieves the Owner of the responsibility to purchase and maintain such insurance except insurance required by Section A.2.3.1.3 and Section A.2.3.3. The Contractor shall comply with all obligations of the Owner under Section A.2.3 except to the extent provided below. The Contractor shall disclose to the Owner the amount of any deductible, and the Owner shall be responsible for losses within the deductible. Upon request, the Contractor shall provide the Owner with a copy of the property insurance policy or policies required. The Owner shall adjust and settle the loss with the insurer and be the trustee of the proceeds of the property insurance in accordance with Article 11 of the General Conditions unless otherwise set forth below:
>
> ***(Where the Contractor's obligation to provide property insurance differs from the Owner's obligations as described under Section A.2.3, indicate such differences in the space below. Additionally, if a party other than the Owner will be responsible for adjusting and settling a loss with the insurer and acting as the trustee of the proceeds of property insurance in accordance with Article 11 of the General Conditions, indicate the responsible party below.)***

This new language is similar in character to former Section 11.3.1.2 of the A201–2007 (and now the new Section 11.2.1 in the A201–2017), which provides a mechanism for the owner to notify the contractor of the owner's intent *not* to purchase property insurance, and, in turn, the contractor's option to procure the property insurance and charge back the premium via change order. However, under Section 11.2.1 of the A201–2017, the initial responsibility for procuring the insurance is on the owner, and the contractor may not know whether this obligation will be shifted to the contractor until after contract execution, thus prompting the change order process for the cost of the insurance. Note that Section 11.2.2 of the A201–2017 also addresses the owner's *failure* to procure required insurance (as opposed to the owner's decision not to procure insurance, as in Section 11.2.1 of the A201–2017) and the contractor's rights to cure in that situation.

The key difference with Section 11.2.1 of the A201–2017 is that this Section A.3.3.2.1 affirmatively places the obligation to procure the property insurance on the contractor rather than the owner. Under this contracting format, Exhibit A will be completed at the outset of the project, so the parties will determine up front whether the owner or the contractor will procure the required property insurance. Thus, this section does not specifically allow the contractor to charge back the cost of the premium of the property insurance via change order (as is the process in Section 11.2.1 of the A201–2017), since presumably the cost will be built into the contractor's Contract Sum.

Note that even under this elective section, the owner remains responsible for procuring property insurance after substantial completion (Section A.2.3.1.3) and property insurance during construction for existing structures, when the construction involves an addition to an existing structure (Section A.2.3.3). The owner also remains responsible for payment of losses that are within the deductible of the property insurance policy. Coordination will be needed with the applicable waiver of subrogation provisions relating to property insurance.

Finally, even though the obligation to procure property insurance is shifted to the contractor under this section, the owner remains responsible for adjusting and settling any loss under the policy, consistent with Section 11.5 of the A201–2017. If the parties agree to change this point of responsibility for adjusting and settling losses with the insurer, then they must address that in this section. The parties should also consider coordinating any such changes with similar changes to Section 11.5 of the A201–2017 in order to avoid potential inconsistencies between the two documents.

§ A.3.3.2.2 Railroad Protective Liability Insurance, with policy limits of not less than ($) per claim and ($) in the aggregate, for Work within fifty (50) feet of railroad property.

This is new language and addresses an area of insurance coverage not previously included in the A201–2007. This coverage would apply to contractors who perform work on or near a railroad right-of-way.

§ A.3.3.2.3 Asbestos Abatement Liability Insurance, with policy limits of not less than ($) per claim and ($) in the aggregate, for liability arising from the encapsulation, removal, handling, storage, transportation, and disposal of asbestos-containing materials.

This is new language and addresses an area of insurance coverage not previously included in the A201–2007. Similar to other risks discussed above, this language requires coverage for asbestos abatement liability and is appropriate where the contractor's work specifically includes work that involves these risks, as such risks may be excluded under a standard CGL insurance policy.

§ A.3.3.2.4 Insurance for physical damage to property while it is in storage and in transit to the construction site on an "all-risks" completed value form.

Section A.2.3.1 addresses the owner's obligation to procure builder's risk insurance on an "all risk" basis, to include, among other things, "materials supplied or installed, by others . . . for the entire Project at the site on a replacement cost basis." Thus, this section does not specifically address coverage for property that is not "at the site," but rather is in storage or in transit to the construction site (i.e., is intended to be used for the Project).[97]

Further, the contractor's CGL insurance policy typically includes an exclusion for "personal property in the care, custody or control of the insured."[98] The gap filler, then, is the addition of an "all risk" property insurance policy that covers materials and personal property, whether they be in storage or in transit before arriving at the construction site.[99]

§ A.3.3.2.5 Property insurance on an "all-risks" completed value form, covering property owned by the Contractor and used on the Project, including scaffolding and other equipment.

As noted above, the contractor's CGL insurance policy typically includes an exclusion for "personal property in the care, custody or control of the insured." Similarly, a builder's risk policy may exclude coverage for the contractor's tools, machinery, plant, and equipment, where that property is not destined to become a permanent part of the project. In addition, the terms of a builder's risk policy may exclude coverage for "temporary works," including scaffolding, formwork, falsework, shoring, fences, and temporary buildings or structures, including office and job site trailers.[100]

This provision provides the parties the option of requiring the contractor to procure property insurance on an "all risk" basis in order to cover this potential gap for scaffolding and other equipment owned by contractor but not intended to become part of the completed project.

97. Zurich American Ins. Co. v. Keating Bldg. Corp., 513 F. Supp. 2d 55 (D.N.J. 2007) (noting that, in interpreting "all risk" builder's risk policy, the policy "insures physical damage to the insured property at the 'Insured Project.' The Policy defines the 'Property Insured' to include all property used to construct the 'Insured Project' and 'Insured Project' as 'the work which the Insured is contractually obligated to perform in accordance with the contract documents'").

98. First Specialty Ins. Corp. v. Main Coast Marine Const., Inc., 497 F. Supp. 2d 11 (D. Ma. 2007) (addressing Exclusion j(4) of CGL insurance policy relating to exclusion of coverage for property damage to "personal property in the care, custody or control of the insured").

99. Black & Veatch Corp. v. Wellington Syndicate, 302 S.W.3d 114, 123–25 (Mo. Ct. App. 2009) (finding that "all risk" builder's risk insurance policy covered "all risks of physical loss or damage to property owned or to be used by Black & Veatch 'while in transit,'" including losses incurred by the ocean transit of materials intended for use on construction project).

100. Lend Lease (U.S.) Const. LMB Inc. v. Zurich Am. Ins. Co., 22 N.Y.S.3d 24 (App. Div. 2015) (noting definition of "Temporary Works" in builder's risk policy and holding that tower crane is not included within definition, but is excluded under language excluding contractor's tools, machinery, plant, and equipment).

§ A.3.4 PERFORMANCE BOND AND PAYMENT BOND

The Contractor shall provide surety bonds, from a company or companies lawfully authorized to issue surety bonds in the jurisdiction where the Project is located, as follows:

(Specify type and penal sum of bonds.)

Type	Penal Sum ($0.00)
Payment Bond	
Performance Bond	

Payment and Performance Bonds shall be AIA Document A312™, Payment Bond and Performance Bond, or contain provisions identical to AIA Document A312™, current as of the date of this Agreement.

This language is essentially an expansion of Article 10 of the A101–2007, with additional language about the bond type, the penal sum (in place of "Bond Amount"), and reference to the AIA Document A312 bond forms.

CHAPTER 12:
Article 12 Uncovering and Correction of Work

Daniel S. Brennan

ARTICLE 12 UNCOVERING AND CORRECTION OF WORK

Article 12 addresses the duties of the contractor to (i) keep the work uncovered (or to affirmatively uncover the work) to facilitate examination by the architect; and (ii) correct work that fails to conform to the Contract Documents. The 2017 edition of the A201 does not include any significant changes from the 2007 version.

§ 12.1 UNCOVERING OF WORK

> **§ 12.1.1** If a portion of the Work is covered contrary to the Architect's request or to requirements specifically expressed in the Contract Documents, it must, if requested in writing by the Architect, be uncovered for the Architect's examination and be replaced at the Contractor's expense without change in the Contract Time.

All construction projects involve efforts from the various participants to complete the work in a timely manner, meet the budget, and satisfy the quality standards required by the Contract Documents. Often these goals are in tension. Schedule and cost pressure could erode quality goals. For example, keeping work uncovered (such as plumbing or mechanical systems in walls or ceilings) to await an architect's examination could disrupt the contractor's progress even if the architect is conducting the examination in accordance with the agreed-upon project schedule. Section 12.1.1 discourages the contractor from proceeding faster than the architect can conduct its examination, as the contractor will be responsible for uncovering and replacing the covered work at its own expense if the work is covered before that examination. Section 12.1.1, therefore, attempts to align incentives so that quality control is not compromised by an overly ambitious contractor.[1] If the contractor is performing ahead of schedule, it might be prudent for the owner to meet with the contractor and architect to address the need to have the architect's personnel available to perform the examination of the work and to remind the contractor of the consequences of improperly covering the work. Of course, if the Contract Documents are silent as to whether the work in question can be covered and the architect does not alert the contractor beforehand that the work must remain uncovered to allow examination, this section suggests by inference that any request to uncover the work should be compensable.[2] That inference, however, is probably overcome by Section 12.1.2, which addresses who bears the cost of uncovering the work if the architect later requests work to be uncovered.

> **§ 12.1.2** If a portion of the Work has been covered that the Architect has not specifically requested to examine prior to its being covered, the Architect may request to see such Work and it shall be uncovered by the Contractor. If such Work is in accordance with the Contract Documents, ~~costs of uncovering and replacement~~the Contractor shall~~, by~~ be entitled to an equitable adjustment to the Contract Sum and Contract Time as may be

1. In the context of a cost reimbursement contract (with or without a guaranteed maximum price), the question of whether the contractor should be entitled to reimbursement for uncovering and replacing work that was improperly covered is often a subject of vigorous negotiation especially as it relates to whether the contractor should be allowed to access funds in a construction contingency.

2. The 2007 AIA Document Commentary to Section 12.1.1 specifically emphasizes the importance of the architect identifying in the Contract Documents the portions of the work that are to be examined before being covered. *AIA Document Commentary A201™ General Conditions of the Contract for Construction,* available at https://www.aiacontracts.org/contract-documents/22076-general-conditions-of-the-contract-for-construction.

appropriate ~~Change Order, be at the Owner's expense~~. If such Work is not in accordance with the Contract Documents, ~~such~~the costs of uncovering the Work, and the cost of correction, shall be at the Contractor's expense ~~unless the condition was caused by the Owner or a separate contractor in which event the Owner shall be responsible for payment of such costs~~.

The changes to Section 12.1.2 clarify the cost and time implications if the architect requests that previously covered work be exposed for examination. Like the 2007 version of this section, the responsibility for the costs and time impacts of uncovering work depends on whether the uncovered work complies with the Contract Documents. The changes to the second sentence of this section clarify that if the uncovered work complies with the Contract Documents, the contractor is entitled to an equitable adjustment to both the Contract Sum and the Contract Time. The 2007 language did not specifically entitle the contractor to an adjustment to the Contract Time. These changes thus recognize the delay that an order to uncover work may have on the schedule.

The final sentence of this section, with slightly different wording, reaffirms that the contractor is to bear the expense of uncovering and replacing the work if it was not in conformance with the Contract Documents. Presumably the contractor must also absorb the adverse impact to the Contract Time caused by the covering of nonconforming work.

§ 12.2 CORRECTION OF WORK

§ 12.2.1 BEFORE ~~OR AFTER~~ SUBSTANTIAL COMPLETION

The Contractor shall promptly correct Work rejected by the Architect or failing to conform to the requirements of the Contract Documents, ~~whether~~ discovered before ~~or after~~ Substantial Completion and whether or not fabricated, installed or completed. Costs of correcting such rejected Work, including additional testing and inspections, the cost of uncovering and replacement, and compensation for the Architect's services and expenses made necessary thereby, shall be at the Contractor's expense.

The 2017 changes draw a clearer distinction between correction of nonconforming work *before* substantial completion and such correction *after* substantial completion. The obligation to correct nonconforming work *after* substantial completion is now addressed solely in Section 12.2.2.

Under Section 12.2.1, there are two triggers for the contractor's duty to correct: (i) work rejected by the architect; and (ii) work that fails to conform to the requirements of the Contract Documents. These two triggers are stated in the disjunctive. From the contractor's standpoint, it may be troublesome that the architect can reject work even if it conforms to the Contract Documents. This language could be interpreted to empower the architect to require performance above and beyond what the Contract Documents require. If that is the case, the contractor might assert that its price and schedule are based on the Contract Documents and not on the architect's opinion of what work should be rejected, thus laying the groundwork for an adjustment to the Contract Sum and Contract Time.

It is also worth noting that Section 4.2.6 of the A201–2017 states that "[t]he Architect has authority to reject Work that does not conform to the Contract Documents." Because this language ties the architect's authority to reject work only to nonconformance with the Contract Documents, it is unclear whether the disjunctive wording of Section 12.2.1 actually creates any additional grounds for rejecting work and requiring correction by the contractor.

The converse is also true. Even if the architect does not reject the work, the contractor still must correct the work if it does not conform to the Contract Documents regardless of how the nonconformity is discovered.[3] This seems sensible given that the architect does not observe the work every day to be in a position to reject the work before it is completed.

§ 12.2.2 AFTER SUBSTANTIAL COMPLETION

> **§ 12.2.2.1** In addition to the Contractor's obligations under Section 3.5, if, within one year after the date of Substantial Completion of the Work or designated portion thereof or after the date for commencement of warranties established under Section 9.9.1, or by terms of ~~an~~any applicable special warranty required by the Contract Documents, any of the Work is found to be not in accordance with the requirements of the Contract Documents, the Contractor shall correct it promptly after receipt of ~~written~~ notice from the Owner to do so, unless the Owner has previously given the Contractor a written acceptance of such condition. The Owner shall give such notice promptly after discovery of the condition. During the one-year period for correction of Work, if the Owner fails to notify the Contractor and give the Contractor an opportunity to make the correction, the Owner waives the rights to require correction by the Contractor and to make a claim for breach of warranty. If the Contractor fails to correct nonconforming Work within a reasonable time during that period after receipt of notice from the Owner or Architect, the Owner may correct it in accordance with Section 2.~~4~~5.

The only substantive change to Section 12.2.2.1 is the deletion of "written" before "notice" in the first sentence. This change does not mean, however, that verbal notice to the contractor to correct nonconforming work would be sufficient, as Section 1.6.1 requires all notices to be in writing.

The one-year correction period and related duties in Section 12.2.2.1 garner more attention than any other section in Article 12 during contract negotiation. Likewise, courts have interpreted Section 12.2.2.1 (or its predecessors) more than any other provision in Article 12. From the owner's standpoint, the contractor's duty to correct work that does not conform to the Contract Documents is an important protection to ensure the quality of the completed project. The contractor and its subcontractors are usually in the best position to repair or replace defective work in view of their involvement with the original construction. For the contractor, the duty to repair or replace defective work within one year after substantial completion has benefits as well. For example, if the owner fails to notify the contractor of nonconforming work and afford the contractor an opportunity to fix it, the contractor may have defenses to a later claim by the owner, such as the owner's failure to mitigate.

One of the most important aspects of Section 12.2.2.1 is the interplay between this section and Section 3.5.1, which sets forth the contractor's general warranty of workmanship and defect-free work. The first sentence of Section 12.2.2.1 ("In

3. See the AIA commentary on the 2007 version of this language. *AIA Document Commentary A201™ General Conditions of the Contract for Construction,* available at https://www.aiacontracts.org/contract-documents/22076-general-conditions-of-the-contract-for-construction.

addition to the Contractor's obligations under Section 3.5") implies that whatever follows does not erode or modify Section 3.5 but amplifies or enhances it. Section 3.5 contains the contractor's general warranty that the materials and equipment will be new and that the work will conform to the Contract Documents and be free from defects.[4] One interpretation of the interplay between the two sections is that Section 12.2.2.1 creates a remedy while Section 3.5 creates the warranty itself. Problems can arise because owners, contractors, and their lawyers sometimes use the term "warranty" to refer to the one-year correction duty created in Section 12.2.2.1. In fact, the warranty is created not by Section 12.2.2.1 but rather by Section 3.5. Section 12.2.5, discussed below, further clarifies that the one-year correction remedy in Section 12.2.2.1 is just that—a remedy and not a limitation on the owner's other rights under the Contract Documents.

The interplay between the two sections may be further complicated by the language in Section 12.2.2.1 that states "[d]uring the one-year period for correction of Work, if the Owner fails to notify the Contractor and give the Contractor an opportunity to make the correction, the Owner *waives the rights to require correction by the Contractor and to make a claim for breach of warranty*" (emphasis added). The owner's failure to give notice within the one-year period deprives the owner of the right to require correction of the work *and* the right to make a claim for breach of warranty. A contractor could assert that any warranty claim—including a claim based on the warranty in Section 3.5—is waived if the owner fails to provide notice and the opportunity to cure within the one-year period. Even assuming such an assertion is correct, the owner would likely be left with a breach of contract claim based on the contractor's alleged failure to build in accordance with the Contract Documents.

Some commentators have observed that this section was not intended to create an exclusive remedy of breach of warranty.[5] Some courts, however, interpreting, in other contexts, the interplay between a contractor's general warranty and a one-year correction period have not always agreed. In *Philco Corp. v. Automatic Sprinkler Corp.*,[6] for example, a company hired a general contractor to build a new manufacturing facility. The fire protection subcontractor installed a 10-inch-diameter pipe five feet below the surface. That pipe ruptured three years after the project was completed, damaging the building and inventory. The plaintiff and its insurers brought an action based on breach of contract and tort theories. Both the federal district court and the Seventh Circuit interpreted certain warranty and correction language with concepts similar but not identical to the A201 to deny recovery to the plaintiffs under all theories, holding that the contractor's exposure for warranty claims was limited to one year.[7] The Seventh Circuit characterized the plaintiff's argument that the one-year correction period was an additional, nonexclusive remedy as "tortured, strained and inconsistent."[8] To avoid any confusion, some commentators recommend deleting the phrase "and to make a claim for breach of warranty" to preserve the owner's right to make a claim for breach of the warranty contained in Section 3.5.[9]

4. See discussion of Section 3.5 in Chapter 3.

5. *See* Stanley P. Sklar, Edward L. Filer, Tina M. Bird, *Implied Duties of Contractors: Wait a Minute, Where Is That in My Contract?*, 21 CONSTR. LAW. 11, 12 (Summer 2001) ("From a review of the language contained in article 12.2.2, it is a fair assumption that this extended obligation is not intended to be an exclusive remedy, barring an owner's common law action for breach of contract or other express or implied warranty action that may be available under the laws of the jurisdiction or the terms of the contract.").

6. 337 F.3d 405 (7th Cir. 1964).

7. *Id.* at 407–08.

8. *Id.* at 408.

9. *See* S. GREGORY JOY, EUGENE J. HEADY (EDS.), ALTERNATIVE CLAUSES TO STANDARD CONSTRUCTION CONTRACTS § 11.14, at 11-76 (4th ed. 2013).

Section 12.2.2.1 creates an affirmative obligation on the contractor to return to the site and correct defective work within the one-year period. What happens if the contractor (or its subcontractors) refuse? Can the owner bring an action for injunctive relief seeking a court order directing the contractor to return to the site and repair the work? That is unlikely especially if the owner can still bring an action for money damages for breach of warranty.[10] Additionally, injunctive relief forcing a contractor to bring workers to a site raises issues of involuntary servitude and court supervision of construction work that weigh heavily against granting such relief.[11]

Section 12.2.2.1 further requires that once notice is provided, the contractor must correct the work promptly. There is little, if any, guidance in the case law about what "promptly" means in the context of Section 12.2.2.1. The meaning of "promptly" will be determined by the particular circumstances.[12] Some courts have, however, addressed the problem of how many opportunities a contractor should have to repair nonconforming work after notice is provided. In *Berkshire Medical Center, Inc. v. U.W. Marx, Inc.*,[13] for example, a hospital experienced bubbling and cracking of vinyl floor tiles in certain of its operating rooms within one year of completion. The hospital promptly notified the general contractor who then alerted the flooring subcontractor. The subcontractor replaced the floor tiles and repaired the bubbling and splitting tiles. The problems, however, persisted. The general contractor and subcontractor continued repairs throughout the one-year correction period and beyond. Finally, the hospital had enough. The hospital hired another set of contractors to repair the entire floor and brought a breach of express warranty claim against the contractor for the cost to repair. While the warranty provision was not AIA language, the provision contained similar obligations to correct defective construction. A jury ruled in favor of the hospital, and that verdict was upheld on appeal. Addressing the contractor's right to cure after completion, the First Circuit observed "while the contractor gets first crack, there has to be some end point. If the contractor refused to do anything, the owner could do the job itself and sue for the cost; the result cannot be otherwise if, after repeated efforts over an extended period . . . , the contractor has attempted to provide a fix and failed to do so."[14]

The language in Section 12.2.2.1 may also serve to limit the measure of damages for a claim based on defective construction. Some owners and contractors have attempted to use diminution in value, instead of cost of repair, as the measure of damages for defective work. Courts interpreting similar provisions have found that contract language that guarantees against defects in workmanship and material and further requires the contractor to correct those defects effectively establish the cost of repair as the appropriate measure of damages.[15] In *Butler & Sidbury, Inc. v. Green Street Baptist Church*,[16] a general contractor that asserted a mechanics' lien for unpaid work faced a counterclaim from the owner for defects in the brickwork and roofing for a church building. The owner prevailed at trial and the

10. *See* 3 Philip L. Bruner, Patrick J. O'Connor, Bruner & O'Connor on Construction Law § 9.62.50 (2016 Update); State *ex rel.* Park County v. Montana Sixth Judicial Dist. Court, Park County, 833 P.2d 210 (Mont. 1992) (declining to order specific performance prohibiting an owner from making a deductive change because the contractor had an adequate remedy at law for damages).

11. *See also* Northern Delaware Indus. Development Corp. v. E. W. Bliss Co., 245 A.2d 431 (Del. Ch. 1968) (declining to order a contractor to hire 300 additional workers to complete a project).

12. Black's Law Dictionary provides "[t]he meaning of the word ['promptly'] depends largely on the facts in each case, for what is 'prompt' in one situation may not be considered such under other circumstances or conditions. To do something 'promptly' is to do it without delay and with reasonable speed." Black's Law Dictionary 1214 (6th ed. 1990).

13. 644 F.3d 71 (1st Cir. 2011).

14. *Id.* at 77. See also Barrack v. Kolea, 651 A.2d 149, 153 (Pa. Super. Ct. 1994) (clauses that require a contractor to correct defective work "are not interpreted to mean, however, that an innocent party's sole remedy would be to allow the breaching party unlimited attempts to cure the defect, to have unlimited bites at the apple, so to speak").

15. *See* Bruner & O'Connor, *supra* note 10.

16. 367 S.E.2d 380 (N.C. Ct. App. 1988).

jury awarded damages of $343,000 for the costs to repair. On appeal, the general contractor argued that the jury was improperly instructed on damages. The trial judge instructed the jury that the measure of damages for the owner's claims should be the cost to repair. This instruction was based on a provision in the construction contract that the general contractor guaranteed its work against defects for one year and promised to "make good" any defects.[17] The general contractor claimed that the jury should have been instructed that if the defects could not be corrected without undoing a substantial part of the completed work then the damages must be based on diminution in value.[18] The appellate court rejected this argument based on the general contractor's failure to propose such an instruction or object to the instruction given by the trial court. The appellate court further concluded that, in any event, the trial court properly instructed the jury that "[w]here a construction contract contains a guarantee against faulty materials or workmanship such as appeared in this contract, the measure of damages is controlled by the contract and the proper measure of damages is the cost of repairs."[19]

§ 12.2.2.2 The one-year period for correction of Work shall be extended with respect to portions of Work first performed after Substantial Completion by the period of time between Substantial Completion and the actual completion of that portion of the Work.

This section, which remains unchanged in the 2017 revision, addresses the situation where certain portions of the work are completed after substantial completion. For those portions of the work, the owner does not want to receive a correction period of less than one year. This section addresses that issue by extending the time for correction for the requisite amount of time to provide the owner with a full year of correction obligations from the contractor for *all* work.[20]

§ 12.2.2.3 The one-year period for correction of Work shall not be extended by corrective Work performed by the Contractor pursuant to this Section 12.2.

This section also remains unchanged from the 2007 edition. Once the contractor has corrected nonconforming work within the one-year period, that corrective work does not trigger *another* year of correction obligations for the contractor. This would create a perverse disincentive for the contractor. If performing corrective work would trigger another year of exposure, the contractor might weigh the risks of performing the corrective work versus ignoring the owner's request to correct. On the other hand, some commentators have argued that all of the work should be in full compliance for one year before the contractor's correction duties expire.[21] The AIA language seems to represent the commonly held view that the contractor's correction of the work should not trigger another year of exposure. When using the A201, most prudent contractors strive, or should strive, to account for only one year of exposure to correct the work either in their bids, their subcontracts, or both.

17. *Id.* at 383.
18. *Id.* at 384.
19. *Id.* at 384–85.
20. See the AIA commentary on the 2007 version of this language. *AIA Document Commentary A201™ General Conditions of the Contract for Construction*, available at www.aia.org.
21. *See* Joy & Heady, *supra* note 9, at 11-77.

Finally, the AIA commentary to the 2007 version of this language notes that if the original one-year correction period expires but the owner later has to repair work that was repaired during the correction period, the owner is not obligated to notify and allow the contractor an opportunity to repair.[22] The owner can undertake self-help in the first instance and then pursue a claim against the contractor.

> **§ 12.2.3** The Contractor shall remove from the site portions of the Work that are not in accordance with the requirements of the Contract Documents and are neither corrected by the Contractor nor accepted by the Owner.

This provision was also not changed in the 2017 edition.

If the work does not satisfy the Contract Documents and has neither been corrected nor accepted by the owner, then the contractor must remove it. This concept is logical because if the work does not satisfy one the three criteria enumerated in this section, then it does not belong as part of the project.

> **§ 12.2.4** The Contractor shall bear the cost of correcting destroyed or damaged construction of the Owner or Separate Contractors, whether completed or partially completed, ~~of the Owner or separate contractors~~ caused by the Contractor's correction or removal of Work that is not in accordance with the requirements of the Contract Documents.

When the contractor has to correct nonconforming work, it may encounter other work performed by the owner or the owner's separate contractors. The correction may necessitate the removal or damage to work performed by these other forces. This section makes it clear that the contractor is responsible for the cost of replacement or correction of that work. One issue the owner must consider is that if the separate contractor's work is still under warranty and the contractor removes, damages, or repairs that work as part of its corrective measures, will the contractor's efforts void the warranty of the separate contractor to the owner? In all likelihood, the separate contractor would have the better argument on this question. As such, the owner might want to confer with the separate contractor about the scope of the contractor's corrective measures that impact the separate contractor's work so as to preserve any warranty from the separate contractor.

> **§ 12.2.5** Nothing contained in this Section 12.2 shall be construed to establish a period of limitation with respect to other obligations the Contractor has under the Contract Documents. Establishment of the one-year period for correction of Work as described in Section 12.2.2 relates only to the specific obligation of the Contractor to correct the Work, and has no relationship to the time within which the obligation to comply with the Contract Documents may be sought to be enforced, nor to the time within which proceedings may be commenced to establish the Contractor's liability with respect to the Contractor's obligations other than specifically to correct the Work.

22. See the AIA commentary on the 2007 version of this language. *AIA Document Commentary A201™ General Conditions of the Contract for Construction,* available at www.aia.org.

Language in construction contracts creating a one-year correction period has, at times, been interpreted to establish a time limitation for an owner to bring an action to enforce a contractor's warranty obligation or breach of contract claim for defective construction.[23] In other cases, a contractor's duty to correct work within one year has been interpreted as an exclusive remedy. For example, in *Independent Consolidated School District No 24 v. Carlstrom*,[24] a school district brought a breach of contract claim against its contractor for damages relating to improper floor settlement, a defect that first appeared more than five and a half years after the work was substantially completed. After an arbitrator awarded the school district damages, the contractor successfully challenged the award based on the claim being untimely in light of the warranty and correction of work language that was not standard AIA language. The relevant language provided as follows:

> [n]either the final certificate nor payment nor any provision in the Contract Documents shall relieve the Contractor of responsibility for faulty materials or workmanship and, unless otherwise specified, he shall remedy any defects due thereto and pay for any damage to other work resulting therefrom, which shall appear within a period of one year from the date of substantial completion. The Owner shall give notice of observed defects with reasonable promptness. All questions arising under this article shall be decided by the Architect subject to arbitration.[25]

The Supreme Court of Minnesota concluded that this provision (and related language pertaining to claims surviving final payment) showed an intent that "claims for damages may be made by the owner after final payment, but those claims are limited to those appearing within one year of substantial completion."[26]

At the other end of the spectrum, courts have actually held that a cause of action against a contractor cannot even accrue before the expiration of the one-year correction period. In *Vogelsang v. McQuestion*,[27] a homeowner sued a contractor for defects based on a breach of contract theory. The defendant asserted that the statute of limitations for the breach of contract action commenced to run upon substantial completion. However, the relevant contract contained a provision, similar to Section 12.2.2.1, requiring the contractor to correct defects that appeared within one year of substantial completion. The court used this clause to hold that the statute of limitations did not even begin to run until the one-year correction period had expired without the contractor having fulfilled its duty to correct.[28]

The current language of Section 12.2.5—which is the same as in the 2007 edition—appears to dispel any confusion about what the one-year period in Section 12.2 means. The one-year period is not a statute of limitations, nor is it an exclusive remedy. The one-year period defines only the time frame within which the contractor is obligated, if properly notified, to correct nonconforming work. It does not bar a breach of contract claim for defective construction.[29]

23. *See* Philco Corp. v. Automatic Sprinkler Corp., 337 F.3d 405 (7th Cir. 1964).
24. 151 N.W.2d 784 (Minn. 1967).
25. *Id.* at 785–86.
26. *Id.* at 787.
27. 518 N.Y.S.2d 345 (Sup. Ct. 1987).
28. *Id.* at 346.
29. *See* Mount Calvary Baptist Church v. Williams Construction Co. of Port Allen, No. 06-1674, 2007 WL 2193771 (La. Ct. App. Aug. 1, 2007) (the owner's breach of contract claim for defective construction under the 1997 A201 was cognizable under the warranty of Section 3.5.1 and was not time-barred by the one-year correction period).

§ 12.3 ACCEPTANCE OF NONCONFORMING WORK

> If the Owner prefers to accept Work that is not in accordance with the requirements of the Contract Documents, the Owner may do so instead of requiring its removal and correction, in which case the Contract Sum will be reduced as appropriate and equitable. Such adjustment shall be effected whether or not final payment has been made.

Under this section, which remains unchanged in the 2017 version, the owner has the option of taking a reduction in the Contract Sum for the value of the nonconforming work. That reduction must be "appropriate and equitable," concepts that leave room for disagreement even among reasonable minds. A curious feature of this section is that the reduction in the Contract Sum can be effected even if final payment has been made. Presumably that would be accomplished through the claims process following an owner's demand for payment and the contractor's refusal to pay. After all, if the owner discovers nonconforming work after final payment but within one year, it has the contractual right to receive corrective work or recovery of money already paid. As a practical matter, it is reasonable to ask how a Contract Sum can be adjusted after final payment. Perhaps the intent is to maintain the owner's right to assert a claim (either for diminution in value or cost to correct) after the contractor has completed the work and the owner has accepted the work.

CHAPTER 13:
Article 13 Miscellaneous Provisions

Nancy Greenwald
Matthew J. Ninneman

ARTICLE 13 MISCELLANEOUS PROVISIONS

Miscellaneous provisions like those contained in Section 13 are often treated as simple boilerplate and are not adequately reviewed and discussed during contract preparation and negotiation. In reality, Article 13 contains provisions that have a significant effect on the rights and responsibilities of the parties to the contract and may determine the outcome of contract disputes. The choice of governing law contained in Section 13.1, for example, controls every aspect of the parties' contractual rights and obligations. Article 13 also continues to impact cost implications, liability, and risk-shifting for obligations to test and inspect the work. So while the intent of the A201–2017 is to set clear, consistent standards for performance, by their nature, the terms of Article 13 contain provisions that are not unique to construction projects, and interpretation of the language of this article as well as the contract as a whole, will largely depend on the choice of laws made in this article. Special care must be taken during contract negotiation to evaluate the impact of these clauses on performance standards and conflict resolution.

§ 13.1 GOVERNING LAW

> The Contract shall be governed by the law of the place where the Project is located ~~except that, if~~, excluding that jurisdiction's choice of law rules. If the parties have selected arbitration as the method of binding dispute resolution, the Federal Arbitration Act shall govern Section 15.4.

The intent of the first sentence of this section is to ensure that the contract will be governed by the law of the place in which the project is located. Among other things, the change to the language of this sentence is designed to remove a potential loophole; that is, to eliminate the possibility that the application of the choice of law rules in the specified jurisdiction could be applied to argue that the governing law of a different jurisdiction should be applied.

Choice of governing law is a particularly important provision because it determines every aspect of the parties' rights and obligations under the contract. These include interpretations of provisions governing standards of care, good faith and fair dealing, indemnification, limitations of liability, payment, liquidated damages, application of the economic loss rule, suspension of work and termination, third-party liabilities, recovery of fees, cost shifting, and applicable statutes of limitations,[1] among others. Although it is not explicitly stated in this section, because the definition of "Claims" in Section 15.1.1 includes "other disputes and matters in question between the Owner and the Contractor arising out of or relating to the Contract," the choice of laws provision extends to noncontract claims and defenses that arise, for example, from tort law, such as claims for fraud, negligence, and negligent misrepresentation.

1. The general provisions of the A201 governing the time limits on claims has been moved to Section 15.1.2. The application of statutes of limitations in arbitration proceedings has been the subject of dispute in recent years. As a result, Section 15.4.1.1 includes language clarifying that claims for arbitration may not be made after the date when the institution of legal or equitable proceedings based on the claim would be barred by the applicable state statute of limitations.

About half of the states have adopted mandatory choice of law statutes for construction projects.[2] Section 13.1 is consistent with such statutes. Most such laws provide that the law of that state will govern contracts for projects located in the state, regardless of what the contract specifies. Some also purport to void clauses requiring that disputes be resolved outside the state. Some are further limited to subcontracts.[3] Still others apply only where one party to the contract is a domestic corporation.[4] Attorneys involved at the point of contract negotiation need to familiarize themselves with the applicable laws of the state in which the project is located.

Section 13.1 does not specifically address the issue of forum selection. Whether the contract calls for litigation or arbitration, the choice of venue can have a significant effect on the process and costs.[5] The unanimous ruling of the United States Supreme Court in *Atlantic Marine Construction Co. v. United States District Court for the Western District of Texas*[6] makes clear that, absent contrary law, federal courts will enforce valid forum-selection clauses "in all but the most exceptional cases."[7] The Court overturned the Fifth Circuit Court of Appeals, which had affirmed the reasoning of the Western District of Texas concluding that venue was proper in the Western District of Texas despite the forum selection provision of the contract at issue, which provided that all disputes between the parties "shall be litigated in the Circuit Court for the City of Norfolk, Virginia, or in the United States District Court for the District of Virginia, Norfolk Division."[8] In doing so, the Supreme Court reviewed the factors taken into account by the Western District of Texas and concluded that these were "private-interest" factors, factors the Court deemed essentially waived by the parties' agreement to the forum selection provision. The Court therefore did not address the question of what "public-interest" factors might overturn an otherwise valid forum-selection clause.

The ruling is significant in that the forum selection clause at issue was contained in a construction subcontract. However, because the case arose from a project located entirely on federal land, and state law did not apply, the opinion does not reference any of the state statutes that explicitly void or make voidable forum-selection clauses of the type at issue in *Atlantic Marine*. As a result, the decision leaves open the question of whether such provisions establish public interest factors that would override the parties' forum selection provisions.

---

2. *See* Arizona, ARIZ. REV. STAT. § 32-1129.05; California, CAL. CIV. PROC. CODE § 410.42; Colorado, COLO. REV. STAT. § 13-21-111.5; Connecticut, CONN. GEN. STAT. ANN. § 42-158m; Florida, FLA. STAT. § 47.025; Illinois, 815 ILL. COMP. STAT. 665/10, Indiana, IND. CODE § 32-28-3-17; Kansas, KAN. STAT. ANN. § 16-121(e); Louisiana, LA. REV. STAT. ANN. § 9:2779; Minnesota, MINN. STAT. § 337.10; Montana, MONT. CODE ANN. § 28-2-2116; Nevada, NEV. REV. STAT. § 108.2453; New Mexico, N.M. STAT. ANN. § 57-28A1; New York, N.Y. GEN. BUS. LAW § 757; North Carolina, N.C. GEN. STAT. § 22B-2; Ohio, OHIO REV. CODE ANN. § 4113.62; Oklahoma, OKLA. STAT. tit. 15, § 15-821 (2014); Oregon, OR. REV. STAT. § 701.640; Pennsylvania, 73 PA. CONS. STAT. § 514 (Pa. Statutes Title 73, Chapter 14A, § 514); Rhode Island, R.I. GEN. LAWS § 6-34.1-1; Tennessee, TENN. CODE ANN. § 66-11-208; TEX. BUS. & COM. CODE ANN. § 272.001; Utah, UTAH CODE ANN. § 13-8-3; Virginia, VA. CODE ANN. § 8.01-262.1; and Wisconsin, WIS. STAT. § 779.135. At least three other states have statutes containing anti-forum selection clauses that apply to, but are not limited to construction contracts—Idaho, IDAHO CODE § 29-110; Nebraska, NEB. REV. STAT. § 25-415; and South Carolina, S.C. CODE ANN. § 15-7-120.

3. *See, e.g.*, California, CAL. CIV. PROC. CODE § 410.42.

4. *See, e.g.*, Florida, FLA. STAT. § 47.025.

5. Lack of clarity on venue creates unnecessary and costly uncertainty that may increase the parties' costs by requiring judicial resolution. *See., e.g.*, *In re* E. Beach Project Phase I, Ltd. & E. Beach Project Holdings, L.L.C., No. 14-11-00393-CV, 2011 Tex. App. LEXIS 5101 (July 7, 2011) ("Accordingly, venue lies in Galveston County if section 13.1.1 of the 2002 contract controls, but it lies in Harris County if section 4.6.1 of the 2005 contract controls."). Factors to be considered include the efficiencies of time savings for fact witnesses and a lower cost of employing local counsel, both for their knowledge of state law and for their knowledge of the local courts, court rules, judges, and customs. Subpoena power may also be geographically constrained by the project location. In the end, forum-selection clauses that make the forum far away from the situs of a construction project can create inefficiencies that make such a selection ill-advised.

6. 134 S. Ct. 568 (2013).

7. *Id.* at 581.

8. *Id.* at 575.

The reference to the Federal Arbitration Act (FAA) was added in the 2007 version of the A201. It was included to promote the AIA documents as a national standard, to avoid doubt as to which arbitration standards apply, and to make the arbitration provision somewhat less prone to differing interpretation in individual states. United States Supreme Court decisions have established that the FAA preempts state regulation that purports to restrict arbitration agreements. In *AT&T Mobility LLC v. Concepcion*,[9] the Court held that the FAA preempted a California law prohibiting class arbitration waivers. Most recently, in *DIRECTV, Inc. v. Imburgia*,[10] the Court reiterated this holding, confirming the validity of a class action arbitration waiver in a contract drafted prior to the court's decision in *Concepcion* that incorporated the same California law that was preempted in *Concepcion*. The Court's holding establishes that the FAA preempts state laws that do not place arbitration contracts on an equal footing with other types of contracts.

Read together, the two provisions of Section 13.1 clarify that while the governing law is the law of the place where the project is located, when arbitration is selected, the FAA rather than the law of the place where the project is located will govern the arbitration process. Though the question of whether state laws prohibiting venue outside the state remain open as to jurisdiction in litigation, a number of lower courts have expressly held that the FAA supersedes such state laws.[11]

§ 13.2 SUCCESSORS AND ASSIGNS

> **§ 13.2.1** The Owner and Contractor respectively bind themselves, their partners, successors, assigns, and legal representatives to covenants, agreements, and obligations contained in the Contract Documents. Except as provided in Section 13.2.2, neither party to the Contract shall assign the Contract as a whole without written consent of the other. If either party attempts to make ~~such~~ an assignment without such consent, that party shall nevertheless remain legally responsible for all obligations under the Contract.

The purpose of a clause prohibiting unilateral assignment of a contract is to ensure that the identities of the parties will remain consistent. Examples of instances when a party may wish to assign a contract include, on the owner's side, a desire to sell all or a portion of the project prior to completion and, on the contractor's side, a desire to assign payments due its own lenders. Section 13.2.1 protects each side from assignment by the other party without written permission. Assignment is particularly important as it impacts obligations for payment and performance, as well as controls the rights of use of the instruments of service. It provides the further protection that if either party ignores this provision, the assignment is ineffective in releasing the party making the assignment from its obligations under the contract. The only exception is for the owner's lender under Section 13.2.2.[12]

9. 563 U.S. 333 (2011).

10. 136 S. Ct. 463 (2015).

11. *See, e.g.*, R.A. Bright Construction, Inc. v. Weis Builders, Inc., 930 N.E.2d 565 (Ill. App. Ct. 2010); LaSalle Group, Inc. v. Electromation of Del. County, Inc., 880 N.E.2d 330 (Ind. Ct. App. 2008); Lodgeworks, L.P. v. C.F. Jordan Construction, LLC, No. 11-1317-KHV, 2012 U.S. Dist. LEXIS 24592 (D. Kan. Feb. 27, 2012); M.A. Mortenson/The Meyne Co. v. Edward E. Gillen Co., No. 03-5135 (PAM/RLE), 2003 U.S. Dist. LEXIS 23175 (D. Minn. Dec. 17, 2003); Aspen Spa Props., LLC v. Int'l Design Concepts, 527 F. Supp. 2d 469 (E.D.N.C. 2007); GEM Mech. Servs. v. DV II, LLC, No. 12-93-M, 2012 U.S. Dist. LEXIS 133591 (D.R.I. Sept. 17, 2012); Ope Int'l LP v. Chet Morrison Contractors, 258 F.3d 443 (5th Cir. 2001); Cleveland Construction, Inc. v. Levco Construction, Inc., 359 S.W.3d 843 (Tex. App. 2012).

12. There are some instances in which restrictions on assignment may not be enforceable; for example, when the assigning party is a debtor in bankruptcy. While the debtor's right is not unqualified, Section 365 of the United States Bankruptcy Code provides that a debtor may assume and assign an executory contract or unexpired lease "notwithstanding a provision in [the] executory contract or unexpired lease of the debtor, or in applicable law, that prohibits, restricts, or conditions the assignment of such contract or lease." 11 U.S.C. § 365(f)(1) (2012).

An assignment in violation of this section may be a breach of contract, and the assigning party may be held to its obligations under the contract, but the resulting assignment still may be enforceable. A detailed discussion of this section is found in *Oliver/Hatcher Construction & Development, Inc. v. Shain Park Associates*,[13] in which the court reversed the lower court's holding that an assignment in violation of this provision is void. The language of this section has also been relied upon to enforce arbitration against an assignee.[14]

> **§ 13.2.2** The Owner may, without consent of the Contractor, assign the Contract to a lender providing construction financing for the Project, if the lender assumes the Owner's rights and obligations under the Contract Documents. The Contractor shall execute all consents reasonably required to facilitate ~~such~~the assignment.

Many construction financing lenders require that the owner assign the terms of its contract with the contractor to the lender such that, if the owner defaults, the lender has the right (but not the obligation) to complete the project by assuming the owner's obligations under the contract. This section is designed to facilitate the owner's construction financing process. While not addressed in this book, great care and consideration should be taken by the contracting parties, and in particular contractors and design professionals, to understand what lender assignments exist or are contemplated by the owner and its lender, as the assignment can significantly affect obligations for payment and performance, lien rights, and rights with respect to the instruments of service.

~~§ 13.3 WRITTEN NOTICE~~

> ~~Written notice shall be deemed to have been duly served if delivered in person to the individual, to a member of the firm or entity, or to an officer of the corporation for which it was intended; or if delivered at, or sent by registered or certified mail or by courier service providing proof of delivery to, the last business address known to the party giving notice.~~

Section 13.3. has been moved to Section 1.6.

~~§ 13.4~~§ 13.3 RIGHTS AND REMEDIES

> **§ 13.~~4~~3.1** Duties and obligations imposed by the Contract Documents and rights and remedies available thereunder shall be in addition to and not a limitation of duties, obligations, rights, and remedies otherwise imposed or available by law.

> **§ 13.~~4~~3.2** No action or failure to act by the Owner, Architect, or Contractor shall constitute a waiver of a right or duty afforded them under the Contract, nor shall such action or failure to act constitute approval of or acquiescence in a breach ~~there-under~~thereunder, except as may be specifically agreed upon in writing.

13. No. 275500, 2008 Mich. App. LEXIS 1084, at *8 (May 22, 2008).

14. McPhee Elec. Ltd., LLC v. Signal Perfection Ltd., Civil No. WDQ-10-0831, 2010 U.S. Dist. LEXIS 84049, at *6 (D. Md. Aug. 16, 2010).

While seemingly innocuous, Section 13.3.2 can have significant impacts on future claims and dispute resolution. Under the common law, a failure or delay in acting can be construed as an acquiescence to the offending conduct and can be used to support doctrines such as estoppel, laches, and waiver. It is also well-established that the parties' course of performance or conduct can be interpreted to modify the terms of a contract, including waiver of certain terms. This provision is, in part, intended to restrict those defenses. Notably, however, it does not apply in all instances, and care should be taken by practitioners in assessing how this provision may impact a particular party's claims and defenses. For example, in *RCR Building Corp. v. Pinnacle Hospitality Partners*,[15] the court rejected the owner's argument that Section 13.4.2 should be applied to excuse its failure to make a timely claim for liquidated damages.

§ 13.~~5~~4 TESTS AND INSPECTIONS

§ 13.~~5~~4.1 Tests, inspections, and approvals of portions of the Work shall be made as required by the Contract Documents and by applicable laws, statutes, ordinances, codes, rules, and regulations or lawful orders of public authorities. Unless otherwise provided, the Contractor shall make arrangements for such tests, inspections, and approvals with an independent testing laboratory or entity acceptable to the Owner, or with the appropriate public authority, and shall bear all related costs of tests, inspections, and approvals. The Contractor shall give the Architect timely notice of when and where tests and inspections are to be made so that the Architect may be present for such procedures. The Owner shall bear costs of ~~(1)~~ tests, inspections, or approvals that do not become requirements until after bids are received or negotiations concluded~~;~~. The Owner shall directly arrange and ~~(2)~~pay for tests, inspections, or approvals where building codes or applicable laws or regulations ~~prohibit the Owner from delegating their cost to the Contractor~~so require.

§ 13.~~5~~4.2 If the Architect, Owner, or public authorities having jurisdiction determine that portions of the Work require additional testing, inspection, or approval not included under Section 13.~~5~~4.1, the Architect will, upon written authorization from the Owner, instruct the Contractor to make arrangements for such additional testing, inspection, or approval, by an entity acceptable to the Owner, and the Contractor shall give timely notice to the Architect of when and where tests and inspections are to be made so that the Architect may be present for such procedures. Such costs, except as provided in Section 13.~~5~~4.3, shall be at the Owner's expense.

§ 13.~~5~~4.3 If ~~such~~ procedures for testing, inspection, or approval under Sections 13.~~5~~4.1 and 13.~~5~~4.2 reveal failure of the portions of the Work to comply with requirements established by the Contract Documents, all costs made necessary by such failure, including those of repeated procedures and compensation for the Architect's services and expenses, shall be at the Contractor's expense.

15. No. M2012-00286-COA-R3-CV, 2012 Tenn. App. LEXIS 787, at *37 (Nov. 15, 2012).

> **§ 13.~~5~~4.4** Required certificates of testing, inspection, or approval shall, unless otherwise required by the Contract Documents, be secured by the Contractor and promptly delivered to the Architect.

> **§ 13.~~5~~4.5** If the Architect is to observe tests, inspections, or approvals required by the Contract Documents, the Architect will do so promptly and, where practicable, at the normal place of testing.

> **§ 13.~~5~~4.6** Tests or inspections conducted pursuant to the Contract Documents shall be made promptly to avoid unreasonable delay in the Work.

No changes were made to Section 13.4 in the 2017 version (except for the numbering). The purpose of this section is to clarify the allocation of responsibilities and costs for testing and inspection required during the course of the project. The default responsibility rests with the contractor under Section 13.4.1 to schedule, conduct, and pay for tests, inspections, and approvals required by the Contract Documents as well as by applicable laws, statutes, ordinances, codes, rules, regulations, and lawful orders of public authorities. Section 13.4.2 shifts the costs of additional tests not covered by Section 13.4.1 to the owner. The responsibility for the work and costs of repairing work that fails required testing rests with the contractor under Section 13.4.3. The remaining sections address procedures. Section 13.4.4 provides that certificates and other documents evidencing testing be promptly delivered by the contractor to the architect. Section 13.4.5 establishes the procedure where the architect is required to observe an inspection. Finally, Section 13.4.6 requires that inspections be performed timely so as not to delay work on the project.

§ 13.~~6~~5 INTEREST

> Payments due and unpaid under the Contract Documents shall bear interest from the date payment is due at ~~such~~the rate ~~as~~ the parties ~~may~~ agree upon in writing or, in the absence thereof, at the legal rate prevailing from time to time at the place where the Project is located.

Section 13.5 sets a default rate for interest for "payments due and unpaid under the Contract Documents." If the parties do not otherwise agree in writing, the provision defaults to the "legal rate prevailing from time to time at the place where the Project is located." Interest, and therefore the choice of interest rates, can become a significant factor in the calculation of damages when large amounts of money are involved. Considerations include whether a floating or a fixed interest rate should be selected and, if a floating interest rate, to which index it should be tied. Unless a different rate is agreed upon, the courts will enforce this provision to apply the applicable statutory rate of interest.[16]

16. *See, e.g.*, Town of Stratford v. A. Secondino & Son, No. FBT CV 06-5004618 S, 2010 Conn. Super. LEXIS 5564, at *35 (June 15, 2010); DePasquale Bldg. & Realty Co. v. R.I. Bd. of Governors for Higher Educ., M.P. No. PC07-6393, 2009 R.I. Super. LEXIS 79, at *53 (June 29, 2009).

~~§ 13.7 TIME LIMITS ON CLAIMS~~

> ~~The Owner and Contractor shall commence all claims and causes of action, whether in contract, tort, breach of warranty or otherwise, against the other arising out of or related to the Contract in accordance with the requirements of the final dispute resolution method selected in the Agreement within the time period specified by applicable law, but in any case not more than 10 years after the date of Substantial Completion of the Work. The Owner and Contractor waive all claims and causes of action not commenced in accordance with this Section 13.7.~~

Section 13.7 has been moved to Section 15.1.2.

CHAPTER 14:
Article 14 Termination or Suspension of the Contract

Michael B. Lapicola
F. Elliotte Quinn IV
C. Michael Shull III

ARTICLE 14 TERMINATION OR SUSPENSION OF THE CONTRACT

Article 14 governs the rights of the owner and the contractor to terminate the contract and suspend performance of the work. Section 14.1 regulates the contractor's right to terminate the contract. Section 14.2 describes when the owner can terminate for cause. Section 14.3 allows the owner to suspend performance of the work for its convenience, and Section 14.4 delineates the conditions under which the owner can terminate the contract for its convenience. The material changes to Article 14 in the 2017 version clarify what the contractor is entitled to be paid when the contractor terminates under Section 14.1 and when the owner terminates for convenience under Section 14.4. Other than that, the 2017 revisions to Article 14 were not significant.

§ 14.1 TERMINATION BY THE CONTRACTOR

§ 14.1.1 The Contractor may terminate the Contract if the Work is stopped for a period of 30 consecutive days through no act or fault of the Contractor ~~or~~, a Subcontractor, a Sub-subcontractor ~~or~~, their agents or employees, or any other persons or entities performing portions of the Work ~~under direct or indirect contract with the Contractor~~, for any of the following reasons:

.1 Issuance of an order of a court or other public authority having jurisdiction that requires all Work to be stopped;

.2 An act of government, such as a declaration of national emergency, that requires all Work to be stopped;

.3 Because the Architect has not issued a Certificate for Payment and has not notified the Contractor of the reason for withholding certification as provided in Section 9.4.1, or because the Owner has not made payment on a Certificate for Payment within the time stated in the Contract Documents; or

.4 The Owner has failed to furnish to the Contractor ~~promptly, upon the Contractor's request,~~ reasonable evidence as required by Section 2.2~~.1~~.

Section 14.1 provides for termination by the contactor upon the occurrence of specific events, but there is an unresolved issue as to whether the events listed in Section 14.1 are the only instances in which the contractor can terminate the contract or whether other actions by an owner might constitute a material breach under common law and permit a contractor to terminate the contract. A majority of commentators agree that Section 14.1 supplements the common law of termination for material breach and establishes specific events that allow a contractor to terminate regardless of whether those events would constitute a material breach under common law.[1] Others view the events specified in Section 14.1 as the only instances in which a contractor may terminate.[2] While there do not appear to be any reported decisions resolving the issue, a contractor likely can still terminate the contract due

1. *See* 2 Philip L. Bruner, Patrick J. O'Connor, Jr., Bruner & O'Connor on Construction Law § 5:255 (2016 Update); American Bar Association Forum on the Construction Industry, The Construction Contracts Book: How to Find Common Ground in Negotiating Design and Construction Clauses 197 (Daniel S. Brennan et al. eds., 2d ed. 2008); 2 Justin Sweet & Jonathan Sweet, Sweet on Construction Industry Contracts: Major AIA Documents 197 (3d ed. 1996).

2. *See* Werner Sabo, Legal Guide to AIA Documents 397–98 (4th ed. 1998).

to a common law material breach because no language in Section 14.1 explicitly limits termination to the events listed there.[3]

Section 14.1.1 provides four specific grounds allowing a contractor to terminate the contract. The first two provisions, Sections 14.1.1.1 and 14.1.1.2, deal with a contractor's inability to perform the work due to a government order or declaration requiring the contractor to stop the work. A decision arising from Hurricane Katrina's impact on New Orleans, *In re Frischhertz Constr. Co.*, is instructive on the parameters of these two grounds for termination and indicates that whether these grounds are satisfied when stopped work could or did resume is judged using the contract's definition of "Work" in Section 1.1.3, consideration of the totality of the circumstances, and a reasonableness standard.[4] There, the contractor was working on a project in New Orleans when Hurricane Katrina struck and the government issued a mandatory evacuation order. Three days after the evacuation order expired, the government issued an order prohibiting anyone from entering the city. The government subsequently issued an order permitting re-entry into the project area, but that order was suspended three days later due to Hurricane Rita. The contractor terminated the contract under Sections 14.1.1.1 and 14.1.1.2 on the basis that the evacuation order and subsequent order prohibiting re-entry were government acts requiring the contractor to stop all work for more than 30 consecutive days. The owner argued that there were gaps in the orders during which the contractor could have returned to the site and recommenced work and that the termination was therefore improper. The court rejected the owner's argument because while some of the contractor's personnel may have been able to access the project during gaps in the orders, it was not actually possible for the contractor to "restart work on the job."[5] Additionally, the court held that the government issued an emergency declaration that lasted for more than 30 consecutive days and the conditions in New Orleans during the period of that emergency declaration were such that work could not recommence.[6] The owner also argued that the termination was improper because the contractor actually returned to the project and performed work. The evidence indicated the project supervisor returned to the project on one instance, inspected a piece of equipment to determine whether it could be removed from the site, and inspected the project for damage. The court employed the contract's definition of "Work" in Section 1.1.3 and concluded that "[e]ven the most strained interpretation" would not permit it to conclude that those actions constituted resumption of work on the project.[7]

The third provision, Section 14.1.1.3, deals with nonpayment for work performed as a result of the architect not issuing a certificate for payment or the owner not making payment on a certificate for payment.[8] To terminate the contract under this provision, the contractor must satisfy a number of hurdles. As set forth in Article 9, the contractor must submit applications for payment to the architect, and the architect then either issues a certificate for payment to the owner or informs the contractor of the reasons for withholding payment. Once a certificate for payment is

---

3. *See* 14 Samuel Williston, Williston on Contracts § 43:5 (Richard A. Lord ed., West 4th ed. 2017 Update); Restatement (Second) of Contracts § 237 (1981); *see also* General Ins. Co. of Am. v. K. Capolino Constr. Corp., 983 F. Supp. 403, 428 (S.D.N.Y. 1997) (relying on both common law of termination for material breach and Section 14.1.1.3 as justifying contractor's termination).

4. *See In re* Frischhertz Constr. Co., No. 05-21605, 2007 WL 2965049 (Bankr. E.D. La. Oct. 8, 2007), *aff'd in part and rev'd in part on other grounds*, 410 F. F. App'x 861 (5th Cir. 2010).

5. *Id.* at *2 (relying on fact that there was not "unlimited access to the job site for all of the people necessary to restart work on the job").

6. *Id.*

7. *Id.* at *3.

8. General Ins. Co. of Am. v. K. Capolino Constr. Corp., 984 F. Supp. 403, 428 (S.D.N.Y. 1997) (finding contractor permitted to terminate under Section 14.1.1.3 where the architect failed to certify uncontested progress payments).

issued, the owner must pay the contractor within the time provided in the contract. If the owner does not make payment or the architect does not issue a certificate for payment, the contractor then must follow the timeline and procedures in Section 9.7 and can then stop work on the project. The procedure involves waiting seven days after making an application for payment to the architect or seven days after the date the contract required the owner to make payment, providing seven days' notice to the owner and architect, and then stopping work. Only if 30 days elapse after the contractor stopped work can the contractor then terminate the contract.

As discussed, while commentators disagree as to whether a contractor can terminate for a common law material breach in addition to the grounds provided in Section 14.1, courts likely would reject a claim that a failure to make payment is a material breach that justifies a termination of the contract earlier than that available by following Section 14.1.1.3's procedure. Section 14.1.1.3 sets out a specific procedure and time frame for terminating the contract for failure to make payment, and contract law generally provides that a contract's conditions precedent to contract termination must be satisfied.[9] Accordingly, while failure to make payment may be a material breach, a court would likely still require the terminating contractor to fulfill the conditions of Section 14.1.1.3 before terminating. Additionally, a failure to fulfill the conditions in Section 14.1.1.3 before terminating the contract for nonpayment could be a breach of contract by the contractor.[10] Courts will permit termination of a contract despite a party's failure to comply with technical details in regard to notice of termination,[11] but courts are unlikely to view the entirety of Section 14.1.1.3 as a technicality and much more likely to view it as a condition precedent to termination with which a contractor must comply.[12]

The fourth provision, Section 14.1.1.4, permits a contractor to terminate the contract when the owner fails to provide "reasonable evidence" of financial arrangements to fund the project. "Reasonable evidence" is a requirement in Section 2.2 under which before the work begins or upon the occurrence of certain events after the work begins, the owner must provide evidence of financial arrangements sufficient for the owner to perform its contractual obligations. An unresolved issue here is the relationship between Section 2.2's prohibition on the owner altering the financial arrangements after providing reasonable evidence to the contractor, and Section 14.1.1.4, which does not specifically permit the contractor to terminate as a result of the owner's modification of its financial arrangements. A material modification of the financial arrangements, however, likely can serve as an indirect route to termination under Section 14.1.1.4 should a contractor need to terminate for that reason. A contractor's belief that an owner materially modified the financial arrangements would likely constitute a reasonable concern regarding the owner's ability to make payments and thereby entitle the contractor to request reasonable evidence under Section 2.2.2. Should the owner then fail to provide reasonable evidence, the contractor would be entitled to terminate the contract under Section 14.1.1.4.

The only notable change to Section 14.1.1 in the 2017 version of the A201 is the deletion of language in Section 14.1.1.4 regarding the owner's failure to "promptly"

9. *See, e.g.*, Southland Metals, Inc. v. Am. Castings, LLC, 800 F.3d 452, 460 (8th Cir. 2015); Keystone Dedicated Logistics, LLC v. JGB Enters., Inc., 77 A.3d 1, 8 (Pa. Super. Ct. 2013); 17B C.J.S. *Contracts* § 606 (2017).

10. *See* Mike Building & Contracting, Inc. v. Just Homes, LLC, 901 N.Y.S.2d 458, 470 (Sup. Ct. 2010) (holding a party breached a construction contract by terminating the contract without satisfying the conditions precedent to termination); Paragon Restoration Group, Inc. v. Cambridge Square Condominiums, No. I 2004-12472, 2006 WL 4094363, at *6 (N.Y. Sup. Ct. May 11, 2006).

11. *See, e.g.*, Luv N'Care Ltd. v. Angel Juvenile Prods., 640 F. App'x 389, 392 (5th Cir. 2016); Sun Secured Fin. LLC v. ARCS Commercial Mortgage Co., 730 F. Supp. 2d 132, 139–40 (D.D.C. 2010).

12. *See Mike Building & Contracting*, 901 N.Y.S.2d at 470; *Paragon Restoration Group*, 2006 WL 4094363, at *6.

provide reasonable evidence to the contractor upon request. This change results from the addition of language in Section 2.2.2 providing that the contractor can stop the work if the owner does not provide reasonable evidence within 14 days after a request.

> **§ 14.1.2** The Contractor ~~or~~, a Subcontractor, a Sub-subcontractor ~~or~~, their agents or employees, or any other persons or entities performing portions of the Work ~~under direct or indirect contract with the Contractor~~, repeated suspensions, delays, or interruptions of the entire Work by the Owner as described in Section 14.3, constitute in the aggregate more than 100 percent of the total number of days scheduled for completion, or 120 days in any 365-day period, whichever is less.

Section 14.1.2 serves as a limitation on the owner's right to suspend the contract for convenience under Section 14.3.1. Under Section 14.3.1, the owner can suspend the work for as long as the owner desires, but under Section 14.1.2, if the suspension extends beyond either 100 percent of the days scheduled for completion or 120 days in a 365-day period, the contractor can terminate the contract. Additionally, as detailed in the discussion of Section 14.3.1, a contractor may be able to terminate a contract as a result of constructive suspension even when the owner has not formally suspended the work.

Section 14.1.2 acts as a limitation on the owner's right to suspend for convenience through that section's language that any suspension, delays, or interruptions be "as described in Section 14.3." Therefore, any suspension, delay, or interruption of the work for reasons other than the owner's suspension for convenience will justify termination only if the suspension, delay, or interruption satisfies Section 14.1.4 and its requirement that work is stopped due to the owner's failure to perform its contractual obligations to allow the work to progress.

Section 14.1.2 also applies only where the suspension, delays, or interruptions are not the result of an action or fault of the contractor or those working directly or indirectly for the contractor. An owner could thus argue that it did not suspend the work for convenience but rather had to suspend the work due to the contractor's actions or inactions. If a court accepts this interpretation of the contractual language and finds factual support, a contractor would not be entitled to terminate the contract and could be liable for breach by failing to perform following an improper termination.

> **§ 14.1.3** If one of the reasons described in Section 14.1.1 or 14.1.2 exists, the Contractor may, upon seven days' ~~written~~ notice to the Owner and Architect, terminate the Contract and recover from the Owner payment for Work executed, ~~including~~as well as reasonable overhead and profit~~;~~ on Work not executed, and costs incurred by reason of such termination~~, and damages~~.

While Sections 14.1.1, 14.1.2, and 14.1.4 govern instances in which a contractor can terminate the contract, Section 14.1.3 provides the mechanism for termination (essentially a notice requirement), and the damages a contractor can recover following termination. Section 14.1.3 states that it applies when termination occurs pursuant to Sections 14.1.1 or 14.1.2, but it also applies when termination occurs pursuant to Section 14.1.4, which allows a contractor to "terminate the Contract and recover from the Owner as provided in Section 14.1.3."

For the notice component of Section 14.1.3, the provision reflects one alteration from previous versions in that the 2017 version of the A201 deletes the requirement that the notice be "written." The required notice, however, still must be in writing because a change to Section 1.6.1 requires all notices to be in writing.

Once the contractor has provided the seven-day notice under Section 14.1.3, nothing in Section 14.1.3 allows an owner to prevent the termination from occurring. Unlike many contractual provisions related to breach or termination, Article 14 does not explicitly provide that the owner can use the seven-day notice period to cure the reason for termination, and no reported decisions exist on whether an implied opportunity to cure exists.[13] At least one commentator, however, has acknowledged that some practitioners have interpreted the seven-day notice period as a right to cure.[14] If Section 14.1.3 does not imply an opportunity to cure, then the presumed purpose of the seven days' notice is to give the owner time to begin making arrangements for the project following termination of the contract.

Courts have reached differing results on the impact of a contractor's failure to satisfy Section 14.1.3's notice requirement. The Massachusetts Supreme Judicial Court held that a contractor's failure to give the contractually required seven days' notice was a substantial breach.[15] An Ohio Court of Appeals, on the other hand, found that a contractor providing a notice stating that the contract terminated as of the date of the notice rather than after the seven-day notice period constituted substantial compliance and did not constitute a material breach.[16]

For the recoverable damages component of Section 14.1.3, the 2017 version of the A201 provides for three types of damages a contractor can recover following a termination by the contractor: (1) payment for work executed, (2) reasonable overhead and profit for work not executed, and (3) costs incurred as a result of termination. The inclusion in the 2017 version of Section 14.1.3 of overhead and profit for work not executed as a recoverable element of damage is a major change from the former version of the A201. The prior version of the A201 permitted the contractor to recover reasonable overhead and profit only for work actually executed. Additionally, while the contractor can now recover overhead and profit for work not executed when the contractor terminates the contract due to the owner's breach, Section 14.4 of the A201–2017 no longer permits the contractor to recover overhead and profit for work not executed when the owner terminates for convenience.

While clarifying the types of damages recoverable following a contractor's termination, what is included within each of those categories of recoverable damages remains open for courts to determine. On the issue of what is recoverable as payment for work executed, courts have held that Section 14.1.3 entitles a contractor to recover retainage withheld from payments for work executed.[17] The U.S. Court of Appeals for the Fifth Circuit reasoned that Section 14.1.3 "unequivocally" entitles a contractor to payment for work executed "irrespective of substantial completion." Retainage is simply compensation due but withheld, and therefore, a contractor is entitled to recover retainage.[18]

---

13. *See* BRUNER & O'CONNOR, *supra* note 1, § 5:263.
14. *Id.*
15. Providence Washington Ins. Co. v. Beck, 255 N.E.2d 600, 601 (Mass. 1970).
16. Gillard v. Green, No. 00-CA-54, 2001 WL 1682940, at *5 (Ohio Ct. App. Dec. 28, 2001). In reaching that result, the court relied on the fact that the contractor remained on the site wrapping up its work after providing the notice and the lack of any evidence that the owner was prejudiced by the lack of seven days' notice.
17. *See In re* Roy Frischhertz Constr. Co., 401 F. App'x 861 (5th Cir. 2010); Iteld v. Four Corners Constr., L.P., 157 So. 3d 702 (La. Ct. App. 2013).
18. *Frischhertz Constr. Co.*, 401 F. App'x at 862.

For the recoverable "reasonable overhead and profit," courts have repeatedly dealt with the issues of the recoverability of overhead and lost profits following a breach of contract, and the law of the applicable jurisdiction should provide guidance here.[19] Additionally, while Section 14.1.3 does not mention Section 7.3.4 (previously numbered 7.3.7) and its specifications as to a contractor's costs, the recoverable profit for unexecuted work likely involves subtracting the costs set out in Section 7.3.4 from the contract price. Additionally, Section 7.3.4 allows a contractor to receive overhead and profit resulting from change orders, and the methods used to calculate those items in relation to change orders can be applied to a contractor's termination.

Some of the costs provided for in Section 7.3.4 may still be incurred despite a termination (e.g., machinery needed for the job that is subject to a rental period extending beyond the termination), and those costs presumably are the direct costs incurred as a result of, and recoverable after, termination.

> **§ 14.1.4** If the Work is stopped for a period of 60 consecutive days through no act or fault of the Contractor ~~or~~, a Subcontractor, a Sub-subcontractor, or their agents or employees or any other persons or entities performing portions of the Work ~~under contract with the Contractor~~ because the Owner has repeatedly failed to fulfill the Owner's obligations under the Contract Documents with respect to matters important to the progress of the Work, the Contractor may, upon seven additional days' ~~written~~ notice to the Owner and the Architect, terminate the Contract and recover from the Owner as provided in Section 14.1.3.

Section 14.1.4 is similar to Section 14.1.2 in that it permits the contractor to terminate the contract due to stoppage of the work, but it differs in that it requires that the work be stopped due to the owner's failure to fulfill certain contractual obligations. In terms of what it means for the work to be "stopped," courts have held that the work can be stopped for purposes of Section 14.1.4 even where the contractor performs punch-list or minor work during the delay period.[20] These decisions indicate that the work will be deemed "stopped" if a contractor cannot perform work critical to the completion of the project for 60 or more consecutive days. It is unclear, however, how these holdings comport with the contract's broad definition of "Work" and Section 14.1.4's requirement that the "Work" has been stopped. Additionally, these decisions are unclear as to what constitutes work "not critical to completion" and whether this means that so long as no work in the project's critical path occurs for 60 consecutive days, the work was stopped, and the contractor can terminate pursuant to Section 14.1.4.

Courts have also addressed what constitutes an owner's repeated failure to fulfill contractual obligations and have held that an owner's failure to provide information or direction necessary for the contractor to perform is grounds for termination under Section 14.1.4.[21] In one decision, an owner failed to fulfill its obligations under Section 14.1.4 where the owner "failed to issue adequate direction to proceed."[22]

---

19. *See* 6 Bruner & O'Connor, *supra* note 1, § 19:111 & 106 (2016 Update); 24 Williston, *supra* note 3, § 64:11 (4th ed. May 2017 Update).

20. *See* General Ins. Co. of Am. v. K. Capolino Constr. Corp., 984 F. Supp. 403, 428 n.43 (S.D.N.Y. 1997); Lott Constructors, Inc. v. Jackson Twp. Bd. of Educ., No. CIV. 92-2402 (CSF), 1992 WL 201313 at *7 (D.N.J. Aug. 12, 1992).

21. *See General Ins. Co. of Am.*, 984 F. Supp. 403; *Lott Constructors, Inc.*, 1992 WL 201313; Gillard v. Green, No. 00-CA-54, 2001 WL 1682940, at *4 (Ohio Ct. App. Dec. 28, 2001).

22. *General Ins. Co. of Am.*, 984 F. Supp. at 428.

In that case, after an issue arose as to details of the work, the contractor sought direction as to how to proceed, but the owner failed to provide the required direction. The court held that this failure justified termination under Section 14.1.4.[23]

In a second decision, a contractor began work on a project but noticed cracking masonry apparently resulting from the movement of structural steel.[24] The contractor notified the architect of the problem, stopped the work, and waited for the owner and architect to issue a directive on how to proceed. The court held the contractor rightfully terminated the contract pursuant to Section 14.1.4 because the owner failed to provide sufficient plans for the work to continue, and failed to timely provide sufficient new plans and a change order or change directive after the deficiency was discovered.[25]

Additionally, that decision indicates that a deficiency in the plans that causes the contractor to be unable to perform the work, even if the deficiency is the result of the architect's actions or omissions, is a failure of the owner to fulfill its contractual obligations and permits the contractor to terminate the contract under Section 14.1.4.[26] As discussed, the masonry cracked as a result of a deficient design for the structural steel in the original plans, and the contractor asserted it rightfully terminated the contract under Section 14.1.4. In response, the owner argued that because the architect, not the owner, prepared the plans, the work stoppage was not the owner's fault. The court rejected the owner's argument on the basis that because the owner is responsible for providing plans sufficient for the contractor to perform the work, because the owner retained the architect to prepare plans, and because there was no privity between the contractor and the architect, the work stoppage resulting from the deficient plans was the owner's fault.[27]

Finally, one item of note in relation to the notice requirement in Section 14.1.4 is that it requires "seven additional days' notice." The use of "additional" here presumably indicates that this notice period is in addition to the 60-day period during which the owner fails to fulfill obligations. In other words, if the owner fails to fulfill its obligations for 53 days, the contractor cannot provide notice on that day and then terminate the contract seven days later. Rather, the contractor must wait for the owner to have not performed its obligations for 60 days, and then provide notice to the owner and wait an additional seven days before terminating the contract.

§ 14.2 TERMINATION BY THE OWNER FOR CAUSE

§ 14.2.1 The Owner may terminate the Contract if the Contractor

.1 repeatedly refuses or fails to supply enough properly skilled workers or proper materials;

.2 fails to make payment to Subcontractors ~~for materials~~ or ~~labor~~suppliers in accordance with the respective agreements between the Contractor and the Subcontractors or Suppliers;

.3 repeatedly disregards applicable laws, statutes, ordinances, codes, rules and regulations, or lawful orders of a public authority; or

.4 otherwise is guilty of substantial breach of a provision of the Contract Documents.

23. *Id. See also Gillard*, 2001 WL 1682940, at *4 (holding that an owner's failure to "make prompt decisions as to materials or processes to be used in completing" the project was a breach of the owner's contractual obligations justifying the contractor's termination).

24. *Lott Constructors, Inc.*, 1992 WL 201313, at *1.

25. *Id.* at *6–7.

26. *See id.* at *6.

27. *Id.*

This section allows the owner to terminate the contract for cause if the contractor fails to perform, giving three specific yet common failings, and a catch-all fourth basis. The decision of whether to terminate the contractor for cause is most often a difficult and multifaceted one. Even the mere consideration of a termination for cause likely indicates serious performance issues that threaten the project and likely its schedule. But the contractor, as well as its surety, if any, are certain to object or challenge the sufficiency of the basis for termination for cause, as well as whether the owner has followed the required protocols for termination. Termination for cause, whether under Section 14.2 or otherwise, is a drastic step, resulting in forfeiture of rights, and thus the "cause" must be a substantial (material) breach of the contractor's obligations under the contract.[28]

The first basis discussed for termination in Section 14.2.1.1 is if the contractor repeatedly refuses or fails to supply enough properly skilled workers or proper materials. This section remains unchanged from the 2007 version, whose only alteration of the 1997 version language was the substitution of "repeatedly" for "persistently," a change that seemed to lower the bar for the owner. The AIA Commentary for the 2007 version made clear that "isolated instances" of an insufficient number of workers or improper materials would not act to satisfy this section.[29] A prudent owner would therefore consider whether it could demonstrate more than mere isolated instances of insufficient manpower, and instead demonstrate not only repetition of this shortcoming by the contractor, but also documentation of the shortcoming and notice to the contractor. Whether or not the contractor has supplied "enough" workers is another determination that the owner must make, in conjunction with obtaining the required certification by the architect discussed below in Section 14.2.2. Absent abandonment of the project by the contractor with work remaining to be performed, the owner and architect might compare the contractor's current progress against the project schedule as a way of examining whether sufficient workers are being supplied.[30]

Section 14.2.1.2 contains the lone revision to 14.2.1 in the 2017 version. Whereas the 2007 version references a failure by the contractor to make payments to subcontractors for materials or labor, the 2017 version references a failure by the contractor to make payments to subcontractors "or suppliers" in accordance with their respective agreements. This revision is made throughout the 2017 version, beginning with Section 1.5.1. The removal of the limitation "for labor and materials" clarifies that failure to make payment to subcontractors satisfies this subsection, regardless of whether the payment was to be for "labor or materials." The revision also eliminates possible arguments by the contractor that a particular first-tier entity is not a "Subcontractor" by parsing that definition.

Unlike 14.2.1.1 and 14.2.1.3, the adjective "repeatedly" is not present in 14.2.1.2. The AIA Commentary states that "such conduct must occur repeatedly" but in doing so highlights the word "repeatedly" only in 14.2.1.1 and 14.2.1.3.[31] The literal language does reference a failure to make payment to subcontractors (plural) indicating that the contractor must fail in making payment to more than just one subcontractor.

Section 14.2.1.3 outlines a basis for termination for cause where the contractor (again *repeatedly*) disregards applicable laws, statutes, ordinances, codes, rules and regulations, or lawful orders of a public authority. Citations by a public authority,

28. *See* 5 Bruner & O'Connor, *supra* note 1, § 18:36.
29. American Institute of Architects, AIA Document A201–2007 Commentary, at *53 (2007).
30. *See* Sabo, *supra* note 2, § 4.81 Termination by the Owner for Cause: ¶ 14.2.1. (5th ed. 2008).
31. American Institute of Architects, AIA Document A201–2007 Commentary, at *53, (2007).

particularly as to safety, might assist the owner in substantiating a termination for cause based upon this section. As the list of legal authorities to be obeyed is comprehensive, from federal law to the most local ordinances, violations of law that concern safety, environmental protection, workforce, labor, and employment laws are all seemingly implicated, as well as any statutes or codes that might pertain specifically to construction, such as Prompt Payment Acts.

Section 14.2.1.4 is the attempted catch-all provision, which finds a basis for termination for cause where the contractor "otherwise is guilty of a substantial breach of a provision of the Contract Documents." On at least one occasion, the failure by the contractor to provide the required surety bonds has been deemed such a material breach under this provision.[32]

§ 14.2.2 When any of the ~~above~~ reasons described in Section 14.2.1 exist, ~~the Owner,~~and upon certification by the ~~Initial Decision Maker~~Architect that sufficient cause exists to justify such action, the Owner may, without prejudice to any other rights or remedies of the Owner and after giving the Contractor and the Contractor's surety, if any, seven days' ~~written~~ notice, terminate employment of the Contractor and may, subject to any prior rights of the surety:

.1 Exclude the Contractor from the site and take possession of all materials, equipment, tools, and construction equipment and machinery thereon owned by the Contractor;

.2 Accept assignment of subcontracts pursuant to Section 5.4; and

.3 Finish the Work by whatever reasonable method the Owner may deem expedient. Upon written request of the Contractor, the Owner shall furnish to the Contractor a detailed accounting of the costs incurred by the Owner in finishing the Work.

Whereas Section 14.2.1 provides the basis for a termination for cause, Section 14.2.2 outlines the protocol for doing so, including two key required actions: (1) the certification by the Initial Decision Maker, and (2) the seven-day notice. The lone revision in Section 14.2.2 is the removal of the word "written" concerning the notice, which is now made unnecessary by the global requirement in Section 1.6.1 that all notices be in writing. While impartiality on the part of the Initial Decision Maker was not made express in the 2007 version, the 2017 version makes it clear in Section 1.1.8 that it "shall not show partiality to the Owner or Contractor." The 1997 version filled this role with the architect, before switching to "Initial Decision Maker" in the 2007 version.

The insistence that the Initial Decision Maker arrive at its conclusion in an impartial manner addresses prior difficulties where an architect (under contract with the owner and being paid by the owner) is asked directly by the owner to certify that termination is warranted despite little or any knowledge of the proffered basis under 14.2.1.1.[33] If found to have been issued as a rubber stamp or without the proprietaries necessary to fulfill the Initial Decision Maker's function, the owner may yet be entitled to terminate for default and recover damages based upon common-law rights of termination.[34]

32. Underwriter's Group, Inc. v. Clear Creek Independent School District, CA No. G-05-334, 2008 WL 112104 (S.D. Tex. Jan. 9, 2008).

33. *See* Ingrassia Construction Co., Inc. v. Vernon Township Board of Education, 784 A.2d 73 (N.J. Super Ct. App. Div. 2001).

34. *Id. See also* Town of Plainfield v. Paden Engineering Co, Inc., 943 N.E.2d 904 (Ind. Ct. App. 2011).

After obtaining the certification that a basis for termination exists, the owner must then notify both the contractor and its surety (if a performance bond has been provided) seven days before terminating the contractor's employment. Given the importance of this step, many practitioners provide the notification via certified mail or some other easily verifiable means of confirming receipt. If a performance bond was issued, it is important for the owner to comply not only with Section 14.2.2 but also with any terms of the performance bond, especially taking care if written on a form other than the AIA A312.

The A312 provides for additional steps the owner must take, including a notice that the owner is considering declaring the contractor in default, as well as a request for a meeting with the contractor and surety. The owner therefore should take care to follow the requirements of both the bond and 14.2.2, even though doing so may create anxiety on a project that is already experiencing a delay in completion and other problems.

The task of complying with the merged protocols of both Section 14.2.2 and the A312 is further complicated by the fact that in 2010, the AIA modified the A312 bond form for the first time since 1984. In the 1984 version, the owner was required not only to provide written notice to the contractor and surety that a default was being considered, but also to request a meeting with all three parties within 15 days. In the 2010 version, the same notice must be given, but the request for a meeting by the owner is not mandatory. If the owner chooses not to request a meeting, then the burden to request the meeting is shifted to the surety, which may request such a meeting within five business days, upon which the owner must attend. Unless the owner agrees otherwise, the meeting must be held within ten business days of the surety's receipt of the owner's notice.

The AIA Commentary on the A312–2010 expressly states that the revision was made because some courts had construed the request for a meeting as a condition precedent to the surety's obligations.[35] The AIA reasoned that it recognized that an owner may believe such a conference to be unproductive and might prefer to move forward with terminating the contractor without this step. The requirement to provide notice to the surety that the owner is considering declaring a contractor default remains in the 2010 version, but a new Section 4 of the A312 states that "Failure on the part of the Owner to comply with the notice requirement in Section 3.1 shall not constitute a failure to comply with a condition precedent to the Surety's obligations, or release the Surety from its obligations, except to the extent the Surety demonstrates actual prejudice." Additionally, the requirement in the 1984 version that the owner not declare a contractor default earlier than 20 days after the contractor and surety have received notice that a default is being considered has been removed entirely. In its Commentary, the AIA states that in many cases the owner may have determined that the contractor had failed to perform to such a degree that terminating the contractor's right to proceed is warranted, and such an additional 20-day waiting period may be unnecessary delay.[36] This revision is well-reasoned, as in many cases the performance may be so irredeemable that putting off termination and the replacement by another capable contractor may be in the best interests of the owner, but also the surety (and perhaps the terminated contractor itself) as further schedule delay and mounting damages might be stemmed earlier.

35. American Institute of Architects, AIA Bond Form Commentary and Comparison—AIA Documents A310–2010 and A312, at *7 (2010).

36. *Id.*

Failure to follow the required termination protocols of the seven-day letter, as well as any conditions of a bond itself, have been deemed a failure to satisfy a condition precedent, resulting in a performance bond surety's release from its obligations under its bond.[37] In *Stonington Water Street Assoc., LLC v. Hodess Building Co, Inc.*, the failure to abide by the conditions precedent such as the seven-day notice was detrimental to the bond claim, despite the fact that the contractor had essentially abandoned the project and the surety in question was already making payments to subcontractors under the contractor's payment bond. In that case the A312–1984 had been used. The basis of the court's finding of failure of a condition precedent, however, was the seven-day notice required under Section 14.2.2, as well as the failure to comply with A312 notices. The holding in *Stonington*, then, cannot be discounted as to conditions precedent merely because of the change to the A312 form in 2010.

Following the protocol provisions, Section 14.2.2 provides the owner with specific rights following termination, subject to the prior rights of the surety, if any. The owner may exclude the contractor from the site and take possession of all materials, equipment, tools, and construction equipment owned by the contractor; accept assignment of subcontracts pursuant to Section 5.4; and finish the work by whatever reasonable method the owner may deem expedient. Similar to the above discussion, the owner's rights listed in Section 14.2.2 must be read in conjunction with any requirements listed in any performance bond, particularly the A312, if the owner is seeking not only to terminate the contractor but also to preserve the owner's rights as to the surety.

§ 14.2.3 When the Owner terminates the Contract for one of the reasons stated in Section 14.2.1, the Contractor shall not be entitled to receive further payment until the Work is finished.

If the owner has an adequate basis under 14.2.1 and terminates the contractor properly, the contractor will not receive further payment until the work is finished. The Commentary to the 2007 version explained that following completion of the work by others, the architect must certify the costs involved and provide an accounting to determine whether further payment is required by the owner or whether the contractor has a debt to the owner. Special attention should be paid to keeping an accurate accounting of such costs given the high likelihood of litigation following termination, so that the owner may allow compliance with Section 14.2.4's requirement for certification by the Initial Decision Maker.

The terms of any performance bond will likely require the owner to make the remainder of funds under the contract available if the performance bond surety accepts its responsibility and provides for a completion contractor. The A312 bond form states that the surety's obligation arise, among other requirements, when "the Owner has agreed to pay the Balance of the Contract Price to the Surety in accordance with the terms of the Construction Contract or to a Contractor selected to perform the Construction Contract in accordance with the terms of the Contract with the Owner."

37. Stonington Water Street Assoc., LLC v. Hodess Building Co, Inc., 792 F. Supp. 2d 253 (D. Conn. 2011). *See also* Bank of Brewton, Inc. v. International Fidelity Ins. Co., 827 So. 2d 747 (Ala. 2002) (holding that while the owner corresponded many times with the surety regarding performance problems, the owner's failure to officially state that it was actually terminating the contractor operated as a failure of a condition precedent to the surety's obligations); MCK Building Assoc. Inc. v. St. Lawrence Univ., 754 N.Y.S.2d 397 (App. Div. 2003).

At least one case has held that pursuant to the language in Section 14.2.3, should a portion of the contract funds remain following completion of the work by another contractor, such funds "become due" only on the date the work is completed, and therefore any pre-judgment interest provision in the contract must be read to begin calculating interest only upon completion, rather than on termination.[38]

> **§ 14.2.4** If the unpaid balance of the Contract Sum exceeds costs of finishing the Work, including compensation for the Architect's services and expenses made necessary thereby, and other damages incurred by the Owner and not expressly waived, such excess shall be paid to the Contractor. If such costs and damages exceed the unpaid balance, the Contractor shall pay the difference to the Owner. The amount to be paid to the Contractor or Owner, as the case may be, shall be certified by the Initial Decision Maker, upon application, and this obligation for payment shall survive termination of the Contract.

The only prior revision to this section in 2007 was the insertion of "and other damages incurred by the Owner and not expressly waived." This edit allowed the owner to also withhold funds, if not waived in Section 15.1.6, to compensate the owner for damages other than the costs to complete the Project. The 2017 version's revision in Section 1.1.8 regarding the Initial Decision Maker's requirement of impartiality, however, does apply to the certification under Section 14.2.4.

§ 14.3 SUSPENSION BY THE OWNER FOR CONVENIENCE

> **§ 14.3.1** The Owner may, without cause, order the Contractor in writing to suspend, delay or interrupt the Work, in whole or in part for such period of time as the Owner may determine.

This section gives the owner the right to suspend, delay, or otherwise interrupt the work, in whole or in part, for the convenience of the owner. Such a clause gives the owner the flexibility it may need to temporarily suspend performance of the work to accommodate unexpected or changing circumstances related to the project—such as changes in financing, user needs, applicable laws, or market conditions. Without a suspension-of-work clause, any material interruption of the work by the owner could be deemed a breach of contract.[39]

To give the owner the flexibility it needs to suspend work without breaching its contract, most standard form construction contracts now include a suspension-of-work provision expressly allowing the owner to suspend performance for its convenience.[40] The addition of such a provision eliminates the contractor's common law breach of contract claim arising from a suspension by the owner, and instead allows the contractor to seek whatever contractual remedy is allowed by the suspension provision.[41]

38. R.S. Deering Mechanical Contractors v. Livesey Company, 468 N.W.2d 758 (Wis. Ct. App. 1991).

39. *See* Guerini Stone Co. v. P.J. Carlin Constr. Co. Inc., 248 U.S. 334 (1919) (work stoppage constituted material breach of contract); Haney v. United States, 676 F.2d 584 (Ct. Cl. 1982) (same); L.L. Hall Constr. Co. v. United States, 379 F.2d 559 (Ct. Cl. 1966) (owner's delay of contractor's performance is a breach of implied duty to not delay the contract).

40. 5 Bruner & O'Connor, *supra* note 1, § 15:85.

41. *See* United States v. Utah Constr. & Min. Co., 384 U.S. 394 (1966) (suspension of work provision converts remedy from breach of contract claim to claim for relief under the contract).

Section 14.3.1 requires a suspension to be directed by the owner in writing. Some jurisdictions, however, in interpreting the federal suspension of work clause, have recognized the doctrine of "constructive suspension" in situations where the contractor's performance of work is effectively suspended or delayed but the owner has not formally suspended the work in writing.[42] In such cases, the contractor is allowed to take advantage of the remedy contained in the suspension of work clause, even though the owner did not direct the contractor in writing to suspend the work, and even where the owner may not acknowledge that the work was suspended.[43]

The federal doctrine of constructive acceleration has not been applied in any reported case involving Section 14.3. Perhaps that is because the federal suspension of work clause contains language not found in the AIA's suspension of work clause—language expressly stating that the contractor is entitled to an adjustment in the contract price if the contractor's performance is delayed for an unreasonable period of time by the actions of the contracting officer. In fact, some jurisdictions have expressly rejected the concept of constructive acceleration in cases where the applicable suspension of work clause required the suspension to be in writing.[44]

Finally, Section 14.3.1 does not limit the period of time that the owner can suspend the work. Nevertheless, the contractor may be entitled to terminate the contract if the total aggregate days of suspension by the owner exceeds "100 percent of the total number of days scheduled for completion, or 120 days in any 365-day period, whichever is less."[45]

§ 14.3.2 The Contract Sum and Contract Time shall be adjusted for increases in the cost and time caused by suspension, delay, or interruption ~~as described in~~under Section 14.3.1. Adjustment of the Contract Sum shall include profit. No adjustment shall be made to the extent

.1 that performance is, was, or would have been, so suspended, delayed, or interrupted, by another cause for which the Contractor is responsible; or

.2 that an equitable adjustment is made or denied under another provision of the Contract.

Section 14.3.2 provides the contractor's remedy when the owner suspends the work. Specifically, the contractor is entitled to a change order adjusting the Contract Sum to account for additional costs (plus profit) incurred as a result of the suspension, as well as an extension of the schedule for any resulting delay. This section does not specify how the "costs" caused by a suspension are to be calculated. Presumably, the parties could rely on the definition of costs found in Section 7.3.4 of the A201, which is used in connection with change orders. Alternatively, for a cost-plus contract, the A133's more detailed definition of the "Cost of the Work" could be used. Similarly, because Section 14.3.2 does not specify how to determine the amount of profit the contractor is entitled to recover, the parties must either use the markup for overhead and profit contained in their agreement or agree on some other reasonable amount of profit. Although not expressly stated in Section 14.3, any disputes regarding the amount of the contractor's adjustment should be resolved in accordance with the claim and dispute provisions in Article 15.[46]

---

42. *See, e.g.*, T.C. Bateson Constr. Co. v. United States, 319 F.2d 135 (Ct. Cl. 1963).
43. *See id.* at 159–60.
44. *See, e.g.*, Bonacorso Constr. Corp. v. Com., 668 N.E.2d 366 (Mass. App. Ct. 1996) (doctrine of constructive suspension is inconsistent with statutory requirement that the suspension be directed in writing).
45. Section 14.1.2.
46. See Section 7.3.5.

Section 14.3.2 identifies two situations in which the contractor is not entitled to a contract adjustment: (1) when the contractor's performance "is, was or would have been so suspended, delayed or interrupted by a cause for which the Contractor is responsible" or (2) when the contract adjustment was already made or denied under another contract provision. The first exception is based on the general legal principle that a compensable delay by the owner is offset by a concurrent inexcusable delay by the contractor.[47]

The second exception recognizes the fact that there are other contract provisions that the contractor may have already relied upon to submit a claim for a suspension, delay, or interruption contemplated by Section 14.3.2. If the contractor already asserted a claim under the change order or claim provisions of the contract, the contractor cannot double dip by asserting a separate claim for the same delay under this section. Moreover, if the contractor's claim has been denied based upon other contract language—such as a failure to provide timely notice as required in Section 15.1.3—Section 14.3.2 does not give the contractor a second bite at the apple.

§ 14.4 TERMINATION BY THE OWNER FOR CONVENIENCE

> **§ 14.4.1** The Owner may, at any time, terminate the Contract for the Owner's convenience and without cause.

This section allows the owner to terminate the contract at any time for any reason. Similar to the suspension for convenience clause in Section 14.3, this termination for convenience clause allows the owner to act unilaterally in a way that would have been considered a breach of contract under the common law.[48] Without such a provision, an attempted termination by the owner without cause would be a material breach of the contract, entitling the contractor to recover damages for wrongful termination or repudiation.[49]

Termination for convenience clauses first arose in the United States at the end of the Civil War, when the federal government needed to end military procurements that were no longer necessary.[50] Termination for convenience clauses are now commonplace in both government contracts and industry construction contract forms. Section 14.4 was first added to the A201 in 1997.

Due to concerns that a one-sided termination for convenience clause lacks adequate consideration, some courts have held that a termination for convenience can be exercised only when circumstances affecting the parties' expectations have changed.[51] But the majority of modern court decisions have rejected the "changed circumstances" test, instead enforcing termination for convenience provisions unless the owner acted in bad faith or abused its discretion.[52]

---

47. *See* Green v. General Services Admin., 96-2 B.C.A. (CCH) ¶ 28306, 1996 WL 220993 (Gen. Services Admin. B.C.A. 1996) (contractor not allowed to recover delay damages from compensable delay that was concurrent with inexcusable delay).

48. *See* Bruner & O'Connor, *supra* note 1, § 5:270.

49. *See, e.g.*, Wilmette Partners v. Hamel, 594 N.E.2d 1177 (Ill. App. Ct. 1992) (owner repudiated contract by wrongfully terminating contract and ordering contractor off the jobsite without adequate cause).

50. Torncello v. United States, 681 F.2d 756, 764 (Ct. Cl. 1982).

51. *See, e.g., id.*

52. *See* Krygoski Constr. Co., Inc. v. United States, 94 F.3d 1537 (Fed. Cir. 1996) (rejecting "change in circumstances" test for "bad faith" standard); A.J. Temple Marble & Tile, Inc. v. Long Island R.R., 659 N.Y.S.2d 412 (Sup. Ct. 1997), *aff'd in part* 682 N.Y.S.2d 422 (App. Div. 1998) (contractor must show that owner acted with "malicious intent" or "animus" in terminating for convenience in order to prevail on wrongful termination claim); Northrop Grumman Corp. v. United States, 46 Fed. Cl. 622 (2000) (termination for convenience constitutes breach of contract where owner abused its discretion or acted in bad faith). *But see* SAK & Associates, Inc. v. Ferguson Constr. Inc., 357 P.3d 671 (Wash. Ct. App. 2015) (termination for convenience clauses are not illusory because the parties' partial performance constitutes consideration; the duty of good faith and fair dealing does not limit the enforceability of unambiguous termination for convenience clauses).

§ 14.4.2 Upon receipt of ~~written~~ notice from the Owner of such termination for the Owner's convenience, the Contractor shall

- **.1** cease operations as directed by the Owner in the notice;
- **.2** take actions necessary, or that the Owner may direct, for the protection and preservation of the Work; and
- **.3** except for Work directed to be performed prior to the effective date of termination stated in the notice, terminate all existing subcontracts and purchase orders and enter into no further subcontracts and purchase orders.

Section 14.4.2 describes the procedures the parties must follow when the owner terminates for convenience. First, the termination notice from the owner must be in writing under Section 1.6.1. Although Section 14.4 does not require the owner to identify the reason for the termination, it may be wise for the owner to articulate a good faith basis for terminating the contract in the written notice of termination in order to refute any allegation that the termination was motivated by bad faith.

Once the owner terminates the contract, the contractor is then obligated to take the following actions: (1) stop all operations as directed by the owner, (2) take actions that are necessary to protect and preserve the existing work, including any actions directed by the owner, and (3) terminate all subcontracts and purchase orders, except to the extent the owner has directed the contractor to continue to perform certain work prior to the effective date of the termination. Although Section 14.4, on its face, contemplates only a total termination, and not a partial termination, Sections 14.4.2.1 and 14.4.2.3 would presumably allow the owner to direct the contractor to continue performing certain portions of the work while terminating the rest of the contract.

Since Section 14.4.2.3 requires the contractor to terminate its subcontracts and purchase orders if the owner terminates for convenience, a contractor may want to negotiate termination for convenience clauses (or strong flow down provisions) in its subcontracts and purchase orders to make sure that the contractor is not subject to damages for wrongful termination against its subcontractors and suppliers. Similarly, an owner may want to require its contractor to include such a clause in its subcontracts and purchase orders so that the contractor does not attempt to pass through wrongful termination damages to the owner.[53]

§ 14.4.3 In case of such termination for the Owner's convenience, the Owner shall pay the Contractor ~~shall be entitled to receive payment~~ for Work properly executed~~, and~~; costs incurred by reason of ~~such~~the termination, ~~along with reasonable overhead and profit on the Work not executed~~including costs attributable to termination of Subcontracts; and the termination fee, if any, set forth in the Agreement.

The AIA's 2017 revised owner-contractor agreements have been modified to encourage the parties to establish a "termination fee" that applies when the owner terminates the agreement for its convenience. Section 14.4.3 has been revised to reference the termination fee negotiated in the agreement, rather than to require

53. *See, e.g.*, Clark-Fitzpatrick, Inc./Franki Foundation Co., v. Gill, 652 A.2d 440 (R.I. 1994) (contractor was permitted to assert pass-through claim to owner for subcontractor's lost profits due to wrongful termination).

payment of "reasonable overhead and profit" on that portion of the work that was not performed before the termination. In addition, this section has been modified to clarify that the costs the contractor is entitled to recover upon a termination for convenience include the costs of terminating subcontracts.

Section 14.4.3 describes the compensation the contractor is entitled to recover from the owner in the event of a termination for convenience. Specifically, the contractor is entitled to recover three categories of damages: (1) payment for work the contractor executed; (2) other costs incurred by the contractor as a result of the termination, including costs attributable to terminating subcontracts; and (3) a termination fee, if the parties negotiated one in the agreement.

The first category of damages is dependent upon the compensation terms in the agreement. For example, if the contractor and owner have entered into a stipulated price contract, the contractor would likely be entitled to recover a percentage of the stipulated price, based upon the percentage of work the contractor has completed. If, on the other hand, the parties have entered into a cost-plus agreement, the contractor would be reimbursed for all costs incurred that meet the definition of the "Cost of the Work" in the agreement, plus a portion of the contractor's fee proportionate to (i) the completed work (for a fixed fee), or (ii) the costs incurred (if the fee is a percentage of the costs).

The second category of damages the contractor is entitled to recover under Section 14.4.3 is costs incurred by reason of the termination for convenience. Such costs might include, for example, material restocking fees and demobilization costs. The 2017 version of Section 14.4.3 expressly added to this category of damages costs incurred by the contractor in terminating subcontracts. Thus, if the contractor is required to pay a termination fee to any subcontractor in the event of a termination for convenience, the contractor can recover that termination fee from the owner.

The final category of damages the contractor is entitled to recover in Section 14.4.3 is a termination fee that the parties are invited to negotiate and set forth in their agreement. In the 1997 and 2007 versions of the A201, rather than a termination fee, the contractor was entitled to recover "reasonable overheard and profit" on the unperformed portions of the work. This change was made in the 2017 version likely because many owners simply deleted, without much discussion or negotiation, the old language allowing the contractor to recover unearned profit. The addition of a termination fee invites the parties to discuss and negotiate the contractor's compensation for losses it would incur upon a termination for convenience. In contrast to Section 14.4.3, the contractor is entitled to recover unearned overhead and profit upon a termination for the owner's default under Section 14.1.3. This difference in remedies seems to suggest that the owner is more culpable under a termination for default than a termination for convenience, even though neither type of termination results from any failure by the contractor. The concept of a negotiated termination fee under Section 14.4.3 appears to be an effort to strike a balance between the owner's need for the flexibility to terminate the agreement for its convenience and the contractor's right to be compensated for at least a portion of the financial benefit it bargained for.

CHAPTER 15:
Article 15 Claims and Disputes

Neal M. Eiseman
Richard F. Greenleaf
Aileen M. Leipprandt

ARTICLE 15 CLAIMS AND DISPUTES

Construction claims and disputes are inevitable, so it is imperative that the owner and contractor understand how and when to make a claim and the process for claim resolution. Article 15 addresses those requirements.

Other sections in the General Conditions identify opportunities for a claim and direct claimants to Article 15 for purposes of proceeding with such claim. Section 3.2.4, for example, provides that if the contractor believes that it is entitled to additional cost or time because of the architect's response to the contractor's notices or RFIs, then the contractor *must* make a claim as provided in Article 15. Similarly, Section 3.7.4 provides that if either party disputes the architect's determination regarding equitable adjustments to the Contract Time or Contract Sum due to concealed or unknown conditions, the parties *may* proceed under Article 15 claim provisions. Section 6.1.3 provides that if the contractor determines that a schedule change as a result of work by the owner's own forces requires a change in the Contract Sum or Contract Time, the contractor *may* make a claim in accordance with Article 15. And Section 7.3.5 and Section 7.3.9 respectively state that if the contractor disagrees with the architect's determination of adjustments in the Contract Time or Contract Sum resulting from a change order, the contractor *may* make a claim in accordance with Article 15.

Article 15 is frequently modified by parties to specify the precise period of limitations (Section 15.1.2), shorten the 21-day notice requirement for claims (Section 15.1.3.1), eliminate the contractor's right to additional costs for a delay (Section 15.1.6.1), impose stringent requirements on claims related to adverse weather (Section 15.1.6.2), modify the mutual waiver of consequential damages (Section 15.1.7), and detail a particular alternative dispute resolution process desired by the parties (Sections 15.3 and 15.4).

§ 15.1 CLAIMS

§ 15.1.1 DEFINITION

A Claim is a demand or assertion by one of the parties seeking, as a matter of right, payment of money, a change in the Contract Time, or other relief with respect to the terms of the Contract. The term "Claim" also includes other disputes and matters in question between the Owner and Contractor arising out of or relating to the Contract. The responsibility to substantiate Claims shall rest with the party making the Claim. This Section 15.1.1 does not require the Owner to file a Claim in order to impose liquidated damages in accordance with the Contract Documents.

The definition of a "Claim" is very broad. This is particularly important when assessing whether a dispute is subject to arbitration.[1]

In 2017, AIA drafters added the last sentence to Section 15.1.1, clarifying that an agreement as to liquidated damages is self-executing and does not require the owner to submit a claim to preserve the owner's rights. Recent cases related to the owner's obligation to timely assert a claim for liquidated damages or face waiver of such claim may have prompted the revision. In *Hedenberg & Co., Inc. v. St. Luke's*

1. See discussion at Section 15.4.1.

Hospital of Duluth,[2] for example, the court determined that the owner's claim for liquidated damages was a "Claim" under the A201 for which the owner was required to give timely notice. Because the owner did not provide timely notice, the court barred the owner's claim.

The court in *RCR Building Corporation v. Pinnacle Hospitality Partners*[3] reached a similar result, albeit on an extreme fact pattern. *RCR Building* involved a payment dispute on a hotel construction project. The court ruled that the owner's claim for liquidated damages was a "Claim" as defined in the A201–1997, requiring the owner to assert the claim within 21 days after the event giving rise to the claim. Because the owner did not initiate the claim or even provide notice of its intent to seek liquidated damages until well after substantial completion, the court ruled the claim was barred. The facts in *RCR* appear to limit its precedential value. The owner there admitted to causing a substantial portion of the delays. The owner admitted to making scope changes after the original completion date. And the owner failed to provide notice of its intent to assert liquidated damages until after the parties had negotiated and agreed upon a final payment amount.

Often, the owner and contractor resolve disputes related to the imposition of liquidated damages during the project closeout, which may occur months after a missed milestone or months after a missed substantial completion date triggering a liquidated damages claim. The 2017 revision to this section clarifies that the owner is not obligated to assert its liquidated damages claim in order to preserve it.

Section 15.1.1 broadly defines a "claim" to include an assertion seeking payment. Consistent with that broad definition, a party has the right to submit a claim where that party disagrees with the architect's decision on adjustments to the Contract Sum for change directives.[4] Likewise a party may submit a claim when the party disputes the architect's decision regarding a certificate for payment.[5] In both of these scenarios, a party must initiate the claim within 21 days after the occurrence giving rise to the claim.[6] When neither the owner nor the contractor disputes the architect's decision on a certificate for payment, but the owner nonetheless does not timely pay the contractor, the General Conditions are not clear on whether the contractor must nonetheless initiate a formal claim within 21 days. One could argue that the contractor's "Application and Certificate for Payment," typically submitted on the AIA G702 form, is a written request for payment and therefore suffices as a "claim." As such, the claim process—whether formal or not—has begun, obviating the need to file a formal claim within 21 days. This interpretation is sensible and reasonable, particularly in light of the 2017 revision to Section 15.1.1 clarifying that an owner need not file a "claim" in order to assess liquidated damages. Regardless, if the owner fails to pay an approved certificate for payment, the only method under the A201 for the contractor to resolve that non-payment is to submit a claim to the Initial Decision Maker (IDM), as the IDM's decision may be a condition precedent to mediation,[7] and mediation is a condition precedent to binding dispute resolution.[8]

2. No. C7-95-1683, 1996 WL 146732, at *2–3 (Minn. Ct. App. Apr. 2, 1996).
3. No. M2012-00286-COA-R3-CV, 2012 WL 5830587 (Tenn. Ct. App. Nov. 15, 2012).
4. See Section 7.3.9.
5. See Section 9.5.2.
6. See Section 15.1.3.1.
7. See Section 15.2.1.
8. See Section 15.3.1.

§ 15.1.2 TIME LIMITS ON CLAIMS

The Owner and Contractor shall commence all Claims and causes of action against the other and arising out of or related to the Contract, whether in contract, tort, breach of warranty or otherwise, in accordance with the requirements of the binding dispute resolution method selected in the Agreement and within the period specified by applicable law, but in any case not more than 10 years after the date of Substantial Completion of the Work. The Owner and Contractor waive all Claims and causes of action not commenced in accordance with this Section 15.1.2.

Generally speaking, the statute of limitations begins to run when a claim "accrues." The rules as to when a claim accrues vary from state to state and according to the type of claim. Many states have "discovery" rules that permit a party to file a claim within a certain period of time after the party discovers a defect. The practical impact of "discovery" accrual is that it can significantly extend liability years after the substantial completion date.

Section 15.1.2 (formerly Section 13.7) specifies an outside date (ten years from the date of substantial completion) after which the ability to bring a claim expires. In essence, Section 15.1.2 acts as a statute of repose. After ten years from the date of substantial completion, any cause of action is extinguished and cannot be revived.

Parties should carefully consider their particular state's law for a determination of the accrual and expiration of the statute of limitations and statute of repose. Parties should then determine if Section 15.1.2 expands or contracts a claimant's legal rights to assert a claim and whether it is advantageous to their interest to modify the provision.

§ 15.1.3 NOTICE OF CLAIMS

§ 15.1.3.1 Claims by either the Owner or Contractor ~~must~~, where the condition giving rise to the Claim is first discovered prior to expiration of the period for correction of the Work set forth in Section 12.2.2, shall be initiated by ~~written~~ notice to the other party and to the Initial Decision Maker with a copy sent to the Architect, if the Architect is not serving as the Initial Decision Maker. Claims by either party ~~must~~under this Section 15.1.3.1 shall be initiated within 21 days after occurrence of the event giving rise to such Claim or within 21 days after the claimant first recognizes the condition giving rise to the Claim, whichever is later.

The time for initiating a notice of a claim by either the contractor or the owner has been revised to distinguish between claims first discovered before the warranty period set forth in Section 12.2.2 expires and claims discovered after the warranty period. Only claims in this earlier period are required to be submitted to the IDM. Consistent with earlier versions of this notice section, notices of claims are required to be initiated within 21 days after occurrence of the event giving rise to the claim or within 21 days after the claimant first recognized the condition giving rise to the claim, whichever is later.[9]

During the earlier period recognized by this section, the IDM presumably is still actively engaged in the project and should be kept informed of issues as they

9. Under Section 3.7.4, the 21-day notice provision does not apply to requests for time and money arising from concealed or unknown conditions.

develop and given the opportunity to make decisions that will keep the project moving forward.

The enforceability of the 21-day time limit for claims has been the subject of much debate. Those against whom a claim is being asserted typically argue that failure to serve notice of the claim within 21 days constitutes a waiver of the claim. Those making the claim after the 21-day period has expired, on the other hand, will assert that failure to comply with the 21-day notice provision does not constitute waiver and that the 21-day period is merely a timing mechanism that precedes arbitration or litigation. Those taking the "non-waiver" position have argued that the previous Section 13.4.2—which states that "no action or failure to act by the owner, architect or contractor shall constitute a waiver of a right or duty afforded them under the contract, nor shall such action or failure to act constitute approval of or acquiescence in a breach there under"—supported the non-waiver argument and that the failure to serve the notice is not a waiver of the claim.[10] The language of 13.4.2 has been preserved in Section 13.3.2 of the A201–2017.

Claimants finding themselves outside the 21-day notice period have overcome the lack of timely notice defense by showing that (1) the party relying on the defense had actual or constructive notice of the underlying events giving rise to the claim; or (2) the lack of notice caused no material prejudice to the party who relies on the defense. Generally, with regard to federal contracts, courts have adopted the rule that strict compliance with a notice of claim provision is not required if the contracting officer had actual or imputed knowledge of the circumstances giving rise to the claim or if notice to the contracting officer would have been useless.[11] With regard to private contracts, however, jurisdictions are split on the issue of whether strict compliance with a notice of claim provision is a prerequisite to filing suit or recovery. Some courts follow the rule that strict compliance is required regardless of the counterpart's actual or constructive notice or lack of prejudice.[12] Other courts recognize these exceptions to the notice requirement and allow for a more liberal approach.[13] To avoid this debate, some drafters modify the standard contract language to state that the contractor must strictly comply with the notice and timing requirements for a claim or else the contractor waives the claim. Similarly,

---

10. In the case of *Weigand Construction Co., Inc. v. Stephens Fabrication, Inc.*, 929 N.E.2d 220 (Ind. Ct. App. 2010), this contractual provision prevented the court from finding that Weigand waived its right to enforce the 21-day notice of claim provision. *See also* Beacon4, LLC v. I & L Investments, LLC, No. E2015-01298-COA-R3-CV, 2016 Tenn. App. LEXIS 637 (Aug. 30, 2016).

11. *See* Hoel-Steffen Const. Co. v. United States, 456 F.2d 760 (Ct. Cl. 1972); R. R. Tyler, AGBCA No. 381, 1977 WL 27125 (Apr. 5, 1977); Mil-Pak Company, Inc., ASBCA No. 19733 (Jan. 26, 1976).

12. *See* Interstate Contracting Corp. v. City of Dallas, Tex., 407 F.3d 708, 727–28 (5th Cir. 2005) (rejecting contractor's claims due to claimant's failure to "strictly comply" with contractual claims and notice requirements); Cameo Homes v. Kraus-Anderson Const. Co., 394 F.3d 1084, 1088 (8th Cir. 2005) (upholding dismissal of contractor claims for which no 21-day written notices were given as required by contract); D.W.H. Painting Co., Inc. v. D.W. Ward Const. Co., Inc., 620 S.E.2d 887, 891–92 (N.C. Ct. App. 2005) (upholding denial of contractor's claims due to its failure to provide other prime contractors timely notice of those claims as required by contract); Village of Greenport v. Manning Plumbing & Heating Corp., No. 26981-09, 2012 WL 1834263 (NY. Sup. Ct. May 9, 2012) (rejecting an owner's claim because it failed to comply with the contractual requirement of timely notice of its claim, which was a condition precedent to suit or recovery).

13. *See* Brinderson Corp. v. Hampton Roads Sanitation Dist., 825 F.2d 41, 45 (4th Cir. 1987)) (holding that lack of timely notice did not bar the contractor's claim because the owner had actual or constructive notice of the conditions underlying the claim and an opportunity to investigate); *In re* Redondo Const. Corp., 678 F.3d 115, 123 (1st Cir. 2012) (holding that strict conformity with a contract's written notice provision is not required as long as the counterparty receives substantially the same information through timely actual notice and suffers no prejudice from the nonconformity); City of Meridian v. Petra Inc., 299 P.3d 232, 244–45 (Idaho 2013) (holding that lack of timely notice did not bar the contractor's claim because the owner had actual notice of the underlying events giving rise to the claim); H.E. Contracting v. Franklin Pierce College, 360 F. Supp. 2d 289, 294–95 (D.N.H. 2005) (holding that lack of timely notice did not bar the contractor's claim because the owner had actual notice of the underlying events giving rise to the claim and was not prejudiced by the contractor's failure to comply with the writing requirement"); Neal & Co., Inc. v. City of Dillingham, 923 P.2d 89, 92–93 (Alaska 1996) (holding that lack of timely notice did not bar the contractor's claim because the owner had actual or constructive notice of the conditions underlying the claim); New Ulm Bldg. Center, Inc. v. Studtmann, 225 N.W.2d 4, 5 (Minn. 1974) (holding that lack of timely notice did not bar the contractor's claim because the owner had actual knowledge of the conditions underlying the claim and therefore waived objection to written notice as required by contract).

some drafters modify the standard contract language to state that strict compliance with the notice and timing requirements is a condition precedent to a claim, and the condition precedent cannot be waived.

> **§ 15.1.3.2** Claims by either the Owner or Contractor, where the condition giving rise to the Claim is first discovered after expiration of the period for correction of the Work set forth in Section 12.2.2, shall be initiated by notice to the other party. In such event, no decision by the Initial Decision Maker is required.

This section is entirely new and drafted to address claims where the condition giving rise to the claim is first discovered after both project completion and expiration of the warranty period. The drafters appear to have recognized that claims arising well after the project has been completed likely would be brought after the IDM had been disengaged. Further, for late discovered claims, there would be no need to expedite a ruling from the IDM or any need for the contractor to work through the IDM process. Rather, this provision recognizes mainly latent defects that arise and are first recognized after the contractor has left the project. Although notice must be given, there is no mandatory time period set forth to govern when such notice must be given, except the statute of limitations provided in Section 15.1.2.

> **§ 15.1.4 CONTINUING CONTRACT PERFORMANCE**
>
> **§ 15.1.4.1** Pending final resolution of a Claim, except as otherwise agreed in writing or as provided in Section 9.7 and Article 14, the Contractor shall proceed diligently with performance of the Contract and the Owner shall continue to make payments in accordance with the Contract Documents. ~~The Architect will prepare Change Orders and issue Certificates for Payment in accordance with the decisions of the Initial Decision Maker.~~
>
> **§ 15.1.4.2** The Contract Sum and Contract Time shall be adjusted in accordance with the Initial Decision Maker's decision, subject to the right of either party to proceed in accordance with this Article 15. The Architect will issue Certificates for Payment in accordance with the decision of the Initial Decision Maker.

Language requiring the contractor to diligently perform its work, and the owner to make payments (previously the subject of a single section, Section 15.1.3) has been revised in 2017 with Sections 15.1.4.1 and 15.1.4.2. Where the 2007 version required the architect to prepare change orders and issue certificates of payment in accordance with the IDM's decision, Section 15.1.4.2 now removes the architect's requirement to prepare change orders in accordance with the IDM's decision and instead requires the architect only to issue certificates of payment in accordance with the decision of the IDM. The purpose of this change is that, by definition, a change order requires agreement by the contractor and owner. While an IDM's decision is binding, it may be difficult or impossible to get a party intent on furthering the claims process to sign off on the change order.

Like many construction contracts, the A201 requires continuing contract performance by both the contractor and the owner despite the existence of claims or the pendency of a dispute. Since arbitration panels and courts typically cannot be convened and react fast enough to handle disputes mid-project, the role of the IDM in making a relatively quick decision that allows the project to proceed is critical. If payments to the contractor were suspended entirely as a result of a claim by the owner, contractors would find it financially difficult to continue working on the project because they would be essentially funding the project. It likewise would be untenable for a contractor to bring a project to a standstill every time there was a disagreement or claim for additional compensation, regardless of whether the amount sought was large or small. An early and practical decision by the IDM may make the difference between the total shutdown of a project, including litigation or arbitration, and continuation of the project followed by an orderly claims process.

§ 15.1.5 CLAIMS FOR ADDITIONAL COST

If the Contractor wishes to make a Claim for an increase in the Contract Sum, ~~written~~ notice as provided ~~herein~~in Section 15.1.3 shall be given before proceeding to execute the portion of the Work that is the subject of the Claim. Prior notice is not required for Claims relating to an emergency endangering life or property arising under Section 10.4.

This revised section (formerly Section 15.1.4) contains only a minor change that appears to delete the requirement that claims be in writing. Section 1.6.2, however, generally requires all notices to be in writing.[14]

§ 15.1.~~5~~6 CLAIMS FOR ADDITIONAL TIME

§ 15.1.~~5~~6.1 If the Contractor wishes to make a Claim for an increase in the Contract Time, ~~written~~ notice as provided ~~herein~~in Section 15.1.3 shall be given. The Contractor's Claim shall include an estimate of cost and of probable effect of delay on progress of the Work. In the case of a continuing delay, only one Claim is necessary.

The change to this section in 2017 is cosmetic, yet helpful as it directs the reader to the interrelated Section 15.1.3, which describes the mechanics and timing of giving notice of a claim. Claim preparation is critical and can be time consuming, especially in complex projects with interrelated activities and impacts. Often the person best suited to assemble the claim is simultaneously involved with project management, making it challenging to devote the necessary resources for properly stating a claim. Regardless, parties operate at their peril if they do not give proper attention to claim documentation, as the language used in the claim and the relief sought may be meticulously scrutinized in a subsequent claim proceeding. An owner may modify Section 15.1.6.1 to require the claimant to provide detailed claim documentation, such as attested affidavits of claim, and that the claimant affirmatively waives the claim for failing to strictly abide by such requirement.

§ 15.1.~~5~~6.2 If adverse weather conditions are the basis for a Claim for additional time, such Claim shall be documented by data substantiating that weather conditions were abnormal for the period of time, could not have been reasonably anticipated, and had an adverse effect on the scheduled construction.

14. See Section 15.1.1 for a discussion about whether nonpayment by the owner on a properly submitted payment application requires the submission of a formal claim.

Generally speaking, weather conditions are not within a contractor's control. But a contractor is expected to anticipate standard weather issues. Accordingly, Section 15.1.6.2 provides that a contractor must establish its "adverse weather" claim with data showing that the condition was abnormal and unforeseeable and actually impacted the construction.

The AIA documents do not define "abnormal," and it would be difficult to do so since conditions vary among geographical locations, the particular nature of the project, and the duration of the project. Weather considerations for constructing a power plant in Las Vegas, for example, are dramatically different from constructing a hotel in Jackson Hole, Wyoming. In assessing the viability of any potential adverse weather claim, consideration should be given to the frequency, duration, and intensity of the "abnormal" weather conditions compared with the typical weather ordinarily encountered at the particular location at the particular time for performance.

Because "abnormal" is not defined, parties often will specify criteria that must be exceeded before conditions will be deemed abnormal (for example, a 25-year storm event; measurable precipitation in excess of three inches in a 24-hour period; temperature that does not exceed 32° Fahrenheit during a 24-hour period). The duration of measurement may be critical—per hour, per day, per month, or percent over a specified period of time. Typically, weather conditions are substantiated by the use of weather data from the National Oceanic and Atmosphere Administration (NOAA) for the weather station nearest to the project and for which adequate data is available.

An illustrative case is *Roger Johnson Construction Co. v. Bossier City*,[15] where a project owner withheld funds from the contractor (RJC), claiming liquidated damages because RJC did not timely complete construction. RJC claimed excusable delay due to severe weather, although the contract did not define severe weather. The court first looked to the definition of "severe weather" in the local office of the NOAA and determined that the parties intended to address a more project-specific condition that would make it unreasonable to permit construction. The court ultimately determined that RJC's lack of sufficient personnel—and not "severe weather"—was the cause of delay.

Likewise, in *McDevitt & Street Co. v. Marriott Corp.*,[16] Marriott withheld payment to the contractor because of the contractor's untimely completion of a hotel. The contractor offered expert testimony based on NOAA records that adverse weather conditions impeded its work and could not have been reasonably anticipated. The court ruled in favor of the owner, finding that the contractor had failed to demonstrate that the particular weather condition it encountered actually could not have been reasonably anticipated.[17]

An unforeseeable, abnormal weather event, by itself, may not entitle a contractor to a claim. The contractor still must establish that the weather event had an adverse impact on the scheduled construction. While this might appear easy to establish, it is deceptively more complicated. If the weather event occurs while nearly all the work is being performed indoors, the event may have no impact on the "scheduled construction." Similarly, an owner may require that the adverse weather days be

15. 330 So. 2d 338 (La. Ct. App. 1976).

16. 716 F. Supp. 906 (E.D. Va. 1989).

17. *See also* Ryll International, LLC v Department of Transportation, 2011 WL 2714000 (U.S. Civilian BCA Jun. 30, 2011) (based upon review of NOAA weather records, adverse weather conditions were not severe enough to warrant a finding of excusable delay); Handex of the Carolinas, Inc. v. County of Haywood, 607 S.E.2d 25, 34–35 (N.C. Ct. App. 2005) (the parties agreed that determination of "abnormal weather conditions" should be made based upon records from the NOAA; however, the parties disagreed as to how such data should be interpreted).

cumulative over the entire duration of the project, such that if an adverse weather day is not used within a given month, it carries forward to the succeeding months. Consequently, even if an abnormal weather event occurs, an owner can apply the cumulative weather days to offset the claim. Likewise, the standard of care requires that the contractor account for weather disrupting construction when developing the project schedule. Often a contractor is required to illustrate anticipated adverse weather days on the critical path of the project schedule. The parties may dispute whether the abnormal weather event caused the project to exceed weather delays that the contractor included or should have included in the project schedule.

Adverse weather conditions can have a significant impact upon the schedule and project costs. Oftentimes, however, weather provisions in a construction agreement are not carefully scrutinized or even negotiated. In its simplicity, Section 15.1.6.2 affords the opportunity for a well-documented claim. The parties should evaluate whether it is in their best interest to clarify criteria for entitlement.

§ 15.1.~~6~~ 7 WAIVER OF CLAIMS FOR CONSEQUENTIAL DAMAGES

> The Contractor and Owner waive Claims against each other for consequential damages arising out of or relating to this Contract. This mutual waiver includes
>
> **.1** damages incurred by the Owner for rental expenses, for losses of use, income, profit, financing, business and reputation, and for loss of management or employee productivity or of the services of such persons; and
>
> **.2** damages incurred by the Contractor for principal office expenses including the compensation of personnel stationed there, for losses of financing, business and reputation, and for loss of profit, except anticipated profit arising directly from the Work.
>
> This mutual waiver is applicable, without limitation, to all consequential damages due to either party's termination in accordance with Article 14. Nothing contained in this Section 15.1.~~6~~7 shall be deemed to preclude ~~an award~~assessment of liquidated damages, when applicable, in accordance with the requirements of the Contract Documents.

Outside of clarifying that an assessment of liquidated damages is not within the ambit of the parties' consequential damage waiver unless the contract provides otherwise, the AIA has not changed this section. This risk-allocation provision tries to take away some of the guesswork regarding what constitutes "consequential damages" by expressly identifying specific types of claims that are deemed indirect because they do not arise from the direct act of a party, but as an indirect consequence or result of the act. By its terms, the waiver does not cover the direct costs incurred to complete work or to correct defective or nonconforming work, most change order disputes, or disputes over additional design-related fees.

The clause first appeared in the 1997 version of the A201 to relieve parties from having financial exposure in excess of the benefits of their contract. The case most cited as giving rise to the clause is *Perini Corp. v. Greate Bay Hotel and Casino, Inc.*,[18] where the arbitrators awarded (and the court upheld) Sands Casino's $14.5 million in lost profits due to a delayed opening of its casino in Atlantic City, New Jersey. Because Perini's construction management contract provided for only a $600,000 fee, Perini contended, among other things, that such an enormous

18. 610 A.2d 364 (N.J. 1992).

potential loss of profit had never been contemplated by the parties and, as such, was not recoverable. In a divided opinion, the majority of the Supreme Court of New Jersey disagreed, applying a *Hadley v. Baxendale* analysis, noting that "[c]ompensatory damages were designed to put the injured party in as good a position as he would have had if performance had been rendered as promised." Simply put, paragraph 15.1.6 is intended to avoid the type of result encountered in *Perini*.

There is very little reported case law interpreting the consequential damages clause or its 2007 predecessor, paragraph 15.1.6.[19] In any event, it is well-settled that such clauses seeking to limit the parties' risk are not unconscionable or against public policy and are routinely enforced by courts.[20]

In practice, the clause benefits contractors over owners because owners are generally more likely than contractors to suffer a significant consequential loss on account of a breach by the other party. Consequential losses sustained by a contractor include intangibles such as loss of bonding capacity and loss of profits on unrelated jobs. An owner, however, is relinquishing the right to recover more tangible losses such as the loss of rent from a tenant due to the delay in the completion of the project. Accordingly, it is more likely that an owner will seek to strike the clause while a contractor will retain it.

§ 15.2 INITIAL DECISION

§ 15.2.1 Claims, excluding those where the condition giving rise to the Claim is first discovered after expiration of the period for correction of the Work set forth in Section 12.2.2 or arising under Sections 10.3, 10.4, ~~11.3.9,~~ and 11.~~3.10~~5, shall be referred to the Initial Decision Maker for initial decision. The Architect will serve as the Initial Decision Maker, unless otherwise indicated in the Agreement. Except for those Claims excluded by this Section 15.2.1, an initial decision shall be required as a condition precedent to mediation of any Claim ~~arising prior to the date final payment is due, unless~~. If an initial decision has not been rendered within 30 days ~~have passed~~ after the Claim has been referred to the Initial Decision Maker ~~with no~~, the party asserting the Claim may demand mediation and binding dispute resolution without a decision having been rendered. Unless the Initial Decision Maker and all affected parties agree, the Initial Decision Maker will not decide disputes between the Contractor and persons or entities other than the Owner.

§ 15.2.2 The Initial Decision Maker will review Claims and within ten days of the receipt of a Claim take one or more of the following actions: (1) request additional supporting data from the claimant or a response with supporting data from the other party, (2) reject the Claim in whole or in part, (3) approve the Claim, (4) suggest a compromise, or

19. *See, e.g.*, Construction Enterprises, Inc. v. Waterstone at Panama City Apartments, Inc., No. 3:10-00711, 2011 WL 4431824 (M.D. Tenn. Sept. 22, 2011) (fees and remediation costs identified in an owner-contractor construction contact, and in light of the language of the parties' separate fee agreement, were deemed to be direct out-of-pocket losses outside the scope of A201 waiver of claims for consequential damages); Brownsburg Municipal Building Corporation v. R.L. Turner Corporation, 933 N.E.2d 905 (Ind. Ct. App. 2010) (whether a contractor's delay damage claim, and each element of the claim, were waived by a AIA contract terms depended upon specific material facts to be determined at trial); Jay Jala, LLC v. DDG Construction, Inc., No. 15-3948, 2016 U.S. Dist. LEXIS 150969 (E.D. Pa. Nov. 1, 2016) (where court ruled that owner's pre-purchased motel insurance premiums, advertising costs and equipment leasing expenses were consequential damages that owner waived, but project completion fee paid to replacement contractor, utility costs incurred due to initial contractor's delay and additional months of construction loan interest were deemed direct damages and recoverable).

20. 6 Philip L. Bruner, Patrick J. O'Connor, Brunner & O'Connor on Construction Law § 19:55, at 275–76 n.3 (2016 Update).

(5) advise the parties that the Initial Decision Maker is unable to resolve the Claim if the Initial Decision Maker lacks sufficient information to evaluate the merits of the Claim or if the Initial Decision Maker concludes that, in the Initial Decision Maker's sole discretion, it would be inappropriate for the Initial Decision Maker to resolve the Claim.

§ 15.2.3 In evaluating Claims, the Initial Decision Maker may, but shall not be obligated to, consult with or seek information from either party or from persons with special knowledge or expertise who may assist the Initial Decision Maker in rendering a decision. The Initial Decision Maker may request the Owner to authorize retention of such persons at the Owner's expense.

§ 15.2.4 If the Initial Decision Maker requests a party to provide a response to a Claim or to furnish additional supporting data, such party shall respond, within ten days after receipt of ~~such~~the request, and shall either (1) provide a response on the requested supporting data, (2) advise the Initial Decision Maker when the response or supporting data will be furnished, or (3) advise the Initial Decision Maker that no supporting data will be furnished. Upon receipt of the response or supporting data, if any, the Initial Decision Maker will either reject or approve the Claim in whole or in part.

§ 15.2.5 The Initial Decision Maker will render an initial decision approving or rejecting the Claim, or indicating that the Initial Decision Maker is unable to resolve the Claim. This initial decision shall (1) be in writing; (2) state the reasons therefor; and (3) notify the parties and the Architect, if the Architect is not serving as the Initial Decision Maker, of any change in the Contract Sum or Contract Time or both. The initial decision shall be final and binding on the parties but subject to mediation and, if the parties fail to resolve their dispute through mediation, to binding dispute resolution.

This section distinguishes between early claims first discovered during the project, which require an IDM decision, and late claims discovered after the correction period, which do not. Consistent with the A201–2007, if the IDM fails to render a decision within 30 days after the claim has been referred to him or her, then the party asserting the claim may demand mediation and binding dispute resolution. This section specifies that only the claimant may move the dispute process forward if the IDM does not render a timely decision, but is not required to do so. Although it would make sense for the claimant to be the one to pursue further steps in the dispute process to achieve resolution, if no decision is rendered by the IDM, there is no specific timing requirement by when the claimant must demand mediation as the next step in the dispute process. Practically speaking, and with the option of the IDM to opt out of rendering a decision, it would seem that an IDM who fails to render a timely decision may have just gotten too busy. A claimant who wants a decision on a claim may be well served to remind the IDM of the duties and deadlines imposed by Section 15.2.

§ 15.2.6 Either party may file for mediation of an initial decision at any time, subject to the terms of Section 15.2.6.1.

§ 15.2.6.1 Either party may, within 30 days from the date of receipt of an initial decision, demand in writing that the other party file for mediation ~~within 60 days of the initial decision~~.

If such a demand is made and the party receiving the demand fails to file for mediation within ~~the time required~~30 days after receipt thereof, then both parties waive their rights to mediate or pursue binding dispute resolution proceedings with respect to the initial decision.

§ 15.2.7 In the event of a Claim against the Contractor, the Owner may, but is not obligated to, notify the surety, if any, of the nature and amount of the Claim. If the Claim relates to a possibility of a Contractor's default, the Owner may, but is not obligated to, notify the surety and request the surety's assistance in resolving the controversy.

§ 15.2.8 If a Claim relates to or is the subject of a mechanic's lien, the party asserting such Claim may proceed in accordance with applicable law to comply with the lien notice or filing deadlines.

The date affixed to the IDM's decision previously triggered time periods for filing a demand for mediation, which commenced the formal dispute process. The drafters recognized that although a decision of the IDM may be signed, without its issuance and receipt by the parties, there is no notice of such decision. This minor revision clarifies that the trigger for further proceedings is actual receipt of it.

§ 15.3 MEDIATION

§ 15.3.1 Claims, disputes, or other matters in controversy arising out of or related to the Contract, except those waived as provided for in Sections 9.10.4, 9.10.5, and 15.1.~~6~~7, shall be subject to mediation as a condition precedent to binding dispute resolution.

This language remains unchanged, requiring mediation as a prerequisite to litigation or arbitration except for disputes relating to the waiver of certain claims after the owner has made final payment and to the parties' mutual waiver of consequential damages.

§ 15.3.2 The parties shall endeavor to resolve their Claims by mediation which, unless the parties mutually agree otherwise, shall be administered by the American Arbitration Association in accordance with its Construction Industry Mediation Procedures in effect on the date of the Agreement. A request for mediation shall be made in writing, delivered to the other party to the Contract, and filed with the person or entity administering the mediation. The request may be made concurrently with the filing of binding dispute resolution proceedings but, in such event, mediation shall proceed in advance of binding dispute resolution proceedings, which shall be stayed pending mediation for a period of 60 days from the date of filing, unless stayed for a longer period by agreement of the parties or court order. If an arbitration is stayed pursuant to this Section 15.3.2, the parties may nonetheless proceed to the selection of the arbitrator(s) and agree upon a schedule for later proceedings.

This paragraph, also unchanged, reaffirms the parties' decision to proceed to mediation and provides that it should be conducted in accordance with the Construction Industry Mediation Procedures of the American Arbitration Association (AAA). It also sets forth the protocol for initiating mediation and states that any binding dispute resolution proceedings are stayed for 60 days except for arbitrator selection and the pre-hearing conference.

> **§ 15.3.3** Either party may, within 30 days from the date that mediation has been concluded without resolution of the dispute or 60 days after mediation has been demanded without resolution of the dispute, demand in writing that the other party file for binding dispute resolution. If such a demand is made and the party receiving the demand fails to file for binding dispute resolution within 60 days after receipt thereof, then both parties waive their rights to binding dispute resolution proceedings with respect to the initial decision.

This provision confirms that if the mediation ends in an impasse or 60 days have passed following a mediation demand with no resolution of the dispute, either party may proceed to file a lawsuit or proceed in arbitration. It contains the same language in 15.2.6.1 that notes when an "initial decision is not necessary."

> ~~§ 15.3.3~~**§ 15.3.4** The parties shall share the mediator's fee and any filing fees equally. The mediation shall be held in the place where the Project is located, unless another location is mutually agreed upon. Agreements reached in mediation shall be enforceable as settlement agreements in any court having jurisdiction thereof.

This provision mirrors the protocols already reflected in the AAA's Construction Industry Mediation Procedures and confirms that when the parties enter into a written settlement agreement at the end of a successful mediation, the agreement may be enforced in court.

§ 15.4 ARBITRATION

> **§ 15.4.1** If the parties have selected arbitration as the method for binding dispute resolution in the Agreement, any Claim subject to, but not resolved by, mediation shall be subject to arbitration which, unless the parties mutually agree otherwise, shall be administered by the American Arbitration Association in accordance with its Construction Industry Arbitration Rules in effect on the date of the Agreement. The Arbitration shall be conducted in the place where the Project is located, unless another location is mutually agreed upon. A demand for arbitration shall be made in writing, delivered to the other party to the Contract, and filed with the person or entity administering the arbitration. The party filing a notice of demand for arbitration must assert in the demand all Claims then known to that party on which arbitration is permitted to be demanded.

Section 15.4.1 compels the parties to submit to an arbitration process administered by the AAA in accordance with its rules. Nevertheless, after a dispute has arisen, parties often agree to conduct a private, non-AAA-administered arbitration. One reason for doing so is to avoid the AAA filing fees and administrative costs.

The AAA rules provide a detailed framework for conducting an arbitration. Parties should understand how those rules impact the litigation process and their substantive claims. For instance, the AAA rules provide that the arbitrator has the power to rule on his or her own jurisdiction, including objections as to the existence, scope, or validity of the arbitration agreement. Some courts have concluded that the contract's express reference to the AAA rules clearly evidences the parties' intent to delegate issues of arbitrability to the arbitrator.[21] The parties should carefully consider whether to modify the A201 to expressly allow for the recovery of attorney fees and confirm that the arbitration rules take into account any substantive contract rights to attorney fees and litigation costs.[22]

A party can be compelled to arbitrate where the agreement specifies arbitration, even though that party did not receive a copy of the A201.[23] Given the broad definition of "Claim" in Section 15.1.1, it is unlikely a party will succeed in avoiding arbitration based on an argument that the dispute is not subject to arbitration.[24]

In 2017, the AIA drafters added the requirement that the arbitration be conducted where the project is located. Such requirement is consistent with a majority of states across the county that make void and unenforceable construction contract clauses that require litigation or arbitration outside the state where the project is located.[25]

§ 15.4.1.1 A demand for arbitration shall be made no earlier than concurrently with the filing of a request for mediation, but in no event shall it be made after the date when the institution of legal or equitable proceedings based on the Claim would be barred by the applicable statute of limitations. For statute of limitations purposes, receipt of a written demand for arbitration by the person or entity administering the arbitration shall constitute the institution of legal or equitable proceedings based on the Claim.

The arbitration provision is intended to work independently from the mediation provision set forth in Section 15.3. However, a demand for mediation and arbitration may be filed simultaneously, particularly where a party seeks to avoid a statute of limitations issue. Concurrent filing also is sometimes done in anticipation of an unsuccessful mediation process and to "move along" the dispute resolution process.[26]

---

21. *See, e.g.*, Haddock v. Quinn, 287 S.W.3d 158, 172 (Tex. App. 2009); Summit Constr. Co., Inc. v. L.L.F.J.A.O., LLC, No. 25621, 2012 WL 473953 (Ohio Ct. App. Feb. 15, 2012) (question of whether settlement agreement with release of claims barred arbitration was arbitrable based upon the unlimited clause requiring arbitration); CMH Mfg. v. Hensel Phelps Constr. Co., Civil Action No. SA-12-CV-1223-XR, 2014 U.S. Dist. LEXIS 28484 (W.D. Tex. Mar. 5, 2014) (contract's adoption of the American Arbitration Association Rules is clear evidence that the parties agreed to arbitrate arbitrability). *But see In re* Brenda DeLuca Trust (Elhannon, LLC), 969 N.Y.S.2d 568 (App. Div. 2013) (whether a condition precedent to arbitration [in this case, submission of claim to the architect] has been satisfied is a determination to be made by the courts).

22. See *Lasco Inc. v. Inman Constr. Corp.*, 467 S.W.3d 467 (Tenn. Ct. App. 2015), for a discussion regarding the interplay between the parties' substantive contract right to demand attorney fees and the AAA rules.

23. Fegley Real Estate, LLC v. Boyle Constr., Inc., No. 2009-C-4007, 2011 Pa. Dist. & Cnty. Dec. LEXIS 366 (Pa. C.P. Oct. 20, 2011).

24. Patriot Constr. & Equip., LLC v. Quad States Constr., LLC, No. 2:13-CV-01529, 2013 U.S. Dist. LEXIS 186747 (W.D. La. Nov. 21, 2013) (when an agreement contains a broad arbitration clause, the clause reaches all aspects of the transaction); Shah v. Monpat Constr., Inc., 884 N.Y.S.2d 116 (App. Div. 2009) (where contract contains a broad arbitration clause, the arbitrator shall resolve any issue as to changes in the substantive provisions of a contract or its termination, unless the modification or termination relates to the arbitration clause itself; the issue of whether the written contract was abandoned and issues related to proper service of the demand for arbitration should be determined by the arbitrator); Luzerne Cty. v. D.A. Nolt, Inc., No. 3:14-cv-00831, 2014 U.S. Dist. LEXIS 123994 (M.D. Pa. Sep. 5, 2014) (arbitrator, not the court, should decide whether the underlying dispute is arbitrable; precise issue was whether contractor waived its claim by accepting final payment); G.T. Leach Builders, LLC v. Sapphire V.P., L.P., 458 S.W.3d 502 (Tex. 2015) (where enforceable arbitration agreement exists, arbitrator shall decide dispute over a contractual deadline in an arbitration agreement, a question of procedural arbitrability).

25. See, for example, ARIZ. REV. STAT. ANN. § 32-1129.05 (West 1956 & Supp. 2014); CONN. GEN. STAT. ANN. § 42-158m (West 1958); IDAHO CODE § 29-110 (LexisNexis 2013); KAN. STAT. ANN. § 16-121(E) (2007); MINN. STAT. ANN. § 337.10 (West 2002); MONT. CODE ANN. § 28-2-2116 (2015); NEV. REV. STAT. ANN. § 108.2453(2) (LexisNexis 2013); N.M. STAT. ANN. § 57-28A-1 (West 1978); N.C. GEN. STAT. § 22B-2 (LexisNexis 1944–1999); OHIO REV. CODE ANN. § 4113.62(D) (LexisNexis 2007); OR. REV. STAT. ANN. § 701.640 (West 2008); PA. STAT. ANN. § 514 (West 2013); TENN. CODE ANN. § 66-11-208 (LexisNexis 2015); and WIS. STAT. ANN. § 779.135(2) (West 2001). See also discussion at Section 13.1.

26. *See* LaSalle Grp., Inc. v. Veterans Enter. Tech. Servs., LLC, No. 11-03517-CV-S-DGK, 2012 U.S. Dist. LEXIS 46151 (W.D. Mo. Apr. 2, 2012) (the question of whether a demand for arbitration is premature where arbitration demand occurred before mediation was for the arbitrator to decide; further, the arbitration provision broadly conveyed an intent to include all disputes between the parties, such that a related Liquidating Agreement was also subject to arbitration); C.E. Franklin, Inc. v. Ray Angelini, Inc., No. 07-2652, 2008 U.S. Dist. LEXIS 34968 (E.D. Pa. Apr. 28, 2008) (arbitrator shall determine whether the procedural prerequisites [in this case, mediation] for arbitration have been satisfied); Golden State Foods Corp. v. Columbia/Okura LLC, No. CV 13-8150 RSWL (VBKx), 2014 U.S. Dist. LEXIS 88304 (C.D. Cal. June 26, 2014) (AIA agreement clearly imposes mediation as a condition precedent to arbitration).

§ 15.4.2 The award rendered by the arbitrator or arbitrators shall be final, and judgment may be entered upon it in accordance with applicable law in any court having jurisdiction thereof.

The finality of arbitration is thought to be one of its principal advantages. Generally speaking, there are very narrow grounds for appealing arbitration awards and successful appeals are rare. Under the Federal Arbitration Act (FAA),[27] a party may petition a court to modify or vacate an arbitration award in four limited circumstances: (1) where the award was procured by corruption, fraud, or undue means; (2) where there was evident partiality or corruption in any arbitrator; (3) where the arbitrator engaged in misconduct as to the evidence or scheduling; (4) where the arbitrator exceeded his or her powers.[28] Parties should consult their particular state's arbitration code for specific provisions related to judicial review of arbitration awards.[29]

In 2013, the AAA created Optional Appellate Arbitration Rules. Those rules permit the parties to appeal to a AAA Appeal Tribunal, so long as the parties agree to appeal of an award by stipulation or in their contract. The only grounds for appeal are (1) an error of law that is material and prejudicial, and (2) determinations of fact that are clearly erroneous. The appellate tribunal must render a decision within 30 days of service of the last brief or oral argument. The appellate tribunal may adopt the underlying award, substitute its own award, or request additional information from the parties. The tribunal may not remand to the original arbitration.

§ 15.4.3 The foregoing agreement to arbitrate and other agreements to arbitrate with an additional person or entity duly consented to by parties to the Agreement, shall be specifically enforceable under applicable law in any court having jurisdiction thereof.

Section 15.4.3 reinforces the well-established tenet that agreements to arbitrate are enforceable. The FAA reflects a liberal policy favoring arbitration agreements and enforcing them according to their terms.[30] Where a contract containing an arbitration clause is not challenged and the contract clearly delegates questions of arbitrability to the arbitrator, then arbitrability is an issue for the arbitrator to decide.[31]

---

27. 9 U.S.C. §§ 9–11 (2012).

28. *See* Hall Street Assoc., LLC v. Mattel Inc., 552 U.S. 576, 582 (2008). Some courts have even ruled that manifest disregard for the law is not grounds to vacate an arbitration award. *See* Med. Shoppe Int'l, Inc. v. Turner Investments, Inc., 614 F.3d 485 (8th Cir. 2010); Metzler Contracting Co. LLC v. Stephens, 774 F. Supp. 2d 1073, 1075 (D. Haw. 2011), *aff'd*, 479 F. App'x 783 (9th Cir. 2012) (homeowners not entitled to vacate the arbitration award under the FAA where the arbitrator did not exceed his powers).

29. *See, e.g.*, Neighbors Const. Co. v. Woodland Park at Soldier Creek, LLC, 284 P.3d 1057 (Kan. Ct. App. 2012) (argument that arbitrator exceeded his power lacked merit; a trial court must presume an arbitration award is valid unless there is proof of one of the specific grounds set forth in the Kansas Uniform Arbitration Act, K.S.A. 5-401 *et seq.*).

30. *See, e.g.*, United States *ex rel.* TGK Enterprises, Inc. v. Clayco, Inc., 978 F. Supp. 2d 540 (E.D.N.C. 2013).

31. Eckert/Wordell Architects, Inc. v. FJM Props. of Willmar, LLC, 756 F3d. 1098 (8th Cir. 2014); Emmanuel African Methodist Episcopal Church v. Reynolds Constr. Co., 718 S.E.2d 201 (N.C. Ct. App. 2011) (court erred in denying contractor's motion for arbitration of contract dispute where contract plainly and unambiguously required disputes to be resolved by arbitration).

Where an agreement to arbitrate exists, parties should take care to timely invoke the right to arbitration. If a party delays or otherwise substantially invokes a judicial process, a party may be deemed to have waived arbitration.[32]

§ 15.4.4 CONSOLIDATION OR JOINDER

§ 15.4.4.1 ~~Either party, at its sole discretion,~~ Subject to the rules of the American Arbitration Association or other applicable arbitration rules, either party may consolidate an arbitration conducted under this Agreement with any other arbitration to which it is a party provided that (1) the arbitration agreement governing the other arbitration permits consolidation, (2) the arbitrations to be consolidated substantially involve common questions of law or fact, and (3) the arbitrations employ materially similar procedural rules and methods for selecting arbitrator(s).

§ 15.4.4.2 ~~Either party, at its sole discretion,~~ Subject to the rules of the American Arbitration Association or other applicable arbitration rules, either party may include by joinder persons or entities substantially involved in a common question of law or fact whose presence is required if complete relief is to be accorded in arbitration, provided that the party sought to be joined consents in writing to such joinder. Consent to arbitration involving an additional person or entity shall not constitute consent to arbitration of any claim, dispute or other matter in question not described in the written consent.

The instructions to the A201–2017 state that this section has been revised to clarify that consolidation and joinder decisions cannot be made at the sole decision of the party seeking consolidation and joinder but, rather, must be made in accordance with the applicable arbitration rules.

Recognizing the reality that most construction arbitrations are administered by arbitration providers, such as the AAA, which includes within its rules specific guidelines and protocols relating to the consolidation of claims and joinder of parties, the 2017 revisions now incorporate by reference those provider rules. The paragraph deletes giving "(e)ither party, at its sole discretion" the right to consolidate or join parties; it now requires the party making the request to follow the arbitration provider's rules. This makes sense because the typical arbitration provision provides that the party shall abide by the arbitration provider's rules.

The American Arbitration Association Construction Industry Rules, amended and effective as of July 1, 2016, provide, in pertinent part:

(a) If the parties are unable to agree to consolidate related arbitrations or to the joinder of parties to an ongoing arbitration, the AAA shall directly appoint a single arbitrator (hereinafter referred to as the R-7 arbitrator) for the limited purpose of deciding whether related arbitrations should be consolidated or parties joined. . . .

(b) To request consolidation of arbitrations, the requesting party must have filed a demand for arbitration, including the applicable arbitration

32. *See, e.g.*, Tuscan Builders, L.P. v. 1437 SH6 L.L.C., 438 S.W.3d 717 (Tex. App. 2014) (trial court properly denied builder's motion to compel arbitration because builder substantially invoked judicial process and delayed a year in invoking arbitration provision).

provision(s) from the parties' contract(s) and must provide a written request for consolidation which provides the supporting reasons for such request. . . .

(c) To request joinder of parties, the requesting party must file with the AAA a written request to join parties to an existing arbitration which provides the names and contact information for such parties, names and contact information for the parties' representatives, if known, and supporting reasons for such request. . . .

(d) Absent agreement of all parties, the R-7 arbitrator appointed under this Rule shall not be a Merits Arbitrator who is appointed to any pending case involved in the consolidation request at issue.

(e) If the R-7 arbitrator determines that separate arbitrations shall be consolidated or that the joinder of additional parties is permissible, that arbitrator may also establish a process for selecting Merits Arbitrators for any ongoing or newly constituted case and, unless agreed otherwise by the parties, the allocation of responsibility for arbitrator compensation among the parties, subject to reapportionment by the arbitrator(s) appointed to the newly constituted case in the final arbitration award.

(f) The AAA may take reasonable administrative action to accomplish the consolidation or joinder as directed by the arbitrator. Pending the determination on a consolidation or joinder request, the AAA shall have the authority to stay the arbitration or arbitrations impacted by the consolidation or joinder request.

(g) The AAA shall maintain a panel of construction attorneys who have experience with consolidation or joinder issues. All arbitrators appointed to hear requests under this Rule shall be appointed from that panel, unless the parties agree otherwise.

Recognizing the importance of having an independent arbitrator decide whether to grant consolidation or joinder, the AAA rules require that an "R-7 arbitrator" (who will not be deciding the merits of the parties' underlying dispute) rule upon the request. Other arbitrator providers, such as JAMS, have different rules that should be consulted, if applicable (see, for example, JAMS Engineering and Construction Arbitration Rules and Procedures, effective November 15, 2014 at Rule 6).

If parties are arbitrating independently and not pursuant to the rules of the AAA or another arbitration provider, then a court will have to decide whether two or more actions should be consolidated or whether additional parties should be joined. In that case, the most important consideration will be whether there are common issues of law and fact involving the two disputes or parties.

For example, Rule 42 of the Federal Rules of Civil Procedure provides:

(a) CONSOLIDATION. If actions before the court involve a common question of law or fact, the court may:
   (1) join for hearing or trial any or all matters at issue in the actions;
   (2) consolidate the actions; or
   (3) issue any other orders to avoid unnecessary cost or delay.

Most state court rules contain similar provisions. In New York, CPLR §602 provides:

602. CONSOLIDATION (a) Generally. When actions involving common questions of law or fact are pending before a court, the court, upon motion, may order the actions consolidated, and may make such other orders

concerning proceedings therein as may tend to avoid unnecessary costs or delay.

§ 15.4.4.3 The Owner and Contractor grant to any person or entity made a party to an arbitration conducted under this Section 15.4, whether by joinder or consolidation, the same rights of joinder and consolidation as those of the Owner and Contractor under this Agreement.

The changes to this paragraph are cosmetic and affirm the right of the party whose claims have been consolidated or added, in turn, to bring additional parties into the proceeding with respect to its claims when they too involve common issues of law and fact. The goal here is to provide one forum to enable the interested parties to resolve all of their pending, related disputes.

APPENDIX A:
AIA® Document A201™ – 2017

AIA® Document A201™ – 2017

General Conditions of the Contract for Construction

for the following PROJECT:
(Name and location or address)

THE OWNER:
(Name, legal status and address)

THE ARCHITECT:
(Name, legal status and address)

This document has important legal consequences. Consultation with an attorney is encouraged with respect to its completion or modification.

For guidance in modifying this document to include supplementary conditions, see AIA Document A503™, Guide for Supplementary Conditions.

TABLE OF ARTICLES

Init.

/

INDEX
(Topics and numbers in bold are Section headings.)

Init.

/

ARTICLE 1 GENERAL PROVISIONS

§ 1.1 Basic Definitions

§ 1.1.1 The Contract Documents

The Contract Documents are enumerated in the Agreement between the Owner and Contractor (hereinafter the Agreement) and consist of the Agreement, Conditions of the Contract (General, Supplementary and other Conditions), Drawings, Specifications, Addenda issued prior to execution of the Contract, other documents listed in the Agreement, and Modifications issued after execution of the Contract. A Modification is (1) a written amendment to the Contract signed by both parties, (2) a Change Order, (3) a Construction Change Directive, or (4) a written order for a minor change in the Work issued by the Architect. Unless specifically enumerated in the Agreement, the Contract Documents do not include the advertisement or invitation to bid, Instructions to Bidders, sample forms, other information furnished by the Owner in anticipation of receiving bids or proposals, the Contractor's bid or proposal, or portions of Addenda relating to bidding or proposal requirements.

§ 1.1.2 The Contract

The Contract Documents form the Contract for Construction. The Contract represents the entire and integrated agreement between the parties hereto and supersedes prior negotiations, representations, or agreements, either written or oral. The Contract may be amended or modified only by a Modification. The Contract Documents shall not be construed to create a contractual relationship of any kind (1) between the Contractor and the Architect or the Architect's consultants, (2) between the Owner and a Subcontractor or a Sub-subcontractor, (3) between the Owner and the Architect or the Architect's consultants, or (4) between any persons or entities other than the Owner and the Contractor. The Architect shall, however, be entitled to performance and enforcement of obligations under the Contract intended to facilitate performance of the Architect's duties.

§ 1.1.3 The Work

The term "Work" means the construction and services required by the Contract Documents, whether completed or partially completed, and includes all other labor, materials, equipment, and services provided or to be provided by the Contractor to fulfill the Contractor's obligations. The Work may constitute the whole or a part of the Project.

§ 1.1.4 The Project

The Project is the total construction of which the Work performed under the Contract Documents may be the whole or a part and which may include construction by the Owner and by Separate Contractors.

§ 1.1.5 The Drawings

The Drawings are the graphic and pictorial portions of the Contract Documents showing the design, location and dimensions of the Work, generally including plans, elevations, sections, details, schedules, and diagrams.

§ 1.1.6 The Specifications

The Specifications are that portion of the Contract Documents consisting of the written requirements for materials, equipment, systems, standards and workmanship for the Work, and performance of related services.

§ 1.1.7 Instruments of Service

Instruments of Service are representations, in any medium of expression now known or later developed, of the tangible and intangible creative work performed by the Architect and the Architect's consultants under their respective professional services agreements. Instruments of Service may include, without limitation, studies, surveys, models, sketches, drawings, specifications, and other similar materials.

§ 1.1.8 Initial Decision Maker

The Initial Decision Maker is the person identified in the Agreement to render initial decisions on Claims in accordance with Section 15.2. The Initial Decision Maker shall not show partiality to the Owner or Contractor and shall not be liable for results of interpretations or decisions rendered in good faith.

§ 1.2 Correlation and Intent of the Contract Documents

§ 1.2.1 The intent of the Contract Documents is to include all items necessary for the proper execution and completion of the Work by the Contractor. The Contract Documents are complementary, and what is required by one shall be as binding as if required by all; performance by the Contractor shall be required only to the extent consistent with the Contract Documents and reasonably inferable from them as being necessary to produce the indicated results.

§ 1.2.1.1 The invalidity of any provision of the Contract Documents shall not invalidate the Contract or its remaining provisions. If it is determined that any provision of the Contract Documents violates any law, or is otherwise invalid or unenforceable, then that provision shall be revised to the extent necessary to make that provision legal and enforceable. In such case the Contract Documents shall be construed, to the fullest extent permitted by law, to give effect to the parties' intentions and purposes in executing the Contract.

§ 1.2.2 Organization of the Specifications into divisions, sections and articles, and arrangement of Drawings shall not control the Contractor in dividing the Work among Subcontractors or in establishing the extent of Work to be performed by any trade.

§ 1.2.3 Unless otherwise stated in the Contract Documents, words that have well-known technical or construction industry meanings are used in the Contract Documents in accordance with such recognized meanings.

§ 1.3 Capitalization
Terms capitalized in these General Conditions include those that are (1) specifically defined, (2) the titles of numbered articles, or (3) the titles of other documents published by the American Institute of Architects.

§ 1.4 Interpretation
In the interest of brevity the Contract Documents frequently omit modifying words such as "all" and "any" and articles such as "the" and "an," but the fact that a modifier or an article is absent from one statement and appears in another is not intended to affect the interpretation of either statement.

§ 1.5 Ownership and Use of Drawings, Specifications, and Other Instruments of Service
§ 1.5.1 The Architect and the Architect's consultants shall be deemed the authors and owners of their respective Instruments of Service, including the Drawings and Specifications, and retain all common law, statutory, and other reserved rights in their Instruments of Service, including copyrights. The Contractor, Subcontractors, Sub-subcontractors, and suppliers shall not own or claim a copyright in the Instruments of Service. Submittal or distribution to meet official regulatory requirements or for other purposes in connection with the Project is not to be construed as publication in derogation of the Architect's or Architect's consultants' reserved rights.

§ 1.5.2 The Contractor, Subcontractors, Sub-subcontractors, and suppliers are authorized to use and reproduce the Instruments of Service provided to them, subject to any protocols established pursuant to Sections 1.7 and 1.8, solely and exclusively for execution of the Work. All copies made under this authorization shall bear the copyright notice, if any, shown on the Instruments of Service. The Contractor, Subcontractors, Sub-subcontractors, and suppliers may not use the Instruments of Service on other projects or for additions to the Project outside the scope of the Work without the specific written consent of the Owner, Architect, and the Architect's consultants.

§ 1.6 Notice
§ 1.6.1 Except as otherwise provided in Section 1.6.2, where the Contract Documents require one party to notify or give notice to the other party, such notice shall be provided in writing to the designated representative of the party to whom the notice is addressed and shall be deemed to have been duly served if delivered in person, by mail, by courier, or by electronic transmission if a method for electronic transmission is set forth in the Agreement.

§ 1.6.2 Notice of Claims as provided in Section 15.1.3 shall be provided in writing and shall be deemed to have been duly served only if delivered to the designated representative of the party to whom the notice is addressed by certified or registered mail, or by courier providing proof of delivery.

§ 1.7 Digital Data Use and Transmission
The parties shall agree upon protocols governing the transmission and use of Instruments of Service or any other information or documentation in digital form. The parties will use AIA Document E203™–2013, Building Information Modeling and Digital Data Exhibit, to establish the protocols for the development, use, transmission, and exchange of digital data.

§ 1.8 Building Information Models Use and Reliance
Any use of, or reliance on, all or a portion of a building information model without agreement to protocols governing the use of, and reliance on, the information contained in the model and without having those protocols set forth in AIA Document E203™–2013, Building Information Modeling and Digital Data Exhibit, and the requisite AIA Document G202™–2013, Project Building Information Modeling Protocol Form, shall be at the using or relying party's sole risk and without liability to the other party and its contractors or consultants, the authors of, or contributors to, the building

information model, and each of their agents and employees.

ARTICLE 2 OWNER

§ 2.1 General

§ 2.1.1 The Owner is the person or entity identified as such in the Agreement and is referred to throughout the Contract Documents as if singular in number. The Owner shall designate in writing a representative who shall have express authority to bind the Owner with respect to all matters requiring the Owner's approval or authorization. Except as otherwise provided in Section 4.2.1, the Architect does not have such authority. The term "Owner" means the Owner or the Owner's authorized representative.

§ 2.1.2 The Owner shall furnish to the Contractor, within fifteen days after receipt of a written request, information necessary and relevant for the Contractor to evaluate, give notice of, or enforce mechanic's lien rights. Such information shall include a correct statement of the record legal title to the property on which the Project is located, usually referred to as the site, and the Owner's interest therein.

§ 2.2 Evidence of the Owner's Financial Arrangements

§ 2.2.1 Prior to commencement of the Work and upon written request by the Contractor, the Owner shall furnish to the Contractor reasonable evidence that the Owner has made financial arrangements to fulfill the Owner's obligations under the Contract. The Contractor shall have no obligation to commence the Work until the Owner provides such evidence. If commencement of the Work is delayed under this Section 2.2.1, the Contract Time shall be extended appropriately.

§ 2.2.2 Following commencement of the Work and upon written request by the Contractor, the Owner shall furnish to the Contractor reasonable evidence that the Owner has made financial arrangements to fulfill the Owner's obligations under the Contract only if (1) the Owner fails to make payments to the Contractor as the Contract Documents require; (2) the Contractor identifies in writing a reasonable concern regarding the Owner's ability to make payment when due; or (3) a change in the Work materially changes the Contract Sum. If the Owner fails to provide such evidence, as required, within fourteen days of the Contractor's request, the Contractor may immediately stop the Work and, in that event, shall notify the Owner that the Work has stopped. However, if the request is made because a change in the Work materially changes the Contract Sum under (3) above, the Contractor may immediately stop only that portion of the Work affected by the change until reasonable evidence is provided. If the Work is stopped under this Section 2.2.2, the Contract Time shall be extended appropriately and the Contract Sum shall be increased by the amount of the Contractor's reasonable costs of shutdown, delay and start-up, plus interest as provided in the Contract Documents.

§ 2.2.3 After the Owner furnishes evidence of financial arrangements under this Section 2.2, the Owner shall not materially vary such financial arrangements without prior notice to the Contractor.

§ 2.2.4 Where the Owner has designated information furnished under this Section 2.2 as "confidential," the Contractor shall keep the information confidential and shall not disclose it to any other person. However, the Contractor may disclose "confidential" information, after seven (7) days' notice to the Owner, where disclosure is required by law, including a subpoena or other form of compulsory legal process issued by a court or governmental entity, or by court or arbitrator(s) order. The Contractor may also disclose "confidential" information to its employees, consultants, sureties, Subcontractors and their employees, Sub-subcontractors, and others who need to know the content of such information solely and exclusively for the Project and who agree to maintain the confidentiality of such information.

§ 2.3 Information and Services Required of the Owner

§ 2.3.1 Except for permits and fees that are the responsibility of the Contractor under the Contract Documents, including those required under Section 3.7.1, the Owner shall secure and pay for necessary approvals, easements, assessments and charges required for construction, use or occupancy of permanent structures or for permanent changes in existing facilities.

§ 2.3.2 The Owner shall retain an architect lawfully licensed to practice architecture, or an entity lawfully practicing architecture, in the jurisdiction where the Project is located. That person or entity is identified as the Architect in the Agreement and is referred to throughout the Contract Documents as if singular in number.

§ 2.3.3 If the employment of the Architect terminates, the Owner shall employ a successor to whom the Contractor has no reasonable objection and whose status under the Contract Documents shall be that of the Architect.

§ 2.3.4 The Owner shall furnish surveys describing physical characteristics, legal limitations and utility locations for the site of the Project, and a legal description of the site. The Contractor shall be entitled to rely on the accuracy of information furnished by the Owner but shall exercise proper precautions relating to the safe performance of the Work.

§ 2.3.5 The Owner shall furnish information or services required of the Owner by the Contract Documents with reasonable promptness. The Owner shall also furnish any other information or services under the Owner's control and relevant to the Contractor's performance of the Work with reasonable promptness after receiving the Contractor's written request for such information or services.

§ 2.3.6 Unless otherwise provided in the Contract Documents, the Owner shall furnish to the Contractor one copy of the Contract Documents for purposes of making reproductions pursuant to Section 1.5.2.

§ 2.4 Owner's Right to Stop the Work
If the Contractor fails to correct Work that is not in accordance with the requirements of the Contract Documents as required by Section 12.2 or repeatedly fails to carry out Work in accordance with the Contract Documents, the Owner may issue a written order to the Contractor to stop the Work, or any portion thereof, until the cause for such order has been eliminated; however, the right of the Owner to stop the Work shall not give rise to a duty on the part of the Owner to exercise this right for the benefit of the Contractor or any other person or entity, except to the extent required by Section 6.1.3.

§ 2.5 Owner's Right to Carry Out the Work
If the Contractor defaults or neglects to carry out the Work in accordance with the Contract Documents and fails within a ten-day period after receipt of notice from the Owner to commence and continue correction of such default or neglect with diligence and promptness, the Owner may, without prejudice to other remedies the Owner may have, correct such default or neglect. Such action by the Owner and amounts charged to the Contractor are both subject to prior approval of the Architect and the Architect may, pursuant to Section 9.5.1, withhold or nullify a Certificate for Payment in whole or in part, to the extent reasonably necessary to reimburse the Owner for the reasonable cost of correcting such deficiencies, including Owner's expenses and compensation for the Architect's additional services made necessary by such default, neglect, or failure. If current and future payments are not sufficient to cover such amounts, the Contractor shall pay the difference to the Owner. If the Contractor disagrees with the actions of the Owner or the Architect, or the amounts claimed as costs to the Owner, the Contractor may file a Claim pursuant to Article 15.

ARTICLE 3 CONTRACTOR
§ 3.1 General
§ 3.1.1 The Contractor is the person or entity identified as such in the Agreement and is referred to throughout the Contract Documents as if singular in number. The Contractor shall be lawfully licensed, if required in the jurisdiction where the Project is located. The Contractor shall designate in writing a representative who shall have express authority to bind the Contractor with respect to all matters under this Contract. The term "Contractor" means the Contractor or the Contractor's authorized representative.

§ 3.1.2 The Contractor shall perform the Work in accordance with the Contract Documents.

§ 3.1.3 The Contractor shall not be relieved of its obligations to perform the Work in accordance with the Contract Documents either by activities or duties of the Architect in the Architect's administration of the Contract, or by tests, inspections or approvals required or performed by persons or entities other than the Contractor.

§ 3.2 Review of Contract Documents and Field Conditions by Contractor
§ 3.2.1 Execution of the Contract by the Contractor is a representation that the Contractor has visited the site, become generally familiar with local conditions under which the Work is to be performed, and correlated personal observations with requirements of the Contract Documents.

§ 3.2.2 Because the Contract Documents are complementary, the Contractor shall, before starting each portion of the Work, carefully study and compare the various Contract Documents relative to that portion of the Work, as well as the information furnished by the Owner pursuant to Section 2.3.4, shall take field measurements of any existing conditions related to that portion of the Work, and shall observe any conditions at the site affecting it. These obligations are for the purpose of facilitating coordination and construction by the Contractor and are not for the purpose of discovering errors, omissions, or inconsistencies in the Contract Documents; however, the Contractor shall promptly report to the Architect any errors, inconsistencies or omissions discovered by or made known to the Contractor as a request for information in such form as the Architect may require. It is recognized that the Contractor's review is made in the Contractor's capacity as a contractor and not as a licensed design professional, unless otherwise specifically provided in the Contract Documents.

§ 3.2.3 The Contractor is not required to ascertain that the Contract Documents are in accordance with applicable laws, statutes, ordinances, codes, rules and regulations, or lawful orders of public authorities, but the Contractor shall promptly report to the Architect any nonconformity discovered by or made known to the Contractor as a request for information in such form as the Architect may require.

§ 3.2.4 If the Contractor believes that additional cost or time is involved because of clarifications or instructions the Architect issues in response to the Contractor's notices or requests for information pursuant to Sections 3.2.2 or 3.2.3, the Contractor shall submit Claims as provided in Article 15. If the Contractor fails to perform the obligations of Sections 3.2.2 or 3.2.3, the Contractor shall pay such costs and damages to the Owner, subject to Section 15.1.7, as would have been avoided if the Contractor had performed such obligations. If the Contractor performs those obligations, the Contractor shall not be liable to the Owner or Architect for damages resulting from errors, inconsistencies or omissions in the Contract Documents, for differences between field measurements or conditions and the Contract Documents, or for nonconformities of the Contract Documents to applicable laws, statutes, ordinances, codes, rules and regulations, and lawful orders of public authorities.

§ 3.3 Supervision and Construction Procedures

§ 3.3.1 The Contractor shall supervise and direct the Work, using the Contractor's best skill and attention. The Contractor shall be solely responsible for, and have control over, construction means, methods, techniques, sequences, and procedures, and for coordinating all portions of the Work under the Contract. If the Contract Documents give specific instructions concerning construction means, methods, techniques, sequences, or procedures, the Contractor shall evaluate the jobsite safety thereof and shall be solely responsible for the jobsite safety of such means, methods, techniques, sequences, or procedures. If the Contractor determines that such means, methods, techniques, sequences or procedures may not be safe, the Contractor shall give timely notice to the Owner and Architect, and shall propose alternative means, methods, techniques, sequences, or procedures. The Architect shall evaluate the proposed alternative solely for conformance with the design intent for the completed construction. Unless the Architect objects to the Contractor's proposed alternative, the Contractor shall perform the Work using its alternative means, methods, techniques, sequences, or procedures.

§ 3.3.2 The Contractor shall be responsible to the Owner for acts and omissions of the Contractor's employees, Subcontractors and their agents and employees, and other persons or entities performing portions of the Work for, or on behalf of, the Contractor or any of its Subcontractors.

§ 3.3.3 The Contractor shall be responsible for inspection of portions of Work already performed to determine that such portions are in proper condition to receive subsequent Work.

§ 3.4 Labor and Materials

§ 3.4.1 Unless otherwise provided in the Contract Documents, the Contractor shall provide and pay for labor, materials, equipment, tools, construction equipment and machinery, water, heat, utilities, transportation, and other facilities and services necessary for proper execution and completion of the Work, whether temporary or permanent and whether or not incorporated or to be incorporated in the Work.

§ 3.4.2 Except in the case of minor changes in the Work approved by the Architect in accordance with Section 3.12.8 or ordered by the Architect in accordance with Section 7.4, the Contractor may make substitutions only with the consent of the Owner, after evaluation by the Architect and in accordance with a Change Order or Construction Change Directive.

§ 3.4.3 The Contractor shall enforce strict discipline and good order among the Contractor's employees and other persons carrying out the Work. The Contractor shall not permit employment of unfit persons or persons not properly skilled in tasks assigned to them.

§ 3.5 Warranty

§ 3.5.1 The Contractor warrants to the Owner and Architect that materials and equipment furnished under the Contract will be of good quality and new unless the Contract Documents require or permit otherwise. The Contractor further warrants that the Work will conform to the requirements of the Contract Documents and will be free from defects, except for those inherent in the quality of the Work the Contract Documents require or permit. Work, materials, or equipment not conforming to these requirements may be considered defective. The Contractor's warranty excludes remedy for damage or defect caused by abuse, alterations to the Work not executed by the Contractor, improper or insufficient maintenance, improper operation, or normal wear and tear and normal usage. If required by the Architect, the Contractor shall furnish satisfactory evidence as to the kind and quality of materials and equipment.

§ 3.5.2 All material, equipment, or other special warranties required by the Contract Documents shall be issued in the name of the Owner, or shall be transferable to the Owner, and shall commence in accordance with Section 9.8.4.

§ 3.6 Taxes
The Contractor shall pay sales, consumer, use and similar taxes for the Work provided by the Contractor that are legally enacted when bids are received or negotiations concluded, whether or not yet effective or merely scheduled to go into effect.

§ 3.7 Permits, Fees, Notices and Compliance with Laws
§ 3.7.1 Unless otherwise provided in the Contract Documents, the Contractor shall secure and pay for the building permit as well as for other permits, fees, licenses, and inspections by government agencies necessary for proper execution and completion of the Work that are customarily secured after execution of the Contract and legally required at the time bids are received or negotiations concluded.

§ 3.7.2 The Contractor shall comply with and give notices required by applicable laws, statutes, ordinances, codes, rules and regulations, and lawful orders of public authorities applicable to performance of the Work.

§ 3.7.3 If the Contractor performs Work knowing it to be contrary to applicable laws, statutes, ordinances, codes, rules and regulations, or lawful orders of public authorities, the Contractor shall assume appropriate responsibility for such Work and shall bear the costs attributable to correction.

§ 3.7.4 Concealed or Unknown Conditions
If the Contractor encounters conditions at the site that are (1) subsurface or otherwise concealed physical conditions that differ materially from those indicated in the Contract Documents or (2) unknown physical conditions of an unusual nature that differ materially from those ordinarily found to exist and generally recognized as inherent in construction activities of the character provided for in the Contract Documents, the Contractor shall promptly provide notice to the Owner and the Architect before conditions are disturbed and in no event later than 14 days after first observance of the conditions. The Architect will promptly investigate such conditions and, if the Architect determines that they differ materially and cause an increase or decrease in the Contractor's cost of, or time required for, performance of any part of the Work, will recommend that an equitable adjustment be made in the Contract Sum or Contract Time, or both. If the Architect determines that the conditions at the site are not materially different from those indicated in the Contract Documents and that no change in the terms of the Contract is justified, the Architect shall promptly notify the Owner and Contractor, stating the reasons. If either party disputes the Architect's determination or recommendation, that party may submit a Claim as provided in Article 15.

§ 3.7.5 If, in the course of the Work, the Contractor encounters human remains or recognizes the existence of burial markers, archaeological sites or wetlands not indicated in the Contract Documents, the Contractor shall immediately suspend any operations that would affect them and shall notify the Owner and Architect. Upon receipt of such notice, the Owner shall promptly take any action necessary to obtain governmental authorization required to resume the operations. The Contractor shall continue to suspend such operations until otherwise instructed by the Owner but shall continue with all other operations that do not affect those remains or features. Requests for adjustments in the Contract Sum and Contract Time arising from the existence of such remains or features may be made as provided in Article 15.

§ 3.8 Allowances
§ 3.8.1 The Contractor shall include in the Contract Sum all allowances stated in the Contract Documents. Items covered by allowances shall be supplied for such amounts and by such persons or entities as the Owner may direct, but the Contractor shall not be required to employ persons or entities to whom the Contractor has reasonable objection.

§ 3.8.2 Unless otherwise provided in the Contract Documents,

.1 allowances shall cover the cost to the Contractor of materials and equipment delivered at the site and all required taxes, less applicable trade discounts;

.2 Contractor's costs for unloading and handling at the site, labor, installation costs, overhead, profit, and other expenses contemplated for stated allowance amounts shall be included in the Contract Sum but not in the allowances; and

.3 whenever costs are more than or less than allowances, the Contract Sum shall be adjusted accordingly by Change Order. The amount of the Change Order shall reflect (1) the difference between actual costs and the allowances under Section 3.8.2.1 and (2) changes in Contractor's costs under Section 3.8.2.2.

§ 3.8.3 Materials and equipment under an allowance shall be selected by the Owner with reasonable promptness.

§ 3.9 Superintendent
§ 3.9.1 The Contractor shall employ a competent superintendent and necessary assistants who shall be in attendance at the Project site during performance of the Work. The superintendent shall represent the Contractor, and communications given to the superintendent shall be as binding as if given to the Contractor.

§ 3.9.2 The Contractor, as soon as practicable after award of the Contract, shall notify the Owner and Architect of the name and qualifications of a proposed superintendent. Within 14 days of receipt of the information, the Architect may notify the Contractor, stating whether the Owner or the Architect (1) has reasonable objection to the proposed superintendent or (2) requires additional time for review. Failure of the Architect to provide notice within the 14-day period shall constitute notice of no reasonable objection.

§ 3.9.3 The Contractor shall not employ a proposed superintendent to whom the Owner or Architect has made reasonable and timely objection. The Contractor shall not change the superintendent without the Owner's consent, which shall not unreasonably be withheld or delayed.

§ 3.10 Contractor's Construction and Submittal Schedules
§ 3.10.1 The Contractor, promptly after being awarded the Contract, shall submit for the Owner's and Architect's information a Contractor's construction schedule for the Work. The schedule shall contain detail appropriate for the Project, including (1) the date of commencement of the Work, interim schedule milestone dates, and the date of Substantial Completion; (2) an apportionment of the Work by construction activity; and (3) the time required for completion of each portion of the Work. The schedule shall provide for the orderly progression of the Work to completion and shall not exceed time limits current under the Contract Documents. The schedule shall be revised at appropriate intervals as required by the conditions of the Work and Project.

§ 3.10.2 The Contractor, promptly after being awarded the Contract and thereafter as necessary to maintain a current submittal schedule, shall submit a submittal schedule for the Architect's approval. The Architect's approval shall not be unreasonably delayed or withheld. The submittal schedule shall (1) be coordinated with the Contractor's construction schedule, and (2) allow the Architect reasonable time to review submittals. If the Contractor fails to submit a submittal schedule, or fails to provide submittals in accordance with the approved submittal schedule, the Contractor shall not be entitled to any increase in Contract Sum or extension of Contract Time based on the time required for review of submittals.

§ 3.10.3 The Contractor shall perform the Work in general accordance with the most recent schedules submitted to the Owner and Architect.

§ 3.11 Documents and Samples at the Site
The Contractor shall make available, at the Project site, the Contract Documents, including Change Orders, Construction Change Directives, and other Modifications, in good order and marked currently to indicate field changes and selections made during construction, and the approved Shop Drawings, Product Data, Samples, and similar required submittals. These shall be in electronic form or paper copy, available to the Architect and Owner, and delivered to the Architect for submittal to the Owner upon completion of the Work as a record of the Work as constructed.

§ 3.12 Shop Drawings, Product Data and Samples
§ 3.12.1 Shop Drawings are drawings, diagrams, schedules, and other data specially prepared for the Work by the Contractor or a Subcontractor, Sub-subcontractor, manufacturer, supplier, or distributor to illustrate some portion of the Work.

§ 3.12.2 Product Data are illustrations, standard schedules, performance charts, instructions, brochures, diagrams, and other information furnished by the Contractor to illustrate materials or equipment for some portion of the Work.

§ 3.12.3 Samples are physical examples that illustrate materials, equipment, or workmanship, and establish standards by which the Work will be judged.

§ 3.12.4 Shop Drawings, Product Data, Samples, and similar submittals are not Contract Documents. Their purpose is to demonstrate how the Contractor proposes to conform to the information given and the design concept expressed in the Contract Documents for those portions of the Work for which the Contract Documents require submittals. Review by the Architect is subject to the limitations of Section 4.2.7. Informational submittals upon which the Architect is not expected to take responsive action may be so identified in the Contract Documents. Submittals that are not required by the Contract Documents may be returned by the Architect without action.

§ 3.12.5 The Contractor shall review for compliance with the Contract Documents, approve, and submit to the Architect, Shop Drawings, Product Data, Samples, and similar submittals required by the Contract Documents, in accordance with the submittal schedule approved by the Architect or, in the absence of an approved submittal schedule, with reasonable promptness and in such sequence as to cause no delay in the Work or in the activities of the Owner or of Separate Contractors.

§ 3.12.6 By submitting Shop Drawings, Product Data, Samples, and similar submittals, the Contractor represents to the Owner and Architect that the Contractor has (1) reviewed and approved them, (2) determined and verified materials, field measurements and field construction criteria related thereto, or will do so, and (3) checked and coordinated the information contained within such submittals with the requirements of the Work and of the Contract Documents.

§ 3.12.7 The Contractor shall perform no portion of the Work for which the Contract Documents require submittal and review of Shop Drawings, Product Data, Samples, or similar submittals, until the respective submittal has been approved by the Architect.

§ 3.12.8 The Work shall be in accordance with approved submittals except that the Contractor shall not be relieved of responsibility for deviations from the requirements of the Contract Documents by the Architect's approval of Shop Drawings, Product Data, Samples, or similar submittals, unless the Contractor has specifically notified the Architect of such deviation at the time of submittal and (1) the Architect has given written approval to the specific deviation as a minor change in the Work, or (2) a Change Order or Construction Change Directive has been issued authorizing the deviation. The Contractor shall not be relieved of responsibility for errors or omissions in Shop Drawings, Product Data, Samples, or similar submittals, by the Architect's approval thereof.

§ 3.12.9 The Contractor shall direct specific attention, in writing or on resubmitted Shop Drawings, Product Data, Samples, or similar submittals, to revisions other than those requested by the Architect on previous submittals. In the absence of such notice, the Architect's approval of a resubmission shall not apply to such revisions.

§ 3.12.10 The Contractor shall not be required to provide professional services that constitute the practice of architecture or engineering unless such services are specifically required by the Contract Documents for a portion of the Work or unless the Contractor needs to provide such services in order to carry out the Contractor's responsibilities for construction means, methods, techniques, sequences, and procedures. The Contractor shall not be required to provide professional services in violation of applicable law.

§ 3.12.10.1 If professional design services or certifications by a design professional related to systems, materials, or equipment are specifically required of the Contractor by the Contract Documents, the Owner and the Architect will specify all performance and design criteria that such services must satisfy. The Contractor shall be entitled to rely upon the adequacy and accuracy of the performance and design criteria provided in the Contract Documents. The Contractor shall cause such services or certifications to be provided by an appropriately licensed design professional, whose signature and seal shall appear on all drawings, calculations, specifications, certifications, Shop Drawings, and other submittals prepared by such professional. Shop Drawings, and other submittals related to the Work, designed or certified by such professional, if prepared by others, shall bear such professional's written approval when submitted to the Architect. The Owner and the Architect shall be entitled to rely upon the adequacy and accuracy of the services, certifications, and approvals performed or provided by such design professionals, provided the Owner and Architect have specified to the Contractor the performance and design criteria that such services must satisfy. Pursuant to this Section 3.12.10, the Architect will review and approve or take other appropriate action on submittals only for the limited purpose of checking for conformance with information given and the design concept expressed in the Contract Documents.

§ 3.12.10.2 If the Contract Documents require the Contractor's design professional to certify that the Work has been performed in accordance with the design criteria, the Contractor shall furnish such certifications to the Architect at the time and in the form specified by the Architect.

§ 3.13 Use of Site
The Contractor shall confine operations at the site to areas permitted by applicable laws, statutes, ordinances, codes, rules and regulations, lawful orders of public authorities, and the Contract Documents and shall not unreasonably encumber the site with materials or equipment.

§ 3.14 Cutting and Patching
§ 3.14.1 The Contractor shall be responsible for cutting, fitting, or patching required to complete the Work or to make its parts fit together properly. All areas requiring cutting, fitting, or patching shall be restored to the condition existing prior to the cutting, fitting, or patching, unless otherwise required by the Contract Documents.

§ 3.14.2 The Contractor shall not damage or endanger a portion of the Work or fully or partially completed construction of the Owner or Separate Contractors by cutting, patching, or otherwise altering such construction, or by excavation. The Contractor shall not cut or otherwise alter construction by the Owner or a Separate Contractor except with written consent of the Owner and of the Separate Contractor. Consent shall not be unreasonably withheld. The Contractor shall not unreasonably withhold, from the Owner or a Separate Contractor, its consent to cutting or otherwise altering the Work.

§ 3.15 Cleaning Up
§ 3.15.1 The Contractor shall keep the premises and surrounding area free from accumulation of waste materials and rubbish caused by operations under the Contract. At completion of the Work, the Contractor shall remove waste materials, rubbish, the Contractor's tools, construction equipment, machinery, and surplus materials from and about the Project.

§ 3.15.2 If the Contractor fails to clean up as provided in the Contract Documents, the Owner may do so and the Owner shall be entitled to reimbursement from the Contractor.

§ 3.16 Access to Work
The Contractor shall provide the Owner and Architect with access to the Work in preparation and progress wherever located.

§ 3.17 Royalties, Patents and Copyrights
The Contractor shall pay all royalties and license fees. The Contractor shall defend suits or claims for infringement of copyrights and patent rights and shall hold the Owner and Architect harmless from loss on account thereof, but shall not be responsible for defense or loss when a particular design, process, or product of a particular manufacturer or manufacturers is required by the Contract Documents, or where the copyright violations are contained in Drawings, Specifications, or other documents prepared by the Owner or Architect. However, if an infringement of a copyright or patent is discovered by, or made known to, the Contractor, the Contractor shall be responsible for the loss unless the information is promptly furnished to the Architect.

§ 3.18 Indemnification
§ 3.18.1 To the fullest extent permitted by law, the Contractor shall indemnify and hold harmless the Owner, Architect, Architect's consultants, and agents and employees of any of them from and against claims, damages, losses, and expenses, including but not limited to attorneys' fees, arising out of or resulting from performance of the Work, provided that such claim, damage, loss, or expense is attributable to bodily injury, sickness, disease or death, or to injury to or destruction of tangible property (other than the Work itself), but only to the extent caused by the negligent acts or omissions of the Contractor, a Subcontractor, anyone directly or indirectly employed by them, or anyone for whose acts they may be liable, regardless of whether or not such claim, damage, loss, or expense is caused in part by a party indemnified hereunder. Such obligation shall not be construed to negate, abridge, or reduce other rights or obligations of indemnity that would otherwise exist as to a party or person described in this Section 3.18.

§ 3.18.2 In claims against any person or entity indemnified under this Section 3.18 by an employee of the Contractor, a Subcontractor, anyone directly or indirectly employed by them, or anyone for whose acts they may be liable, the indemnification obligation under Section 3.18.1 shall not be limited by a limitation on amount or type of damages, compensation, or benefits payable by or for the Contractor or a Subcontractor under workers' compensation acts, disability benefit acts, or other employee benefit acts.

ARTICLE 4 ARCHITECT
§ 4.1 General
§ 4.1.1 The Architect is the person or entity retained by the Owner pursuant to Section 2.3.2 and identified as such in the Agreement.

§ 4.1.2 Duties, responsibilities, and limitations of authority of the Architect as set forth in the Contract Documents shall not be restricted, modified, or extended without written consent of the Owner, Contractor, and Architect. Consent shall not be unreasonably withheld.

§ 4.2 Administration of the Contract
§ 4.2.1 The Architect will provide administration of the Contract as described in the Contract Documents and will be an Owner's representative during construction until the date the Architect issues the final Certificate for Payment. The Architect will have authority to act on behalf of the Owner only to the extent provided in the Contract Documents.

§ 4.2.2 The Architect will visit the site at intervals appropriate to the stage of construction, or as otherwise agreed with the Owner, to become generally familiar with the progress and quality of the portion of the Work completed, and to determine in general if the Work observed is being performed in a manner indicating that the Work, when fully completed, will be in accordance with the Contract Documents. However, the Architect will not be required to make exhaustive or continuous on-site inspections to check the quality or quantity of the Work. The Architect will not have control over, charge of, or responsibility for the construction means, methods, techniques, sequences or procedures, or for the safety precautions and programs in connection with the Work, since these are solely the Contractor's rights and responsibilities under the Contract Documents.

§ 4.2.3 On the basis of the site visits, the Architect will keep the Owner reasonably informed about the progress and quality of the portion of the Work completed, and promptly report to the Owner (1) known deviations from the Contract Documents, (2) known deviations from the most recent construction schedule submitted by the Contractor, and (3) defects and deficiencies observed in the Work. The Architect will not be responsible for the Contractor's failure to perform the Work in accordance with the requirements of the Contract Documents. The Architect will not have control over or charge of, and will not be responsible for acts or omissions of, the Contractor, Subcontractors, or their agents or employees, or any other persons or entities performing portions of the Work.

§ 4.2.4 Communications
The Owner and Contractor shall include the Architect in all communications that relate to or affect the Architect's services or professional responsibilities. The Owner shall promptly notify the Architect of the substance of any direct communications between the Owner and the Contractor otherwise relating to the Project. Communications by and with the Architect's consultants shall be through the Architect. Communications by and with Subcontractors and suppliers shall be through the Contractor. Communications by and with Separate Contractors shall be through the Owner. The Contract Documents may specify other communication protocols.

§ 4.2.5 Based on the Architect's evaluations of the Contractor's Applications for Payment, the Architect will review and certify the amounts due the Contractor and will issue Certificates for Payment in such amounts.

§ 4.2.6 The Architect has authority to reject Work that does not conform to the Contract Documents. Whenever the Architect considers it necessary or advisable, the Architect will have authority to require inspection or testing of the Work in accordance with Sections 13.4.2 and 13.4.3, whether or not the Work is fabricated, installed or completed. However, neither this authority of the Architect nor a decision made in good faith either to exercise or not to exercise such authority shall give rise to a duty or responsibility of the Architect to the Contractor, Subcontractors, suppliers, their agents or employees, or other persons or entities performing portions of the Work.

§ 4.2.7 The Architect will review and approve, or take other appropriate action upon, the Contractor's submittals such as Shop Drawings, Product Data, and Samples, but only for the limited purpose of checking for conformance with information given and the design concept expressed in the Contract Documents. The Architect's action will be taken in accordance with the submittal schedule approved by the Architect or, in the absence of an approved submittal schedule, with reasonable promptness while allowing sufficient time in the Architect's professional judgment to permit adequate review. Review of such submittals is not conducted for the purpose of determining the accuracy and completeness of other details such as dimensions and quantities, or for substantiating instructions for installation or performance of equipment or systems, all of which remain the responsibility of the Contractor as required by the Contract Documents. The Architect's review of the Contractor's submittals shall not relieve the Contractor of the obligations under

Init.
/

Sections 3.3, 3.5, and 3.12. The Architect's review shall not constitute approval of safety precautions or of any construction means, methods, techniques, sequences, or procedures. The Architect's approval of a specific item shall not indicate approval of an assembly of which the item is a component.

§ 4.2.8 The Architect will prepare Change Orders and Construction Change Directives, and may order minor changes in the Work as provided in Section 7.4. The Architect will investigate and make determinations and recommendations regarding concealed and unknown conditions as provided in Section 3.7.4.

§ 4.2.9 The Architect will conduct inspections to determine the date or dates of Substantial Completion and the date of final completion; issue Certificates of Substantial Completion pursuant to Section 9.8; receive and forward to the Owner, for the Owner's review and records, written warranties and related documents required by the Contract and assembled by the Contractor pursuant to Section 9.10; and issue a final Certificate for Payment pursuant to Section 9.10.

§ 4.2.10 If the Owner and Architect agree, the Architect will provide one or more Project representatives to assist in carrying out the Architect's responsibilities at the site. The Owner shall notify the Contractor of any change in the duties, responsibilities and limitations of authority of the Project representatives.

§ 4.2.11 The Architect will interpret and decide matters concerning performance under, and requirements of, the Contract Documents on written request of either the Owner or Contractor. The Architect's response to such requests will be made in writing within any time limits agreed upon or otherwise with reasonable promptness.

§ 4.2.12 Interpretations and decisions of the Architect will be consistent with the intent of, and reasonably inferable from, the Contract Documents and will be in writing or in the form of drawings. When making such interpretations and decisions, the Architect will endeavor to secure faithful performance by both Owner and Contractor, will not show partiality to either, and will not be liable for results of interpretations or decisions rendered in good faith.

§ 4.2.13 The Architect's decisions on matters relating to aesthetic effect will be final if consistent with the intent expressed in the Contract Documents.

§ 4.2.14 The Architect will review and respond to requests for information about the Contract Documents. The Architect's response to such requests will be made in writing within any time limits agreed upon or otherwise with reasonable promptness. If appropriate, the Architect will prepare and issue supplemental Drawings and Specifications in response to the requests for information.

ARTICLE 5 SUBCONTRACTORS

§ 5.1 Definitions

§ 5.1.1 A Subcontractor is a person or entity who has a direct contract with the Contractor to perform a portion of the Work at the site. The term "Subcontractor" is referred to throughout the Contract Documents as if singular in number and means a Subcontractor or an authorized representative of the Subcontractor. The term "Subcontractor" does not include a Separate Contractor or the subcontractors of a Separate Contractor.

§ 5.1.2 A Sub-subcontractor is a person or entity who has a direct or indirect contract with a Subcontractor to perform a portion of the Work at the site. The term "Sub-subcontractor" is referred to throughout the Contract Documents as if singular in number and means a Sub-subcontractor or an authorized representative of the Sub-subcontractor.

§ 5.2 Award of Subcontracts and Other Contracts for Portions of the Work

§ 5.2.1 Unless otherwise stated in the Contract Documents, the Contractor, as soon as practicable after award of the Contract, shall notify the Owner and Architect of the persons or entities proposed for each principal portion of the Work, including those who are to furnish materials or equipment fabricated to a special design. Within 14 days of receipt of the information, the Architect may notify the Contractor whether the Owner or the Architect (1) has reasonable objection to any such proposed person or entity or (2) requires additional time for review. Failure of the Architect to provide notice within the 14-day period shall constitute notice of no reasonable objection.

§ 5.2.2 The Contractor shall not contract with a proposed person or entity to whom the Owner or Architect has made reasonable and timely objection. The Contractor shall not be required to contract with anyone to whom the Contractor has made reasonable objection.

§ 5.2.3 If the Owner or Architect has reasonable objection to a person or entity proposed by the Contractor, the Contractor shall propose another to whom the Owner or Architect has no reasonable objection. If the proposed but rejected Subcontractor was reasonably capable of performing the Work, the Contract Sum and Contract Time shall be increased or decreased by the difference, if any, occasioned by such change, and an appropriate Change Order shall be issued before commencement of the substitute Subcontractor's Work. However, no increase in the Contract Sum or Contract Time shall be allowed for such change unless the Contractor has acted promptly and responsively in submitting names as required.

§ 5.2.4 The Contractor shall not substitute a Subcontractor, person, or entity for one previously selected if the Owner or Architect makes reasonable objection to such substitution.

§ 5.3 Subcontractual Relations
By appropriate written agreement, the Contractor shall require each Subcontractor, to the extent of the Work to be performed by the Subcontractor, to be bound to the Contractor by terms of the Contract Documents, and to assume toward the Contractor all the obligations and responsibilities, including the responsibility for safety of the Subcontractor's Work that the Contractor, by these Contract Documents, assumes toward the Owner and Architect. Each subcontract agreement shall preserve and protect the rights of the Owner and Architect under the Contract Documents with respect to the Work to be performed by the Subcontractor so that subcontracting thereof will not prejudice such rights, and shall allow to the Subcontractor, unless specifically provided otherwise in the subcontract agreement, the benefit of all rights, remedies, and redress against the Contractor that the Contractor, by the Contract Documents, has against the Owner. Where appropriate, the Contractor shall require each Subcontractor to enter into similar agreements with Sub-subcontractors. The Contractor shall make available to each proposed Subcontractor, prior to the execution of the subcontract agreement, copies of the Contract Documents to which the Subcontractor will be bound, and, upon written request of the Subcontractor, identify to the Subcontractor terms and conditions of the proposed subcontract agreement that may be at variance with the Contract Documents. Subcontractors will similarly make copies of applicable portions of such documents available to their respective proposed Sub-subcontractors.

§ 5.4 Contingent Assignment of Subcontracts
§ 5.4.1 Each subcontract agreement for a portion of the Work is assigned by the Contractor to the Owner, provided that

- **.1** assignment is effective only after termination of the Contract by the Owner for cause pursuant to Section 14.2 and only for those subcontract agreements that the Owner accepts by notifying the Subcontractor and Contractor; and
- **.2** assignment is subject to the prior rights of the surety, if any, obligated under bond relating to the Contract.

When the Owner accepts the assignment of a subcontract agreement, the Owner assumes the Contractor's rights and obligations under the subcontract.

§ 5.4.2 Upon such assignment, if the Work has been suspended for more than 30 days, the Subcontractor's compensation shall be equitably adjusted for increases in cost resulting from the suspension.

§ 5.4.3 Upon assignment to the Owner under this Section 5.4, the Owner may further assign the subcontract to a successor contractor or other entity. If the Owner assigns the subcontract to a successor contractor or other entity, the Owner shall nevertheless remain legally responsible for all of the successor contractor's obligations under the subcontract.

ARTICLE 6 CONSTRUCTION BY OWNER OR BY SEPARATE CONTRACTORS
§ 6.1 Owner's Right to Perform Construction and to Award Separate Contracts
§ 6.1.1 The term "Separate Contractor(s)" shall mean other contractors retained by the Owner under separate agreements. The Owner reserves the right to perform construction or operations related to the Project with the Owner's own forces, and with Separate Contractors retained under Conditions of the Contract substantially similar to those of this Contract, including those provisions of the Conditions of the Contract related to insurance and waiver of subrogation.

§ 6.1.2 When separate contracts are awarded for different portions of the Project or other construction or operations on the site, the term "Contractor" in the Contract Documents in each case shall mean the Contractor who executes each separate Owner-Contractor Agreement.

§ 6.1.3 The Owner shall provide for coordination of the activities of the Owner's own forces and of each Separate Contractor with the Work of the Contractor, who shall cooperate with them. The Contractor shall participate with any Separate Contractors and the Owner in reviewing their construction schedules. The Contractor shall make any revisions to its construction schedule deemed necessary after a joint review and mutual agreement. The construction schedules shall then constitute the schedules to be used by the Contractor, Separate Contractors, and the Owner until subsequently revised.

§ 6.1.4 Unless otherwise provided in the Contract Documents, when the Owner performs construction or operations related to the Project with the Owner's own forces or with Separate Contractors, the Owner or its Separate Contractors shall have the same obligations and rights that the Contractor has under the Conditions of the Contract, including, without excluding others, those stated in Article 3, this Article 6, and Articles 10, 11, and 12.

§ 6.2 Mutual Responsibility

§ 6.2.1 The Contractor shall afford the Owner and Separate Contractors reasonable opportunity for introduction and storage of their materials and equipment and performance of their activities, and shall connect and coordinate the Contractor's construction and operations with theirs as required by the Contract Documents.

§ 6.2.2 If part of the Contractor's Work depends for proper execution or results upon construction or operations by the Owner or a Separate Contractor, the Contractor shall, prior to proceeding with that portion of the Work, promptly notify the Architect of apparent discrepancies or defects in the construction or operations by the Owner or Separate Contractor that would render it unsuitable for proper execution and results of the Contractor's Work. Failure of the Contractor to notify the Architect of apparent discrepancies or defects prior to proceeding with the Work shall constitute an acknowledgment that the Owner's or Separate Contractor's completed or partially completed construction is fit and proper to receive the Contractor's Work. The Contractor shall not be responsible for discrepancies or defects in the construction or operations by the Owner or Separate Contractor that are not apparent.

§ 6.2.3 The Contractor shall reimburse the Owner for costs the Owner incurs that are payable to a Separate Contractor because of the Contractor's delays, improperly timed activities or defective construction. The Owner shall be responsible to the Contractor for costs the Contractor incurs because of a Separate Contractor's delays, improperly timed activities, damage to the Work or defective construction.

§ 6.2.4 The Contractor shall promptly remedy damage that the Contractor wrongfully causes to completed or partially completed construction or to property of the Owner or Separate Contractor as provided in Section 10.2.5.

§ 6.2.5 The Owner and each Separate Contractor shall have the same responsibilities for cutting and patching as are described for the Contractor in Section 3.14.

§ 6.3 Owner's Right to Clean Up

If a dispute arises among the Contractor, Separate Contractors, and the Owner as to the responsibility under their respective contracts for maintaining the premises and surrounding area free from waste materials and rubbish, the Owner may clean up and the Architect will allocate the cost among those responsible.

ARTICLE 7 CHANGES IN THE WORK

§ 7.1 General

§ 7.1.1 Changes in the Work may be accomplished after execution of the Contract, and without invalidating the Contract, by Change Order, Construction Change Directive or order for a minor change in the Work, subject to the limitations stated in this Article 7 and elsewhere in the Contract Documents.

§ 7.1.2 A Change Order shall be based upon agreement among the Owner, Contractor, and Architect. A Construction Change Directive requires agreement by the Owner and Architect and may or may not be agreed to by the Contractor. An order for a minor change in the Work may be issued by the Architect alone.

§ 7.1.3 Changes in the Work shall be performed under applicable provisions of the Contract Documents. The Contractor shall proceed promptly with changes in the Work, unless otherwise provided in the Change Order, Construction Change Directive, or order for a minor change in the Work.

§ 7.2 Change Orders
§ 7.2.1 A Change Order is a written instrument prepared by the Architect and signed by the Owner, Contractor, and Architect stating their agreement upon all of the following:

- **.1** The change in the Work;
- **.2** The amount of the adjustment, if any, in the Contract Sum; and
- **.3** The extent of the adjustment, if any, in the Contract Time.

§ 7.3 Construction Change Directives
§ 7.3.1 A Construction Change Directive is a written order prepared by the Architect and signed by the Owner and Architect, directing a change in the Work prior to agreement on adjustment, if any, in the Contract Sum or Contract Time, or both. The Owner may by Construction Change Directive, without invalidating the Contract, order changes in the Work within the general scope of the Contract consisting of additions, deletions, or other revisions, the Contract Sum and Contract Time being adjusted accordingly.

§ 7.3.2 A Construction Change Directive shall be used in the absence of total agreement on the terms of a Change Order.

§ 7.3.3 If the Construction Change Directive provides for an adjustment to the Contract Sum, the adjustment shall be based on one of the following methods:

- **.1** Mutual acceptance of a lump sum properly itemized and supported by sufficient substantiating data to permit evaluation;
- **.2** Unit prices stated in the Contract Documents or subsequently agreed upon;
- **.3** Cost to be determined in a manner agreed upon by the parties and a mutually acceptable fixed or percentage fee; or
- **.4** As provided in Section 7.3.4.

§ 7.3.4 If the Contractor does not respond promptly or disagrees with the method for adjustment in the Contract Sum, the Architect shall determine the adjustment on the basis of reasonable expenditures and savings of those performing the Work attributable to the change, including, in case of an increase in the Contract Sum, an amount for overhead and profit as set forth in the Agreement, or if no such amount is set forth in the Agreement, a reasonable amount. In such case, and also under Section 7.3.3.3, the Contractor shall keep and present, in such form as the Architect may prescribe, an itemized accounting together with appropriate supporting data. Unless otherwise provided in the Contract Documents, costs for the purposes of this Section 7.3.4 shall be limited to the following:

- **.1** Costs of labor, including applicable payroll taxes, fringe benefits required by agreement or custom, workers' compensation insurance, and other employee costs approved by the Architect;
- **.2** Costs of materials, supplies, and equipment, including cost of transportation, whether incorporated or consumed;
- **.3** Rental costs of machinery and equipment, exclusive of hand tools, whether rented from the Contractor or others;
- **.4** Costs of premiums for all bonds and insurance, permit fees, and sales, use, or similar taxes, directly related to the change; and
- **.5** Costs of supervision and field office personnel directly attributable to the change.

§ 7.3.5 If the Contractor disagrees with the adjustment in the Contract Time, the Contractor may make a Claim in accordance with applicable provisions of Article 15.

§ 7.3.6 Upon receipt of a Construction Change Directive, the Contractor shall promptly proceed with the change in the Work involved and advise the Architect of the Contractor's agreement or disagreement with the method, if any, provided in the Construction Change Directive for determining the proposed adjustment in the Contract Sum or Contract Time.

§ 7.3.7 A Construction Change Directive signed by the Contractor indicates the Contractor's agreement therewith, including adjustment in Contract Sum and Contract Time or the method for determining them. Such agreement shall be effective immediately and shall be recorded as a Change Order.

§ 7.3.8 The amount of credit to be allowed by the Contractor to the Owner for a deletion or change that results in a net decrease in the Contract Sum shall be actual net cost as confirmed by the Architect. When both additions and credits covering related Work or substitutions are involved in a change, the allowance for overhead and profit shall be figured on the basis of net increase, if any, with respect to that change.

§ 7.3.9 Pending final determination of the total cost of a Construction Change Directive to the Owner, the Contractor may request payment for Work completed under the Construction Change Directive in Applications for Payment. The Architect will make an interim determination for purposes of monthly certification for payment for those costs and certify for payment the amount that the Architect determines, in the Architect's professional judgment, to be reasonably justified. The Architect's interim determination of cost shall adjust the Contract Sum on the same basis as a Change Order, subject to the right of either party to disagree and assert a Claim in accordance with Article 15.

§ 7.3.10 When the Owner and Contractor agree with a determination made by the Architect concerning the adjustments in the Contract Sum and Contract Time, or otherwise reach agreement upon the adjustments, such agreement shall be effective immediately and the Architect will prepare a Change Order. Change Orders may be issued for all or any part of a Construction Change Directive.

§ 7.4 Minor Changes in the Work
The Architect may order minor changes in the Work that are consistent with the intent of the Contract Documents and do not involve an adjustment in the Contract Sum or an extension of the Contract Time. The Architect's order for minor changes shall be in writing. If the Contractor believes that the proposed minor change in the Work will affect the Contract Sum or Contract Time, the Contractor shall notify the Architect and shall not proceed to implement the change in the Work. If the Contractor performs the Work set forth in the Architect's order for a minor change without prior notice to the Architect that such change will affect the Contract Sum or Contract Time, the Contractor waives any adjustment to the Contract Sum or extension of the Contract Time.

ARTICLE 8 TIME
§ 8.1 Definitions
§ 8.1.1 Unless otherwise provided, Contract Time is the period of time, including authorized adjustments, allotted in the Contract Documents for Substantial Completion of the Work.

§ 8.1.2 The date of commencement of the Work is the date established in the Agreement.

§ 8.1.3 The date of Substantial Completion is the date certified by the Architect in accordance with Section 9.8.

§ 8.1.4 The term "day" as used in the Contract Documents shall mean calendar day unless otherwise specifically defined.

§ 8.2 Progress and Completion
§ 8.2.1 Time limits stated in the Contract Documents are of the essence of the Contract. By executing the Agreement, the Contractor confirms that the Contract Time is a reasonable period for performing the Work.

§ 8.2.2 The Contractor shall not knowingly, except by agreement or instruction of the Owner in writing, commence the Work prior to the effective date of insurance required to be furnished by the Contractor and Owner.

§ 8.2.3 The Contractor shall proceed expeditiously with adequate forces and shall achieve Substantial Completion within the Contract Time.

§ 8.3 Delays and Extensions of Time
§ 8.3.1 If the Contractor is delayed at any time in the commencement or progress of the Work by (1) an act or neglect of the Owner or Architect, of an employee of either, or of a Separate Contractor; (2) by changes ordered in the Work; (3) by labor disputes, fire, unusual delay in deliveries, unavoidable casualties, adverse weather conditions documented in accordance with Section 15.1.6.2, or other causes beyond the Contractor's control; (4) by delay authorized by the Owner pending mediation and binding dispute resolution; or (5) by other causes that the Contractor asserts, and the Architect determines, justify delay, then the Contract Time shall be extended for such reasonable time as the Architect may determine.

§ 8.3.2 Claims relating to time shall be made in accordance with applicable provisions of Article 15.

§ 8.3.3 This Section 8.3 does not preclude recovery of damages for delay by either party under other provisions of the Contract Documents.

ARTICLE 9 PAYMENTS AND COMPLETION

§ 9.1 Contract Sum

§ 9.1.1 The Contract Sum is stated in the Agreement and, including authorized adjustments, is the total amount payable by the Owner to the Contractor for performance of the Work under the Contract Documents.

§ 9.1.2 If unit prices are stated in the Contract Documents or subsequently agreed upon, and if quantities originally contemplated are materially changed so that application of such unit prices to the actual quantities causes substantial inequity to the Owner or Contractor, the applicable unit prices shall be equitably adjusted.

§ 9.2 Schedule of Values

Where the Contract is based on a stipulated sum or Guaranteed Maximum Price, the Contractor shall submit a schedule of values to the Architect before the first Application for Payment, allocating the entire Contract Sum to the various portions of the Work. The schedule of values shall be prepared in the form, and supported by the data to substantiate its accuracy, required by the Architect. This schedule, unless objected to by the Architect, shall be used as a basis for reviewing the Contractor's Applications for Payment. Any changes to the schedule of values shall be submitted to the Architect and supported by such data to substantiate its accuracy as the Architect may require, and unless objected to by the Architect, shall be used as a basis for reviewing the Contractor's subsequent Applications for Payment.

§ 9.3 Applications for Payment

§ 9.3.1 At least ten days before the date established for each progress payment, the Contractor shall submit to the Architect an itemized Application for Payment prepared in accordance with the schedule of values, if required under Section 9.2, for completed portions of the Work. The application shall be notarized, if required, and supported by all data substantiating the Contractor's right to payment that the Owner or Architect require, such as copies of requisitions, and releases and waivers of liens from Subcontractors and suppliers, and shall reflect retainage if provided for in the Contract Documents.

§ 9.3.1.1 As provided in Section 7.3.9, such applications may include requests for payment on account of changes in the Work that have been properly authorized by Construction Change Directives, or by interim determinations of the Architect, but not yet included in Change Orders.

§ 9.3.1.2 Applications for Payment shall not include requests for payment for portions of the Work for which the Contractor does not intend to pay a Subcontractor or supplier, unless such Work has been performed by others whom the Contractor intends to pay.

§ 9.3.2 Unless otherwise provided in the Contract Documents, payments shall be made on account of materials and equipment delivered and suitably stored at the site for subsequent incorporation in the Work. If approved in advance by the Owner, payment may similarly be made for materials and equipment suitably stored off the site at a location agreed upon in writing. Payment for materials and equipment stored on or off the site shall be conditioned upon compliance by the Contractor with procedures satisfactory to the Owner to establish the Owner's title to such materials and equipment or otherwise protect the Owner's interest, and shall include the costs of applicable insurance, storage, and transportation to the site, for such materials and equipment stored off the site.

§ 9.3.3 The Contractor warrants that title to all Work covered by an Application for Payment will pass to the Owner no later than the time of payment. The Contractor further warrants that upon submittal of an Application for Payment all Work for which Certificates for Payment have been previously issued and payments received from the Owner shall, to the best of the Contractor's knowledge, information, and belief, be free and clear of liens, claims, security interests, or encumbrances, in favor of the Contractor, Subcontractors, suppliers, or other persons or entities that provided labor, materials, and equipment relating to the Work.

§ 9.4 Certificates for Payment

§ 9.4.1 The Architect will, within seven days after receipt of the Contractor's Application for Payment, either (1) issue to the Owner a Certificate for Payment in the full amount of the Application for Payment, with a copy to the Contractor; or (2) issue to the Owner a Certificate for Payment for such amount as the Architect determines is properly due, and notify the Contractor and Owner of the Architect's reasons for withholding certification in part as provided in Section 9.5.1; or (3) withhold certification of the entire Application for Payment, and notify the Contractor and Owner of the Architect's reason for withholding certification in whole as provided in Section 9.5.1.

§ 9.4.2 The issuance of a Certificate for Payment will constitute a representation by the Architect to the Owner, based on the Architect's evaluation of the Work and the data in the Application for Payment, that, to the best of the Architect's knowledge, information, and belief, the Work has progressed to the point indicated, the quality of the Work is in accordance with the Contract Documents, and that the Contractor is entitled to payment in the amount certified. The foregoing representations are subject to an evaluation of the Work for conformance with the Contract Documents upon Substantial Completion, to results of subsequent tests and inspections, to correction of minor deviations from the Contract Documents prior to completion, and to specific qualifications expressed by the Architect. However, the issuance of a Certificate for Payment will not be a representation that the Architect has (1) made exhaustive or continuous on-site inspections to check the quality or quantity of the Work; (2) reviewed construction means, methods, techniques, sequences, or procedures; (3) reviewed copies of requisitions received from Subcontractors and suppliers and other data requested by the Owner to substantiate the Contractor's right to payment; or (4) made examination to ascertain how or for what purpose the Contractor has used money previously paid on account of the Contract Sum.

§ 9.5 Decisions to Withhold Certification
§ 9.5.1 The Architect may withhold a Certificate for Payment in whole or in part, to the extent reasonably necessary to protect the Owner, if in the Architect's opinion the representations to the Owner required by Section 9.4.2 cannot be made. If the Architect is unable to certify payment in the amount of the Application, the Architect will notify the Contractor and Owner as provided in Section 9.4.1. If the Contractor and Architect cannot agree on a revised amount, the Architect will promptly issue a Certificate for Payment for the amount for which the Architect is able to make such representations to the Owner. The Architect may also withhold a Certificate for Payment or, because of subsequently discovered evidence, may nullify the whole or a part of a Certificate for Payment previously issued, to such extent as may be necessary in the Architect's opinion to protect the Owner from loss for which the Contractor is responsible, including loss resulting from acts and omissions described in Section 3.3.2, because of

.1 defective Work not remedied;
.2 third party claims filed or reasonable evidence indicating probable filing of such claims, unless security acceptable to the Owner is provided by the Contractor;
.3 failure of the Contractor to make payments properly to Subcontractors or suppliers for labor, materials or equipment;
.4 reasonable evidence that the Work cannot be completed for the unpaid balance of the Contract Sum;
.5 damage to the Owner or a Separate Contractor;
.6 reasonable evidence that the Work will not be completed within the Contract Time, and that the unpaid balance would not be adequate to cover actual or liquidated damages for the anticipated delay; or
.7 repeated failure to carry out the Work in accordance with the Contract Documents.

§ 9.5.2 When either party disputes the Architect's decision regarding a Certificate for Payment under Section 9.5.1, in whole or in part, that party may submit a Claim in accordance with Article 15.

§ 9.5.3 When the reasons for withholding certification are removed, certification will be made for amounts previously withheld.

§ 9.5.4 If the Architect withholds certification for payment under Section 9.5.1.3, the Owner may, at its sole option, issue joint checks to the Contractor and to any Subcontractor or supplier to whom the Contractor failed to make payment for Work properly performed or material or equipment suitably delivered. If the Owner makes payments by joint check, the Owner shall notify the Architect and the Contractor shall reflect such payment on its next Application for Payment.

§ 9.6 Progress Payments
§ 9.6.1 After the Architect has issued a Certificate for Payment, the Owner shall make payment in the manner and within the time provided in the Contract Documents, and shall so notify the Architect.

§ 9.6.2 The Contractor shall pay each Subcontractor, no later than seven days after receipt of payment from the Owner, the amount to which the Subcontractor is entitled, reflecting percentages actually retained from payments to the Contractor on account of the Subcontractor's portion of the Work. The Contractor shall, by appropriate agreement with each Subcontractor, require each Subcontractor to make payments to Sub-subcontractors in a similar manner.

§ 9.6.3 The Architect will, on request, furnish to a Subcontractor, if practicable, information regarding percentages of completion or amounts applied for by the Contractor and action taken thereon by the Architect and Owner on account of portions of the Work done by such Subcontractor.

Init.

/

§ 9.6.4 The Owner has the right to request written evidence from the Contractor that the Contractor has properly paid Subcontractors and suppliers amounts paid by the Owner to the Contractor for subcontracted Work. If the Contractor fails to furnish such evidence within seven days, the Owner shall have the right to contact Subcontractors and suppliers to ascertain whether they have been properly paid. Neither the Owner nor Architect shall have an obligation to pay, or to see to the payment of money to, a Subcontractor or supplier, except as may otherwise be required by law.

§ 9.6.5 The Contractor's payments to suppliers shall be treated in a manner similar to that provided in Sections 9.6.2, 9.6.3 and 9.6.4.

§ 9.6.6 A Certificate for Payment, a progress payment, or partial or entire use or occupancy of the Project by the Owner shall not constitute acceptance of Work not in accordance with the Contract Documents.

§ 9.6.7 Unless the Contractor provides the Owner with a payment bond in the full penal sum of the Contract Sum, payments received by the Contractor for Work properly performed by Subcontractors or provided by suppliers shall be held by the Contractor for those Subcontractors or suppliers who performed Work or furnished materials, or both, under contract with the Contractor for which payment was made by the Owner. Nothing contained herein shall require money to be placed in a separate account and not commingled with money of the Contractor, create any fiduciary liability or tort liability on the part of the Contractor for breach of trust, or entitle any person or entity to an award of punitive damages against the Contractor for breach of the requirements of this provision.

§ 9.6.8 Provided the Owner has fulfilled its payment obligations under the Contract Documents, the Contractor shall defend and indemnify the Owner from all loss, liability, damage or expense, including reasonable attorney's fees and litigation expenses, arising out of any lien claim or other claim for payment by any Subcontractor or supplier of any tier. Upon receipt of notice of a lien claim or other claim for payment, the Owner shall notify the Contractor. If approved by the applicable court, when required, the Contractor may substitute a surety bond for the property against which the lien or other claim for payment has been asserted.

§ 9.7 Failure of Payment
If the Architect does not issue a Certificate for Payment, through no fault of the Contractor, within seven days after receipt of the Contractor's Application for Payment, or if the Owner does not pay the Contractor within seven days after the date established in the Contract Documents, the amount certified by the Architect or awarded by binding dispute resolution, then the Contractor may, upon seven additional days' notice to the Owner and Architect, stop the Work until payment of the amount owing has been received. The Contract Time shall be extended appropriately and the Contract Sum shall be increased by the amount of the Contractor's reasonable costs of shutdown, delay and start-up, plus interest as provided for in the Contract Documents.

§ 9.8 Substantial Completion
§ 9.8.1 Substantial Completion is the stage in the progress of the Work when the Work or designated portion thereof is sufficiently complete in accordance with the Contract Documents so that the Owner can occupy or utilize the Work for its intended use.

§ 9.8.2 When the Contractor considers that the Work, or a portion thereof which the Owner agrees to accept separately, is substantially complete, the Contractor shall prepare and submit to the Architect a comprehensive list of items to be completed or corrected prior to final payment. Failure to include an item on such list does not alter the responsibility of the Contractor to complete all Work in accordance with the Contract Documents.

§ 9.8.3 Upon receipt of the Contractor's list, the Architect will make an inspection to determine whether the Work or designated portion thereof is substantially complete. If the Architect's inspection discloses any item, whether or not included on the Contractor's list, which is not sufficiently complete in accordance with the Contract Documents so that the Owner can occupy or utilize the Work or designated portion thereof for its intended use, the Contractor shall, before issuance of the Certificate of Substantial Completion, complete or correct such item upon notification by the Architect. In such case, the Contractor shall then submit a request for another inspection by the Architect to determine Substantial Completion.

§ 9.8.4 When the Work or designated portion thereof is substantially complete, the Architect will prepare a Certificate of Substantial Completion that shall establish the date of Substantial Completion, establish responsibilities of the Owner and Contractor for security, maintenance, heat, utilities, damage to the Work and insurance; and fix the time within which the Contractor shall finish all items on the list accompanying the Certificate. Warranties required by the Contract Documents shall commence on the date of Substantial Completion of the Work or designated portion thereof unless otherwise provided in the Certificate of Substantial Completion.

§ 9.8.5 The Certificate of Substantial Completion shall be submitted to the Owner and Contractor for their written acceptance of responsibilities assigned to them in the Certificate. Upon such acceptance, and consent of surety if any, the Owner shall make payment of retainage applying to the Work or designated portion thereof. Such payment shall be adjusted for Work that is incomplete or not in accordance with the requirements of the Contract Documents.

§ 9.9 Partial Occupancy or Use

§ 9.9.1 The Owner may occupy or use any completed or partially completed portion of the Work at any stage when such portion is designated by separate agreement with the Contractor, provided such occupancy or use is consented to by the insurer and authorized by public authorities having jurisdiction over the Project. Such partial occupancy or use may commence whether or not the portion is substantially complete, provided the Owner and Contractor have accepted in writing the responsibilities assigned to each of them for payments, retainage, if any, security, maintenance, heat, utilities, damage to the Work and insurance, and have agreed in writing concerning the period for correction of the Work and commencement of warranties required by the Contract Documents. When the Contractor considers a portion substantially complete, the Contractor shall prepare and submit a list to the Architect as provided under Section 9.8.2. Consent of the Contractor to partial occupancy or use shall not be unreasonably withheld. The stage of the progress of the Work shall be determined by written agreement between the Owner and Contractor or, if no agreement is reached, by decision of the Architect.

§ 9.9.2 Immediately prior to such partial occupancy or use, the Owner, Contractor, and Architect shall jointly inspect the area to be occupied or portion of the Work to be used in order to determine and record the condition of the Work.

§ 9.9.3 Unless otherwise agreed upon, partial occupancy or use of a portion or portions of the Work shall not constitute acceptance of Work not complying with the requirements of the Contract Documents.

§ 9.10 Final Completion and Final Payment

§ 9.10.1 Upon receipt of the Contractor's notice that the Work is ready for final inspection and acceptance and upon receipt of a final Application for Payment, the Architect will promptly make such inspection. When the Architect finds the Work acceptable under the Contract Documents and the Contract fully performed, the Architect will promptly issue a final Certificate for Payment stating that to the best of the Architect's knowledge, information and belief, and on the basis of the Architect's on-site visits and inspections, the Work has been completed in accordance with the Contract Documents and that the entire balance found to be due the Contractor and noted in the final Certificate is due and payable. The Architect's final Certificate for Payment will constitute a further representation that conditions listed in Section 9.10.2 as precedent to the Contractor's being entitled to final payment have been fulfilled.

§ 9.10.2 Neither final payment nor any remaining retained percentage shall become due until the Contractor submits to the Architect (1) an affidavit that payrolls, bills for materials and equipment, and other indebtedness connected with the Work for which the Owner or the Owner's property might be responsible or encumbered (less amounts withheld by Owner) have been paid or otherwise satisfied, (2) a certificate evidencing that insurance required by the Contract Documents to remain in force after final payment is currently in effect, (3) a written statement that the Contractor knows of no reason that the insurance will not be renewable to cover the period required by the Contract Documents, (4) consent of surety, if any, to final payment, (5) documentation of any special warranties, such as manufacturers' warranties or specific Subcontractor warranties, and (6) if required by the Owner, other data establishing payment or satisfaction of obligations, such as receipts and releases and waivers of liens, claims, security interests, or encumbrances arising out of the Contract, to the extent and in such form as may be designated by the Owner. If a Subcontractor refuses to furnish a release or waiver required by the Owner, the Contractor may furnish a bond satisfactory to the Owner to indemnify the Owner against such lien, claim, security interest, or encumbrance. If a lien, claim, security interest, or encumbrance remains unsatisfied after payments are made, the Contractor shall refund to the Owner all money that the Owner may be compelled to pay in discharging the lien, claim, security interest, or encumbrance, including all costs and reasonable attorneys' fees.

§ 9.10.3 If, after Substantial Completion of the Work, final completion thereof is materially delayed through no fault of the Contractor or by issuance of Change Orders affecting final completion, and the Architect so confirms, the Owner shall, upon application by the Contractor and certification by the Architect, and without terminating the Contract, make payment of the balance due for that portion of the Work fully completed, corrected, and accepted. If the remaining balance for Work not fully completed or corrected is less than retainage stipulated in the Contract Documents, and if bonds have been furnished, the written consent of the surety to payment of the balance due for that portion of the Work fully completed and accepted shall be submitted by the Contractor to the Architect prior to certification of such payment. Such payment shall be made under terms and conditions governing final payment, except that it shall not constitute a waiver of Claims.

§ 9.10.4 The making of final payment shall constitute a waiver of Claims by the Owner except those arising from

.1 liens, Claims, security interests, or encumbrances arising out of the Contract and unsettled;
.2 failure of the Work to comply with the requirements of the Contract Documents;
.3 terms of special warranties required by the Contract Documents; or
.4 audits performed by the Owner, if permitted by the Contract Documents, after final payment.

§ 9.10.5 Acceptance of final payment by the Contractor, a Subcontractor, or a supplier, shall constitute a waiver of claims by that payee except those previously made in writing and identified by that payee as unsettled at the time of final Application for Payment.

ARTICLE 10 PROTECTION OF PERSONS AND PROPERTY
§ 10.1 Safety Precautions and Programs
The Contractor shall be responsible for initiating, maintaining, and supervising all safety precautions and programs in connection with the performance of the Contract.

§ 10.2 Safety of Persons and Property
§ 10.2.1 The Contractor shall take reasonable precautions for safety of, and shall provide reasonable protection to prevent damage, injury, or loss to

.1 employees on the Work and other persons who may be affected thereby;
.2 the Work and materials and equipment to be incorporated therein, whether in storage on or off the site, under care, custody, or control of the Contractor, a Subcontractor, or a Sub-subcontractor; and
.3 other property at the site or adjacent thereto, such as trees, shrubs, lawns, walks, pavements, roadways, structures, and utilities not designated for removal, relocation, or replacement in the course of construction.

§ 10.2.2 The Contractor shall comply with, and give notices required by applicable laws, statutes, ordinances, codes, rules and regulations, and lawful orders of public authorities, bearing on safety of persons or property or their protection from damage, injury, or loss.

§ 10.2.3 The Contractor shall implement, erect, and maintain, as required by existing conditions and performance of the Contract, reasonable safeguards for safety and protection, including posting danger signs and other warnings against hazards; promulgating safety regulations; and notifying the owners and users of adjacent sites and utilities of the safeguards.

§ 10.2.4 When use or storage of explosives or other hazardous materials or equipment, or unusual methods are necessary for execution of the Work, the Contractor shall exercise utmost care and carry on such activities under supervision of properly qualified personnel.

§ 10.2.5 The Contractor shall promptly remedy damage and loss (other than damage or loss insured under property insurance required by the Contract Documents) to property referred to in Sections 10.2.1.2 and 10.2.1.3 caused in whole or in part by the Contractor, a Subcontractor, a Sub-subcontractor, or anyone directly or indirectly employed by any of them, or by anyone for whose acts they may be liable and for which the Contractor is responsible under Sections 10.2.1.2 and 10.2.1.3. The Contractor may make a Claim for the cost to remedy the damage or loss to the extent such damage or loss is attributable to acts or omissions of the Owner or Architect or anyone directly or indirectly employed by either of them, or by anyone for whose acts either of them may be liable, and not attributable to the fault or negligence of the Contractor. The foregoing obligations of the Contractor are in addition to the Contractor's obligations under Section 3.18.

§ 10.2.6 The Contractor shall designate a responsible member of the Contractor's organization at the site whose duty shall be the prevention of accidents. This person shall be the Contractor's superintendent unless otherwise designated by the Contractor in writing to the Owner and Architect.

§ 10.2.7 The Contractor shall not permit any part of the construction or site to be loaded so as to cause damage or create an unsafe condition.

§ 10.2.8 Injury or Damage to Person or Property
If either party suffers injury or damage to person or property because of an act or omission of the other party, or of others for whose acts such party is legally responsible, notice of the injury or damage, whether or not insured, shall be given to the other party within a reasonable time not exceeding 21 days after discovery. The notice shall provide sufficient detail to enable the other party to investigate the matter.

§ 10.3 Hazardous Materials and Substances
§ 10.3.1 The Contractor is responsible for compliance with any requirements included in the Contract Documents regarding hazardous materials or substances. If the Contractor encounters a hazardous material or substance not addressed in the Contract Documents and if reasonable precautions will be inadequate to prevent foreseeable bodily injury or death to persons resulting from a material or substance, including but not limited to asbestos or polychlorinated biphenyl (PCB), encountered on the site by the Contractor, the Contractor shall, upon recognizing the condition, immediately stop Work in the affected area and notify the Owner and Architect of the condition.

§ 10.3.2 Upon receipt of the Contractor's notice, the Owner shall obtain the services of a licensed laboratory to verify the presence or absence of the material or substance reported by the Contractor and, in the event such material or substance is found to be present, to cause it to be rendered harmless. Unless otherwise required by the Contract Documents, the Owner shall furnish in writing to the Contractor and Architect the names and qualifications of persons or entities who are to perform tests verifying the presence or absence of the material or substance or who are to perform the task of removal or safe containment of the material or substance. The Contractor and the Architect will promptly reply to the Owner in writing stating whether or not either has reasonable objection to the persons or entities proposed by the Owner. If either the Contractor or Architect has an objection to a person or entity proposed by the Owner, the Owner shall propose another to whom the Contractor and the Architect have no reasonable objection. When the material or substance has been rendered harmless, Work in the affected area shall resume upon written agreement of the Owner and Contractor. By Change Order, the Contract Time shall be extended appropriately and the Contract Sum shall be increased by the amount of the Contractor's reasonable additional costs of shutdown, delay, and start-up.

§ 10.3.3 To the fullest extent permitted by law, the Owner shall indemnify and hold harmless the Contractor, Subcontractors, Architect, Architect's consultants, and agents and employees of any of them from and against claims, damages, losses, and expenses, including but not limited to attorneys' fees, arising out of or resulting from performance of the Work in the affected area if in fact the material or substance presents the risk of bodily injury or death as described in Section 10.3.1 and has not been rendered harmless, provided that such claim, damage, loss, or expense is attributable to bodily injury, sickness, disease or death, or to injury to or destruction of tangible property (other than the Work itself), except to the extent that such damage, loss, or expense is due to the fault or negligence of the party seeking indemnity.

§ 10.3.4 The Owner shall not be responsible under this Section 10.3 for hazardous materials or substances the Contractor brings to the site unless such materials or substances are required by the Contract Documents. The Owner shall be responsible for hazardous materials or substances required by the Contract Documents, except to the extent of the Contractor's fault or negligence in the use and handling of such materials or substances.

§ 10.3.5 The Contractor shall reimburse the Owner for the cost and expense the Owner incurs (1) for remediation of hazardous materials or substances the Contractor brings to the site and negligently handles, or (2) where the Contractor fails to perform its obligations under Section 10.3.1, except to the extent that the cost and expense are due to the Owner's fault or negligence.

§ 10.3.6 If, without negligence on the part of the Contractor, the Contractor is held liable by a government agency for the cost of remediation of a hazardous material or substance solely by reason of performing Work as required by the Contract Documents, the Owner shall reimburse the Contractor for all cost and expense thereby incurred.

§ 10.4 Emergencies
In an emergency affecting safety of persons or property, the Contractor shall act, at the Contractor's discretion, to prevent threatened damage, injury, or loss. Additional compensation or extension of time claimed by the Contractor on account of an emergency shall be determined as provided in Article 15 and Article 7.

ARTICLE 11 INSURANCE AND BONDS
§ 11.1 Contractor's Insurance and Bonds
§ 11.1.1 The Contractor shall purchase and maintain insurance of the types and limits of liability, containing the endorsements, and subject to the terms and conditions, as described in the Agreement or elsewhere in the Contract Documents. The Contractor shall purchase and maintain the required insurance from an insurance company or insurance companies lawfully authorized to issue insurance in the jurisdiction where the Project is located. The Owner, Architect, and Architect's consultants shall be named as additional insureds under the Contractor's commercial general liability policy or as otherwise described in the Contract Documents.

§ 11.1.2 The Contractor shall provide surety bonds of the types, for such penal sums, and subject to such terms and conditions as required by the Contract Documents. The Contractor shall purchase and maintain the required bonds from a company or companies lawfully authorized to issue surety bonds in the jurisdiction where the Project is located.

§ 11.1.3 Upon the request of any person or entity appearing to be a potential beneficiary of bonds covering payment of obligations arising under the Contract, the Contractor shall promptly furnish a copy of the bonds or shall authorize a copy to be furnished.

§ 11.1.4 Notice of Cancellation or Expiration of Contractor's Required Insurance. Within three (3) business days of the date the Contractor becomes aware of an impending or actual cancellation or expiration of any insurance required by the Contract Documents, the Contractor shall provide notice to the Owner of such impending or actual cancellation or expiration. Upon receipt of notice from the Contractor, the Owner shall, unless the lapse in coverage arises from an act or omission of the Owner, have the right to stop the Work until the lapse in coverage has been cured by the procurement of replacement coverage by the Contractor. The furnishing of notice by the Contractor shall not relieve the Contractor of any contractual obligation to provide any required coverage.

§ 11.2 Owner's Insurance
§ 11.2.1 The Owner shall purchase and maintain insurance of the types and limits of liability, containing the endorsements, and subject to the terms and conditions, as described in the Agreement or elsewhere in the Contract Documents. The Owner shall purchase and maintain the required insurance from an insurance company or insurance companies lawfully authorized to issue insurance in the jurisdiction where the Project is located.

§ 11.2.2 Failure to Purchase Required Property Insurance. If the Owner fails to purchase and maintain the required property insurance, with all of the coverages and in the amounts described in the Agreement or elsewhere in the Contract Documents, the Owner shall inform the Contractor in writing prior to commencement of the Work. Upon receipt of notice from the Owner, the Contractor may delay commencement of the Work and may obtain insurance that will protect the interests of the Contractor, Subcontractors, and Sub-Subcontractors in the Work. When the failure to provide coverage has been cured or resolved, the Contract Sum and Contract Time shall be equitably adjusted. In the event the Owner fails to procure coverage, the Owner waives all rights against the Contractor, Subcontractors, and Sub-subcontractors to the extent the loss to the Owner would have been covered by the insurance to have been procured by the Owner. The cost of the insurance shall be charged to the Owner by a Change Order. If the Owner does not provide written notice, and the Contractor is damaged by the failure or neglect of the Owner to purchase or maintain the required insurance, the Owner shall reimburse the Contractor for all reasonable costs and damages attributable thereto.

§ 11.2.3 Notice of Cancellation or Expiration of Owner's Required Property Insurance. Within three (3) business days of the date the Owner becomes aware of an impending or actual cancellation or expiration of any property insurance required by the Contract Documents, the Owner shall provide notice to the Contractor of such impending or actual cancellation or expiration. Unless the lapse in coverage arises from an act or omission of the Contractor: (1) the Contractor, upon receipt of notice from the Owner, shall have the right to stop the Work until the lapse in coverage has been cured by the procurement of replacement coverage by either the Owner or the Contractor; (2) the Contract Time and Contract Sum shall be equitably adjusted; and (3) the Owner waives all rights against the Contractor, Subcontractors, and Sub-subcontractors to the extent any loss to the Owner would have been covered by the insurance had it not expired or been cancelled. If the Contractor purchases replacement coverage, the cost of the insurance shall be charged to the Owner by an appropriate Change Order. The furnishing of notice by the Owner shall not relieve the Owner of any contractual obligation to provide required insurance.

§ 11.3 Waivers of Subrogation
§ 11.3.1 The Owner and Contractor waive all rights against (1) each other and any of their subcontractors, sub-subcontractors, agents, and employees, each of the other; (2) the Architect and Architect's consultants; and (3) Separate

Contractors, if any, and any of their subcontractors, sub-subcontractors, agents, and employees, for damages caused by fire, or other causes of loss, to the extent those losses are covered by property insurance required by the Agreement or other property insurance applicable to the Project, except such rights as they have to proceeds of such insurance. The Owner or Contractor, as appropriate, shall require similar written waivers in favor of the individuals and entities identified above from the Architect, Architect's consultants, Separate Contractors, subcontractors, and sub-subcontractors. The policies of insurance purchased and maintained by each person or entity agreeing to waive claims pursuant to this section 11.3.1 shall not prohibit this waiver of subrogation. This waiver of subrogation shall be effective as to a person or entity (1) even though that person or entity would otherwise have a duty of indemnification, contractual or otherwise, (2) even though that person or entity did not pay the insurance premium directly or indirectly, or (3) whether or not the person or entity had an insurable interest in the damaged property.

§ 11.3.2 If during the Project construction period the Owner insures properties, real or personal or both, at or adjacent to the site by property insurance under policies separate from those insuring the Project, or if after final payment property insurance is to be provided on the completed Project through a policy or policies other than those insuring the Project during the construction period, to the extent permissible by such policies, the Owner waives all rights in accordance with the terms of Section 11.3.1 for damages caused by fire or other causes of loss covered by this separate property insurance.

§ 11.4 Loss of Use, Business Interruption, and Delay in Completion Insurance
The Owner, at the Owner's option, may purchase and maintain insurance that will protect the Owner against loss of use of the Owner's property, or the inability to conduct normal operations, due to fire or other causes of loss. The Owner waives all rights of action against the Contractor and Architect for loss of use of the Owner's property, due to fire or other hazards however caused.

§11.5 Adjustment and Settlement of Insured Loss
§ 11.5.1 A loss insured under the property insurance required by the Agreement shall be adjusted by the Owner as fiduciary and made payable to the Owner as fiduciary for the insureds, as their interests may appear, subject to requirements of any applicable mortgagee clause and of Section 11.5.2. The Owner shall pay the Architect and Contractor their just shares of insurance proceeds received by the Owner, and by appropriate agreements the Architect and Contractor shall make payments to their consultants and Subcontractors in similar manner.

§ 11.5.2 Prior to settlement of an insured loss, the Owner shall notify the Contractor of the terms of the proposed settlement as well as the proposed allocation of the insurance proceeds. The Contractor shall have 14 days from receipt of notice to object to the proposed settlement or allocation of the proceeds. If the Contractor does not object, the Owner shall settle the loss and the Contractor shall be bound by the settlement and allocation. Upon receipt, the Owner shall deposit the insurance proceeds in a separate account and make the appropriate distributions. Thereafter, if no other agreement is made or the Owner does not terminate the Contract for convenience, the Owner and Contractor shall execute a Change Order for reconstruction of the damaged or destroyed Work in the amount allocated for that purpose. If the Contractor timely objects to either the terms of the proposed settlement or the allocation of the proceeds, the Owner may proceed to settle the insured loss, and any dispute between the Owner and Contractor arising out of the settlement or allocation of the proceeds shall be resolved pursuant to Article 15. Pending resolution of any dispute, the Owner may issue a Construction Change Directive for the reconstruction of the damaged or destroyed Work.

ARTICLE 12 UNCOVERING AND CORRECTION OF WORK
§ 12.1 Uncovering of Work
§ 12.1.1 If a portion of the Work is covered contrary to the Architect's request or to requirements specifically expressed in the Contract Documents, it must, if requested in writing by the Architect, be uncovered for the Architect's examination and be replaced at the Contractor's expense without change in the Contract Time.

§ 12.1.2 If a portion of the Work has been covered that the Architect has not specifically requested to examine prior to its being covered, the Architect may request to see such Work and it shall be uncovered by the Contractor. If such Work is in accordance with the Contract Documents, the Contractor shall be entitled to an equitable adjustment to the Contract Sum and Contract Time as may be appropriate. If such Work is not in accordance with the Contract Documents, the costs of uncovering the Work, and the cost of correction, shall be at the Contractor's expense.

§ 12.2 Correction of Work
§ 12.2.1 Before Substantial Completion
The Contractor shall promptly correct Work rejected by the Architect or failing to conform to the requirements of the

Contract Documents, discovered before Substantial Completion and whether or not fabricated, installed or completed. Costs of correcting such rejected Work, including additional testing and inspections, the cost of uncovering and replacement, and compensation for the Architect's services and expenses made necessary thereby, shall be at the Contractor's expense.

§ 12.2.2 After Substantial Completion

§ 12.2.2.1 In addition to the Contractor's obligations under Section 3.5, if, within one year after the date of Substantial Completion of the Work or designated portion thereof or after the date for commencement of warranties established under Section 9.9.1, or by terms of any applicable special warranty required by the Contract Documents, any of the Work is found to be not in accordance with the requirements of the Contract Documents, the Contractor shall correct it promptly after receipt of notice from the Owner to do so, unless the Owner has previously given the Contractor a written acceptance of such condition. The Owner shall give such notice promptly after discovery of the condition. During the one-year period for correction of Work, if the Owner fails to notify the Contractor and give the Contractor an opportunity to make the correction, the Owner waives the rights to require correction by the Contractor and to make a claim for breach of warranty. If the Contractor fails to correct nonconforming Work within a reasonable time during that period after receipt of notice from the Owner or Architect, the Owner may correct it in accordance with Section 2.5.

§ 12.2.2.2 The one-year period for correction of Work shall be extended with respect to portions of Work first performed after Substantial Completion by the period of time between Substantial Completion and the actual completion of that portion of the Work.

§ 12.2.2.3 The one-year period for correction of Work shall not be extended by corrective Work performed by the Contractor pursuant to this Section 12.2.

§ 12.2.3 The Contractor shall remove from the site portions of the Work that are not in accordance with the requirements of the Contract Documents and are neither corrected by the Contractor nor accepted by the Owner.

§ 12.2.4 The Contractor shall bear the cost of correcting destroyed or damaged construction of the Owner or Separate Contractors, whether completed or partially completed, caused by the Contractor's correction or removal of Work that is not in accordance with the requirements of the Contract Documents.

§ 12.2.5 Nothing contained in this Section 12.2 shall be construed to establish a period of limitation with respect to other obligations the Contractor has under the Contract Documents. Establishment of the one-year period for correction of Work as described in Section 12.2.2 relates only to the specific obligation of the Contractor to correct the Work, and has no relationship to the time within which the obligation to comply with the Contract Documents may be sought to be enforced, nor to the time within which proceedings may be commenced to establish the Contractor's liability with respect to the Contractor's obligations other than specifically to correct the Work.

§ 12.3 Acceptance of Nonconforming Work

If the Owner prefers to accept Work that is not in accordance with the requirements of the Contract Documents, the Owner may do so instead of requiring its removal and correction, in which case the Contract Sum will be reduced as appropriate and equitable. Such adjustment shall be effected whether or not final payment has been made.

ARTICLE 13 MISCELLANEOUS PROVISIONS

§ 13.1 Governing Law

The Contract shall be governed by the law of the place where the Project is located, excluding that jurisdiction's choice of law rules. If the parties have selected arbitration as the method of binding dispute resolution, the Federal Arbitration Act shall govern Section 15.4.

§ 13.2 Successors and Assigns

§ 13.2.1 The Owner and Contractor respectively bind themselves, their partners, successors, assigns, and legal representatives to covenants, agreements, and obligations contained in the Contract Documents. Except as provided in Section 13.2.2, neither party to the Contract shall assign the Contract as a whole without written consent of the other. If either party attempts to make an assignment without such consent, that party shall nevertheless remain legally responsible for all obligations under the Contract.

§ 13.2.2 The Owner may, without consent of the Contractor, assign the Contract to a lender providing construction financing for the Project, if the lender assumes the Owner's rights and obligations under the Contract Documents. The

Contractor shall execute all consents reasonably required to facilitate the assignment.

§ 13.3 Rights and Remedies
§ 13.3.1 Duties and obligations imposed by the Contract Documents and rights and remedies available thereunder shall be in addition to and not a limitation of duties, obligations, rights, and remedies otherwise imposed or available by law.

§ 13.3.2 No action or failure to act by the Owner, Architect, or Contractor shall constitute a waiver of a right or duty afforded them under the Contract, nor shall such action or failure to act constitute approval of or acquiescence in a breach thereunder, except as may be specifically agreed upon in writing.

§ 13.4 Tests and Inspections
§ 13.4.1 Tests, inspections, and approvals of portions of the Work shall be made as required by the Contract Documents and by applicable laws, statutes, ordinances, codes, rules, and regulations or lawful orders of public authorities. Unless otherwise provided, the Contractor shall make arrangements for such tests, inspections, and approvals with an independent testing laboratory or entity acceptable to the Owner, or with the appropriate public authority, and shall bear all related costs of tests, inspections, and approvals. The Contractor shall give the Architect timely notice of when and where tests and inspections are to be made so that the Architect may be present for such procedures. The Owner shall bear costs of tests, inspections, or approvals that do not become requirements until after bids are received or negotiations concluded. The Owner shall directly arrange and pay for tests, inspections, or approvals where building codes or applicable laws or regulations so require.

§ 13.4.2 If the Architect, Owner, or public authorities having jurisdiction determine that portions of the Work require additional testing, inspection, or approval not included under Section 13.4.1, the Architect will, upon written authorization from the Owner, instruct the Contractor to make arrangements for such additional testing, inspection, or approval, by an entity acceptable to the Owner, and the Contractor shall give timely notice to the Architect of when and where tests and inspections are to be made so that the Architect may be present for such procedures. Such costs, except as provided in Section 13.4.3, shall be at the Owner's expense.

§ 13.4.3 If procedures for testing, inspection, or approval under Sections 13.4.1 and 13.4.2 reveal failure of the portions of the Work to comply with requirements established by the Contract Documents, all costs made necessary by such failure, including those of repeated procedures and compensation for the Architect's services and expenses, shall be at the Contractor's expense.

§ 13.4.4 Required certificates of testing, inspection, or approval shall, unless otherwise required by the Contract Documents, be secured by the Contractor and promptly delivered to the Architect.

§ 13.4.5 If the Architect is to observe tests, inspections, or approvals required by the Contract Documents, the Architect will do so promptly and, where practicable, at the normal place of testing.

§ 13.4.6 Tests or inspections conducted pursuant to the Contract Documents shall be made promptly to avoid unreasonable delay in the Work.

§ 13.5 Interest
Payments due and unpaid under the Contract Documents shall bear interest from the date payment is due at the rate the parties agree upon in writing or, in the absence thereof, at the legal rate prevailing from time to time at the place where the Project is located.

ARTICLE 14 TERMINATION OR SUSPENSION OF THE CONTRACT
§ 14.1 Termination by the Contractor
§ 14.1.1 The Contractor may terminate the Contract if the Work is stopped for a period of 30 consecutive days through no act or fault of the Contractor, a Subcontractor, a Sub-subcontractor, their agents or employees, or any other persons or entities performing portions of the Work, for any of the following reasons:

.1 Issuance of an order of a court or other public authority having jurisdiction that requires all Work to be stopped;
.2 An act of government, such as a declaration of national emergency, that requires all Work to be stopped;
.3 Because the Architect has not issued a Certificate for Payment and has not notified the Contractor of the reason for withholding certification as provided in Section 9.4.1, or because the Owner has not made payment on a Certificate for Payment within the time stated in the Contract Documents; or

.4 The Owner has failed to furnish to the Contractor reasonable evidence as required by Section 2.2.

§ 14.1.2 The Contractor may terminate the Contract if, through no act or fault of the Contractor, a Subcontractor, a Sub-subcontractor, their agents or employees, or any other persons or entities performing portions of the Work, repeated suspensions, delays, or interruptions of the entire Work by the Owner as described in Section 14.3, constitute in the aggregate more than 100 percent of the total number of days scheduled for completion, or 120 days in any 365-day period, whichever is less.

§ 14.1.3 If one of the reasons described in Section 14.1.1 or 14.1.2 exists, the Contractor may, upon seven days' notice to the Owner and Architect, terminate the Contract and recover from the Owner payment for Work executed, as well as reasonable overhead and profit on Work not executed, and costs incurred by reason of such termination.

§ 14.1.4 If the Work is stopped for a period of 60 consecutive days through no act or fault of the Contractor, a Subcontractor, a Sub-subcontractor, or their agents or employees or any other persons or entities performing portions of the Work because the Owner has repeatedly failed to fulfill the Owner's obligations under the Contract Documents with respect to matters important to the progress of the Work, the Contractor may, upon seven additional days' notice to the Owner and the Architect, terminate the Contract and recover from the Owner as provided in Section 14.1.3.

§ 14.2 Termination by the Owner for Cause

§ 14.2.1 The Owner may terminate the Contract if the Contractor

.1 repeatedly refuses or fails to supply enough properly skilled workers or proper materials;

.2 fails to make payment to Subcontractors or suppliers in accordance with the respective agreements between the Contractor and the Subcontractors or Suppliers;

.3 repeatedly disregards applicable laws, statutes, ordinances, codes, rules and regulations, or lawful orders of a public authority; or

.4 otherwise is guilty of substantial breach of a provision of the Contract Documents.

§ 14.2.2 When any of the reasons described in Section 14.2.1 exist, and upon certification by the Architect that sufficient cause exists to justify such action, the Owner may, without prejudice to any other rights or remedies of the Owner and after giving the Contractor and the Contractor's surety, if any, seven days' notice, terminate employment of the Contractor and may, subject to any prior rights of the surety:

.1 Exclude the Contractor from the site and take possession of all materials, equipment, tools, and construction equipment and machinery thereon owned by the Contractor;

.2 Accept assignment of subcontracts pursuant to Section 5.4; and

.3 Finish the Work by whatever reasonable method the Owner may deem expedient. Upon written request of the Contractor, the Owner shall furnish to the Contractor a detailed accounting of the costs incurred by the Owner in finishing the Work.

§ 14.2.3 When the Owner terminates the Contract for one of the reasons stated in Section 14.2.1, the Contractor shall not be entitled to receive further payment until the Work is finished.

§ 14.2.4 If the unpaid balance of the Contract Sum exceeds costs of finishing the Work, including compensation for the Architect's services and expenses made necessary thereby, and other damages incurred by the Owner and not expressly waived, such excess shall be paid to the Contractor. If such costs and damages exceed the unpaid balance, the Contractor shall pay the difference to the Owner. The amount to be paid to the Contractor or Owner, as the case may be, shall be certified by the Initial Decision Maker, upon application, and this obligation for payment shall survive termination of the Contract.

§ 14.3 Suspension by the Owner for Convenience

§ 14.3.1 The Owner may, without cause, order the Contractor in writing to suspend, delay or interrupt the Work, in whole or in part for such period of time as the Owner may determine.

§ 14.3.2 The Contract Sum and Contract Time shall be adjusted for increases in the cost and time caused by suspension, delay, or interruption under Section 14.3.1. Adjustment of the Contract Sum shall include profit. No adjustment shall be made to the extent

.1 that performance is, was, or would have been, so suspended, delayed, or interrupted, by another cause for which the Contractor is responsible; or

.2 that an equitable adjustment is made or denied under another provision of the Contract.

§ 14.4 Termination by the Owner for Convenience
§ 14.4.1 The Owner may, at any time, terminate the Contract for the Owner's convenience and without cause.

§ 14.4.2 Upon receipt of notice from the Owner of such termination for the Owner's convenience, the Contractor shall

- **.1** cease operations as directed by the Owner in the notice;
- **.2** take actions necessary, or that the Owner may direct, for the protection and preservation of the Work; and
- **.3** except for Work directed to be performed prior to the effective date of termination stated in the notice, terminate all existing subcontracts and purchase orders and enter into no further subcontracts and purchase orders.

§ 14.4.3 In case of such termination for the Owner's convenience, the Owner shall pay the Contractor for Work properly executed; costs incurred by reason of the termination, including costs attributable to termination of Subcontracts; and the termination fee, if any, set forth in the Agreement.

ARTICLE 15 CLAIMS AND DISPUTES
§ 15.1 Claims
§ 15.1.1 Definition
A Claim is a demand or assertion by one of the parties seeking, as a matter of right, payment of money, a change in the Contract Time, or other relief with respect to the terms of the Contract. The term "Claim" also includes other disputes and matters in question between the Owner and Contractor arising out of or relating to the Contract. The responsibility to substantiate Claims shall rest with the party making the Claim. This Section 15.1.1 does not require the Owner to file a Claim in order to impose liquidated damages in accordance with the Contract Documents.

§ 15.1.2 Time Limits on Claims
The Owner and Contractor shall commence all Claims and causes of action against the other and arising out of or related to the Contract, whether in contract, tort, breach of warranty or otherwise, in accordance with the requirements of the binding dispute resolution method selected in the Agreement and within the period specified by applicable law, but in any case not more than 10 years after the date of Substantial Completion of the Work. The Owner and Contractor waive all Claims and causes of action not commenced in accordance with this Section 15.1.2.

§ 15.1.3 Notice of Claims
§ 15.1.3.1 Claims by either the Owner or Contractor, where the condition giving rise to the Claim is first discovered prior to expiration of the period for correction of the Work set forth in Section 12.2.2, shall be initiated by notice to the other party and to the Initial Decision Maker with a copy sent to the Architect, if the Architect is not serving as the Initial Decision Maker. Claims by either party under this Section 15.1.3.1 shall be initiated within 21 days after occurrence of the event giving rise to such Claim or within 21 days after the claimant first recognizes the condition giving rise to the Claim, whichever is later.

§ 15.1.3.2 Claims by either the Owner or Contractor, where the condition giving rise to the Claim is first discovered after expiration of the period for correction of the Work set forth in Section 12.2.2, shall be initiated by notice to the other party. In such event, no decision by the Initial Decision Maker is required.

§ 15.1.4 Continuing Contract Performance
§ 15.1.4.1 Pending final resolution of a Claim, except as otherwise agreed in writing or as provided in Section 9.7 and Article 14, the Contractor shall proceed diligently with performance of the Contract and the Owner shall continue to make payments in accordance with the Contract Documents.

§ 15.1.4.2 The Contract Sum and Contract Time shall be adjusted in accordance with the Initial Decision Maker's decision, subject to the right of either party to proceed in accordance with this Article 15. The Architect will issue Certificates for Payment in accordance with the decision of the Initial Decision Maker.

§ 15.1.5 Claims for Additional Cost
If the Contractor wishes to make a Claim for an increase in the Contract Sum, notice as provided in Section 15.1.3 shall be given before proceeding to execute the portion of the Work that is the subject of the Claim. Prior notice is not required for Claims relating to an emergency endangering life or property arising under Section 10.4.

§ 15.1.6 Claims for Additional Time
§ 15.1.6.1 If the Contractor wishes to make a Claim for an increase in the Contract Time, notice as provided in Section 15.1.3 shall be given. The Contractor's Claim shall include an estimate of cost and of probable effect of delay on progress of the Work. In the case of a continuing delay, only one Claim is necessary.

§ 15.1.6.2 If adverse weather conditions are the basis for a Claim for additional time, such Claim shall be documented by data substantiating that weather conditions were abnormal for the period of time, could not have been reasonably anticipated, and had an adverse effect on the scheduled construction.

§ 15.1.7 Waiver of Claims for Consequential Damages
The Contractor and Owner waive Claims against each other for consequential damages arising out of or relating to this Contract. This mutual waiver includes

.1 damages incurred by the Owner for rental expenses, for losses of use, income, profit, financing, business and reputation, and for loss of management or employee productivity or of the services of such persons; and

.2 damages incurred by the Contractor for principal office expenses including the compensation of personnel stationed there, for losses of financing, business and reputation, and for loss of profit, except anticipated profit arising directly from the Work.

This mutual waiver is applicable, without limitation, to all consequential damages due to either party's termination in accordance with Article 14. Nothing contained in this Section 15.1.7 shall be deemed to preclude assessment of liquidated damages, when applicable, in accordance with the requirements of the Contract Documents.

§ 15.2 Initial Decision
§ 15.2.1 Claims, excluding those where the condition giving rise to the Claim is first discovered after expiration of the period for correction of the Work set forth in Section 12.2.2 or arising under Sections 10.3, 10.4, and 11.5, shall be referred to the Initial Decision Maker for initial decision. The Architect will serve as the Initial Decision Maker, unless otherwise indicated in the Agreement. Except for those Claims excluded by this Section 15.2.1, an initial decision shall be required as a condition precedent to mediation of any Claim. If an initial decision has not been rendered within 30 days after the Claim has been referred to the Initial Decision Maker, the party asserting the Claim may demand mediation and binding dispute resolution without a decision having been rendered. Unless the Initial Decision Maker and all affected parties agree, the Initial Decision Maker will not decide disputes between the Contractor and persons or entities other than the Owner.

§ 15.2.2 The Initial Decision Maker will review Claims and within ten days of the receipt of a Claim take one or more of the following actions: (1) request additional supporting data from the claimant or a response with supporting data from the other party, (2) reject the Claim in whole or in part, (3) approve the Claim, (4) suggest a compromise, or (5) advise the parties that the Initial Decision Maker is unable to resolve the Claim if the Initial Decision Maker lacks sufficient information to evaluate the merits of the Claim or if the Initial Decision Maker concludes that, in the Initial Decision Maker's sole discretion, it would be inappropriate for the Initial Decision Maker to resolve the Claim.

§ 15.2.3 In evaluating Claims, the Initial Decision Maker may, but shall not be obligated to, consult with or seek information from either party or from persons with special knowledge or expertise who may assist the Initial Decision Maker in rendering a decision. The Initial Decision Maker may request the Owner to authorize retention of such persons at the Owner's expense.

§ 15.2.4 If the Initial Decision Maker requests a party to provide a response to a Claim or to furnish additional supporting data, such party shall respond, within ten days after receipt of the request, and shall either (1) provide a response on the requested supporting data, (2) advise the Initial Decision Maker when the response or supporting data will be furnished, or (3) advise the Initial Decision Maker that no supporting data will be furnished. Upon receipt of the response or supporting data, if any, the Initial Decision Maker will either reject or approve the Claim in whole or in part.

§ 15.2.5 The Initial Decision Maker will render an initial decision approving or rejecting the Claim, or indicating that the Initial Decision Maker is unable to resolve the Claim. This initial decision shall (1) be in writing; (2) state the reasons therefor; and (3) notify the parties and the Architect, if the Architect is not serving as the Initial Decision Maker, of any change in the Contract Sum or Contract Time or both. The initial decision shall be final and binding on the parties but subject to mediation and, if the parties fail to resolve their dispute through mediation, to binding dispute resolution.

§ 15.2.6 Either party may file for mediation of an initial decision at any time, subject to the terms of Section 15.2.6.1.

§ 15.2.6.1 Either party may, within 30 days from the date of receipt of an initial decision, demand in writing that the other party file for mediation. If such a demand is made and the party receiving the demand fails to file for mediation within 30 days after receipt thereof, then both parties waive their rights to mediate or pursue binding dispute resolution proceedings with respect to the initial decision.

§ 15.2.7 In the event of a Claim against the Contractor, the Owner may, but is not obligated to, notify the surety, if any, of the nature and amount of the Claim. If the Claim relates to a possibility of a Contractor's default, the Owner may, but is not obligated to, notify the surety and request the surety's assistance in resolving the controversy.

§ 15.2.8 If a Claim relates to or is the subject of a mechanic's lien, the party asserting such Claim may proceed in accordance with applicable law to comply with the lien notice or filing deadlines.

§ 15.3 Mediation
§ 15.3.1 Claims, disputes, or other matters in controversy arising out of or related to the Contract, except those waived as provided for in Sections 9.10.4, 9.10.5, and 15.1.7, shall be subject to mediation as a condition precedent to binding dispute resolution.

§ 15.3.2 The parties shall endeavor to resolve their Claims by mediation which, unless the parties mutually agree otherwise, shall be administered by the American Arbitration Association in accordance with its Construction Industry Mediation Procedures in effect on the date of the Agreement. A request for mediation shall be made in writing, delivered to the other party to the Contract, and filed with the person or entity administering the mediation. The request may be made concurrently with the filing of binding dispute resolution proceedings but, in such event, mediation shall proceed in advance of binding dispute resolution proceedings, which shall be stayed pending mediation for a period of 60 days from the date of filing, unless stayed for a longer period by agreement of the parties or court order. If an arbitration is stayed pursuant to this Section 15.3.2, the parties may nonetheless proceed to the selection of the arbitrator(s) and agree upon a schedule for later proceedings.

§ 15.3.3 Either party may, within 30 days from the date that mediation has been concluded without resolution of the dispute or 60 days after mediation has been demanded without resolution of the dispute, demand in writing that the other party file for binding dispute resolution. If such a demand is made and the party receiving the demand fails to file for binding dispute resolution within 60 days after receipt thereof, then both parties waive their rights to binding dispute resolution proceedings with respect to the initial decision.

§ 15.3.4 The parties shall share the mediator's fee and any filing fees equally. The mediation shall be held in the place where the Project is located, unless another location is mutually agreed upon. Agreements reached in mediation shall be enforceable as settlement agreements in any court having jurisdiction thereof.

§ 15.4 Arbitration
§ 15.4.1 If the parties have selected arbitration as the method for binding dispute resolution in the Agreement, any Claim subject to, but not resolved by, mediation shall be subject to arbitration which, unless the parties mutually agree otherwise, shall be administered by the American Arbitration Association in accordance with its Construction Industry Arbitration Rules in effect on the date of the Agreement. The Arbitration shall be conducted in the place where the Project is located, unless another location is mutually agreed upon. A demand for arbitration shall be made in writing, delivered to the other party to the Contract, and filed with the person or entity administering the arbitration. The party filing a notice of demand for arbitration must assert in the demand all Claims then known to that party on which arbitration is permitted to be demanded.

§ 15.4.1.1 A demand for arbitration shall be made no earlier than concurrently with the filing of a request for mediation, but in no event shall it be made after the date when the institution of legal or equitable proceedings based on the Claim would be barred by the applicable statute of limitations. For statute of limitations purposes, receipt of a written demand for arbitration by the person or entity administering the arbitration shall constitute the institution of legal or equitable proceedings based on the Claim.

§ 15.4.2 The award rendered by the arbitrator or arbitrators shall be final, and judgment may be entered upon it in accordance with applicable law in any court having jurisdiction thereof.

Init.
/

§ 15.4.3 The foregoing agreement to arbitrate and other agreements to arbitrate with an additional person or entity duly consented to by parties to the Agreement, shall be specifically enforceable under applicable law in any court having jurisdiction thereof.

§ 15.4.4 Consolidation or Joinder
§ 15.4.4.1 Subject to the rules of the American Arbitration Association or other applicable arbitration rules, either party may consolidate an arbitration conducted under this Agreement with any other arbitration to which it is a party provided that (1) the arbitration agreement governing the other arbitration permits consolidation, (2) the arbitrations to be consolidated substantially involve common questions of law or fact, and (3) the arbitrations employ materially similar procedural rules and methods for selecting arbitrator(s).

§ 15.4.4.2 Subject to the rules of the American Arbitration Association or other applicable arbitration rules, either party may include by joinder persons or entities substantially involved in a common question of law or fact whose presence is required if complete relief is to be accorded in arbitration, provided that the party sought to be joined consents in writing to such joinder. Consent to arbitration involving an additional person or entity shall not constitute consent to arbitration of any claim, dispute or other matter in question not described in the written consent.

§ 15.4.4.3 The Owner and Contractor grant to any person or entity made a party to an arbitration conducted under this Section 15.4, whether by joinder or consolidation, the same rights of joinder and consolidation as those of the Owner and Contractor under this Agreement.

APPENDIX B:
AIA® Document Comparative

AIA® Document Comparative

A201™ – 2017 Compared to A201™ – 2007

Additions to A201–2007 are underlined. Deletions from A201–2007 are in ~~strikethrough~~ text.

Title

General Conditions of the Contract for Construction

Cover Page

for the following PROJECT:
(Name and location or address)

THE OWNER:
(Name and address)

THE ARCHITECT:
(Name and address)

Table of Articles

Article 1 General Provisions

§ 1.1 Basic Definitions
§ 1.1.1 The Contract Documents
The Contract Documents are enumerated in the Agreement between the Owner and Contractor (hereinafter the Agreement) and consist of the Agreement, Conditions of the Contract (General, Supplementary and other Conditions), Drawings, Specifications, Addenda issued prior to execution of the Contract, other documents listed in the Agreement, and Modifications issued after execution of the Contract. A Modification is (1) a written amendment to the Contract signed by both parties, (2) a Change Order, (3) a Construction Change Directive, or (4) a written order for a minor change in the Work issued by the Architect. Unless specifically enumerated in the Agreement, the Contract Documents do not include the advertisement or invitation to bid, Instructions to Bidders, sample forms, other information furnished by the Owner in anticipation of receiving bids or proposals, the Contractor's bid or proposal, or portions of Addenda relating to bidding or proposal requirements.

§ 1.1.2 The Contract
The Contract Documents form the Contract for Construction. The Contract represents the entire and integrated agreement between the parties hereto and supersedes prior negotiations, representations, or agreements, either written or oral. The Contract may be amended or modified only by a Modification. The Contract Documents shall

not be construed to create a contractual relationship of any kind (1) between the Contractor and the Architect or the Architect's consultants, (2) between the Owner and a Subcontractor or a Sub-subcontractor, (3) between the Owner and the Architect or the Architect's consultants, or (4) between any persons or entities other than the Owner and the Contractor. The Architect shall, however, be entitled to performance and enforcement of obligations under the Contract intended to facilitate performance of the Architect's duties.

§ 1.1.3 The Work
The term "Work" means the construction and services required by the Contract Documents, whether completed or partially completed, and includes all other labor, materials, equipment, and services provided or to be provided by the Contractor to fulfill the Contractor's obligations. The Work may constitute the whole or a part of the Project.

§ 1.1.4 The Project
The Project is the total construction of which the Work performed under the Contract Documents may be the whole or a part and which may include construction by the Owner and by ~~separate contractors~~Separate Contractors.

§ 1.1.5 The Drawings
The Drawings are the graphic and pictorial portions of the Contract Documents showing the design, location and dimensions of the Work, generally including plans, elevations, sections, details, schedules, and diagrams.

§ 1.1.6 The Specifications
The Specifications are that portion of the Contract Documents consisting of the written requirements for materials, equipment, systems, standards and workmanship for the Work, and performance of related services.

§ 1.1.7 Instruments of Service
Instruments of Service are representations, in any medium of expression now known or later developed, of the tangible and intangible creative work performed by the Architect and the Architect's consultants under their respective professional services agreements. Instruments of Service may include, without limitation, studies, surveys, models, sketches, drawings, specifications, and other similar materials.

§ 1.1.8 Initial Decision Maker
The Initial Decision Maker is the person identified in the Agreement to render initial decisions on Claims in accordance with Section ~~15.2 and certify termination of the Agreement under Section 14.2.2.~~ 15.2. The Initial Decision Maker shall not show partiality to the Owner or Contractor and shall not be liable for results of interpretations or decisions rendered in good faith.

§ 1.2 Correlation and Intent of the Contract Documents
§ 1.2.1 The intent of the Contract Documents is to include all items necessary for the proper execution and completion of the Work by the Contractor. The Contract Documents are complementary, and what is required by one shall be as binding as if required by all; performance by the Contractor shall be required only to the extent consistent with the Contract Documents and reasonably inferable from them as being necessary to produce the indicated results.

§ 1.2.1.1 The invalidity of any provision of the Contract Documents shall not invalidate the Contract or its remaining provisions. If it is determined that any provision of the Contract Documents violates any law, or is otherwise invalid or unenforceable, then that provision shall be revised to the extent necessary to make that provision legal and enforceable. In such case the Contract Documents shall be construed, to the fullest extent permitted by law, to give effect to the parties' intentions and purposes in executing the Contract.

§ 1.2.2 Organization of the Specifications into divisions, sections and articles, and arrangement of Drawings shall not control the Contractor in dividing the Work among Subcontractors or in establishing the extent of Work to be performed by any trade.

§ 1.2.3 Unless otherwise stated in the Contract Documents, words that have well-known technical or construction industry meanings are used in the Contract Documents in accordance with such recognized meanings.

§ 1.3 Capitalization
Terms capitalized in these General Conditions include those that are (1) specifically defined, (2) the titles of numbered articles, or (3) the titles of other documents published by the American Institute of Architects.

§ 1.4 Interpretation
In the interest of brevity the Contract Documents frequently omit modifying words such as "all" and "any" and articles such as "the" and "an," but the fact that a modifier or an article is absent from one statement and appears in another is not intended to affect the interpretation of either statement.

§ 1.5 Ownership and Use of Drawings, Specifications, and Other Instruments of Service
§ 1.5.1 The Architect and the Architect's consultants shall be deemed the authors and owners of their respective Instruments of Service, including the Drawings and Specifications, and ~~will~~ retain all common law, statutory, and other reserved rights in their Instruments of Service, including copyrights. The Contractor, Subcontractors, Sub-subcontractors, and ~~material or equipment~~ suppliers shall not own or claim a copyright in the Instruments of Service. Submittal or distribution to meet official regulatory requirements or for other purposes in connection with ~~this~~the Project is not to be construed as publication in derogation of the Architect's or Architect's consultants' reserved rights.

§ 1.5.2 The Contractor, Subcontractors, Sub-subcontractors, and ~~material or equipment~~ suppliers are authorized to use and reproduce the Instruments of Service provided to them, subject to any protocols established pursuant to Sections 1.7 and 1.8, solely and exclusively for execution of the Work. All copies made under this authorization shall bear the copyright notice, if any, shown on the Instruments of Service. The Contractor, Subcontractors, Sub-subcontractors, and ~~material or equipment~~ suppliers may not use the Instruments of Service on other projects or for additions to ~~this~~the Project outside the scope of the Work without the specific written consent of the Owner, Architect, and the Architect's consultants.

~~**§ 1.6 TRANSMISSION OF DATA IN DIGITAL FORM**~~
§ 1.6 Notice
§ 1.6.1 Except as otherwise provided in Section 1.6.2, where the Contract Documents require one party to notify or give notice to the other party, such notice shall be provided in writing to the designated representative of the party to whom the notice is addressed and shall be deemed to have been duly served if delivered in person, by mail, by courier, or by electronic transmission if a method for electronic transmission is set forth in the Agreement.

§ 1.6.2 Notice of Claims as provided in Section 15.1.3 shall be provided in writing and shall be deemed to have been duly served only if delivered to the designated representative of the party to whom the notice is addressed by certified or registered mail, or by courier providing proof of delivery.

> ***Note:*** *Section 1.6 formerly appeared at Section 13.3.*

§ 1.7 Digital Data Use and Transmission
The parties ~~intend to transmit~~shall agree upon protocols governing the transmission and use of Instruments of Service or any other information or documentation in digital form~~, they shall endeavor~~. The parties will use AIA Document E203™–2013, Building Information Modeling and Digital Data Exhibit, to establish ~~necessary protocols governing such transmissions, unless otherwise already provided in the Agreement or the Contract Documents~~the protocols for the development, use, transmission, and exchange of digital data.

§ 1.8 Building Information Models Use and Reliance
Any use of, or reliance on, all or a portion of a building information model without agreement to protocols governing the use of, and reliance on, the information contained in the model and without having those protocols set forth in AIA Document E203™–2013, Building Information Modeling and Digital Data Exhibit, and the requisite AIA Document G202™–2013, Project Building Information Modeling Protocol Form, shall be at the using or relying party's sole risk and without liability to the other party and its contractors or consultants, the authors of, or contributors to, the building information model, and each of their agents and employees.

Article 2 Owner

§ 2.1 General
§ 2.1.1 The Owner is the person or entity identified as such in the Agreement and is referred to throughout the Contract Documents as if singular in number. The Owner shall designate in writing a representative who shall have express authority to bind the Owner with respect to all matters requiring the Owner's approval or authorization. Except as otherwise provided in Section 4.2.1, the Architect does not have such authority. The term "Owner" means the Owner or the Owner's authorized representative.

§ 2.1.2 The Owner shall furnish to the Contractor, within fifteen days after receipt of a written request, information necessary and relevant for the Contractor to evaluate, give notice of, or enforce mechanic's lien rights. Such information shall include a correct statement of the record legal title to the property on which the Project is located, usually referred to as the site, and the Owner's interest therein.

~~**§ 2.2 INFORMATION AND SERVICES REQUIRED OF THE OWNER**~~
§ 2.2 Evidence of the Owner's Financial Arrangements
§ 2.2.1 Prior to commencement of the Work~~,~~ and upon written request by the Contractor ~~may request in writing that~~, the Owner ~~provide~~shall furnish to the Contractor reasonable evidence that the Owner has made financial arrangements to fulfill the Owner's obligations under the Contract. ~~Thereafter, the~~The Contractor ~~may only request~~ shall have no obligation to commence the Work until the Owner provides such evidence. If commencement of the Work is delayed under this Section 2.2.1, the Contract Time shall be extended appropriately.

§ 2.2.2 Following commencement of the Work and upon written request by the Contractor, the Owner shall furnish to the Contractor reasonable evidence that the Owner has made financial arrangements to fulfill the Owner's obligations under the Contract only if (1) the Owner fails to make payments to the Contractor as the Contract Documents require; (2~~) a change in the Work materially changes the Contract Sum; or (3~~) the Contractor identifies in writing a reasonable concern regarding the Owner's ability to make payment when due; or (3) a change in the Work materially changes the Contract Sum. If the Owner fails to provide such evidence, as required, within fourteen days of the Contractor's request, the Contractor may immediately stop the Work and, in that event, shall notify the Owner that the Work has stopped. However, if the request is made because a change in the Work materially changes the Contract Sum under (3) above, the Contractor may immediately stop only that portion of the Work affected by the change until reasonable evidence is provided. If the Work is stopped under this Section 2.2.2, the Contract Time shall be extended appropriately and the Contract Sum shall be increased by the amount of the Contractor's reasonable costs of shutdown, delay and start-up, plus interest as provided in the Contract Documents.

§ 2.2.3 ~~The Owner shall furnish such evidence as a condition precedent to commencement or continuation of the Work or the portion of the Work affected by a material change. After the Owner furnishes the evidence~~ After the Owner furnishes evidence of financial arrangements under this Section 2.2, the Owner shall not materially vary such financial arrangements without prior notice to the Contractor.

~~§ 2.2.2~~§ 2.2.4 Where the Owner has designated information furnished under this Section 2.2 as "confidential," the Contractor shall keep the information confidential and shall not disclose it to any other person. However, the Contractor may disclose "confidential" information, after seven (7) days' notice to the Owner, where disclosure is required by law, including a subpoena or other form of compulsory legal process issued by a court or governmental entity, or by court or arbitrator(s) order. The Contractor may also disclose "confidential" information to its employees, consultants, sureties, Subcontractors and their employees, Sub-subcontractors, and others who need to know the content of such information solely and exclusively for the Project and who agree to maintain the confidentiality of such information.

§ 2.3 Information and Services Required of the Owner
§ 2.3.1 Except for permits and fees that are the responsibility of the Contractor under the Contract Documents, including those required under Section 3.7.1, the Owner shall secure and pay for necessary approvals, easements, assessments and charges required for construction, use or occupancy of permanent structures or for permanent changes in existing facilities.

§ 2.3.2 The Owner shall retain an architect lawfully licensed to practice architecture, or an entity lawfully practicing architecture, in the jurisdiction where the Project is located. That person or entity is identified as the Architect in the Agreement and is referred to throughout the Contract Documents as if singular in number.

> ***Note:*** *Portions of Section 2.3.2 formerly appeared at Section 4.1.1.*

§ 2.3.3 If the employment of the Architect terminates, the Owner shall employ a successor to whom the Contractor has no reasonable objection and whose status under the Contract Documents shall be that of the Architect.

> ***Note:*** *Portions of Section 2.3.3 formerly appeared at Section 4.1.3.*

§ § 2.2.32.3.4 The Owner shall furnish surveys describing physical characteristics, legal limitations and utility locations for the site of the Project, and a legal description of the site. The Contractor shall be entitled to rely on the accuracy of information furnished by the Owner but shall exercise proper precautions relating to the safe performance of the Work.

§ 2.2.43.5 The Owner shall furnish information or services required of the Owner by the Contract Documents with reasonable promptness. The Owner shall also furnish any other information or services under the Owner's control and relevant to the Contractor's performance of the Work with reasonable promptness after receiving the Contractor's written request for such information or services.

§ 2.2.53.6 Unless otherwise provided in the Contract Documents, the Owner shall furnish to the Contractor one copy of the Contract Documents for purposes of making reproductions pursuant to Section 1.5.2.

§ 2.4 Owner's Right to Stop the Work
If the Contractor fails to correct Work that is not in accordance with the requirements of the Contract Documents as required by Section 12.2 or repeatedly fails to carry out Work in accordance with the Contract Documents, the Owner may issue a written order to the Contractor to stop the Work, or any portion thereof, until the cause for such order has been eliminated; however, the right of the Owner to stop the Work shall not give rise to a duty on the part of the Owner to exercise this right for the benefit of the Contractor or any other person or entity, except to the extent required by Section 6.1.3.

§ 2.5 Owner's Right to Carry Out the Work
If the Contractor defaults or neglects to carry out the Work in accordance with the Contract Documents and fails within a ten-day period after receipt of written notice from the Owner to commence and continue correction of such default or neglect with diligence and promptness, the Owner may, without prejudice to other remedies the Owner may have, correct such deficiencies. In such case an appropriate Change Order shall be issued deducting from payments then or thereafter due the Contractordefault or neglect. Such action by the Owner and amounts charged to the Contractor are both subject to prior approval of the Architect and the Architect may, pursuant to Section 9.5.1, withhold or nullify a Certificate for Payment in whole or in part, to the extent reasonably necessary to reimburse the Owner for the reasonable cost of correcting such deficiencies, including Owner's expenses and compensation for the Architect's additional services made necessary by such default, neglect, or failure. Such action by the OwnerIf current and amounts charged to the Contractor are both subject to prior approval of the Architect. If future payments then or thereafter due the Contractor are not sufficient to cover such amounts, the Contractor shall pay the difference to the Owner. If the Contractor disagrees with the actions of the Owner or the Architect, or the amounts claimed as costs to the Owner, the Contractor may file a Claim pursuant to Article 15.

Article 3 Contractor

§ 3.1 General
§ 3.1.1 The Contractor is the person or entity identified as such in the Agreement and is referred to throughout the Contract Documents as if singular in number. The Contractor shall be lawfully licensed, if required in the jurisdiction where the Project is located. The Contractor shall designate in writing a representative who shall have express authority to bind the Contractor with respect to all matters under this Contract. The term "Contractor" means the Contractor or the Contractor's authorized representative.

§ 3.1.2 The Contractor shall perform the Work in accordance with the Contract Documents.

§ 3.1.3 The Contractor shall not be relieved of its obligations to perform the Work in accordance with the Contract Documents either by activities or duties of the Architect in the Architect's administration of the Contract, or by tests, inspections or approvals required or performed by persons or entities other than the Contractor.

§ 3.2 Review of Contract Documents and Field Conditions by Contractor
§ 3.2.1 Execution of the Contract by the Contractor is a representation that the Contractor has visited the site, become generally familiar with local conditions under which the Work is to be performed, and correlated personal observations with requirements of the Contract Documents.

§ 3.2.2 Because the Contract Documents are complementary, the Contractor shall, before starting each portion of the Work, carefully study and compare the various Contract Documents relative to that portion of the Work, as well as the information furnished by the Owner pursuant to Section 2. 2.3.4, shall take field measurements of any existing conditions related to that portion of the Work, and shall observe any conditions at the site affecting it. These obligations are for the purpose of facilitating coordination and construction by the Contractor and are not for the purpose of discovering errors, omissions, or inconsistencies in the Contract Documents; however, the Contractor shall promptly report to the Architect any errors, inconsistencies or omissions discovered by or made known to the Contractor as a request for information in such form as the Architect may require. It is recognized that the Contractor's review is made in the Contractor's capacity as a contractor and not as a licensed design professional, unless otherwise specifically provided in the Contract Documents.

§ 3.2.3 The Contractor is not required to ascertain that the Contract Documents are in accordance with applicable laws, statutes, ordinances, codes, rules and regulations, or lawful orders of public authorities, but the Contractor shall promptly report to the Architect any nonconformity discovered by or made known to the Contractor as a request for information in such form as the Architect may require.

§ 3.2.4 If the Contractor believes that additional cost or time is involved because of clarifications or instructions the Architect issues in response to the Contractor's notices or requests for information pursuant to Sections 3.2.2 or 3.2.3, the Contractor shall ~~make~~submit Claims as provided in Article 15. If the Contractor fails to perform the obligations of Sections 3.2.2 or 3.2.3, the Contractor shall pay such costs and damages to the Owner, subject to Section 15.1.7, as would have been avoided if the Contractor had performed such obligations. If the Contractor performs those obligations, the Contractor shall not be liable to the Owner or Architect for damages resulting from errors, inconsistencies or omissions in the Contract Documents, for differences between field measurements or conditions and the Contract Documents, or for nonconformities of the Contract Documents to applicable laws, statutes, ordinances, codes, rules and regulations, and lawful orders of public authorities.

§ 3.3 Supervision and Construction Procedures

§ 3.3.1 The Contractor shall supervise and direct the Work, using the Contractor's best skill and attention. The Contractor shall be solely responsible for, and have control over, construction means, methods, techniques, sequences, and procedures, and for coordinating all portions of the Work under the Contract~~, unless the Contract Documents give other specific instructions concerning these matters~~. If the Contract Documents give specific instructions concerning construction means, methods, techniques, sequences, or procedures, the Contractor shall evaluate the jobsite safety thereof and~~, except as stated below,~~ shall be ~~fully and~~ solely responsible for the jobsite safety of such means, methods, techniques, sequences, or procedures. If the Contractor determines that such means, methods, techniques, sequences or procedures may not be safe, the Contractor shall give timely ~~written~~ notice to the Owner and Architect, and shall ~~not proceed with that portion of the Work without further written instructions from the Architect. If the Contractor is then instructed to proceed with the required~~propose alternative means, methods, techniques, sequences, or procedures~~ without acceptance of changes~~. The Architect shall evaluate the proposed ~~by the Contractor, the Owner shall be~~alternative solely ~~responsible~~ for ~~any loss or damage arising solely from those Owner-required~~conformance with the design intent for the completed construction. Unless the Architect objects to the Contractor's proposed alternative, the Contractor shall perform the Work using its alternative means, methods, techniques, sequences, or procedures.

§ 3.3.2 The Contractor shall be responsible to the Owner for acts and omissions of the Contractor's employees, Subcontractors and their agents and employees, and other persons or entities performing portions of the Work for, or on behalf of, the Contractor or any of its Subcontractors.

§ 3.3.3 The Contractor shall be responsible for inspection of portions of Work already performed to determine that such portions are in proper condition to receive subsequent Work.

§ 3.4 Labor and Materials

§ 3.4.1 Unless otherwise provided in the Contract Documents, the Contractor shall provide and pay for labor, materials, equipment, tools, construction equipment and machinery, water, heat, utilities, transportation, and other facilities and services necessary for proper execution and completion of the Work, whether temporary or permanent and whether or not incorporated or to be incorporated in the Work.

§ 3.4.2 Except in the case of minor changes in the Work ~~authorized~~approved by the Architect in accordance with ~~Sections~~ Section 3.12.8 or ordered by the Architect in accordance with Section 7.4, the Contractor may make substitutions only with the consent of the Owner, after evaluation by the Architect and in accordance with a Change Order or Construction Change Directive.

§ 3.4.3 The Contractor shall enforce strict discipline and good order among the Contractor's employees and other persons carrying out the Work. The Contractor shall not permit employment of unfit persons or persons not properly skilled in tasks assigned to them.

§ 3.5 Warranty

§ 3.5.1 The Contractor warrants to the Owner and Architect that materials and equipment furnished under the Contract will be of good quality and new unless the Contract Documents require or permit otherwise. The Contractor further warrants that the Work will conform to the requirements of the Contract Documents and will be free from defects, except for those inherent in the quality of the Work the Contract Documents require or permit. Work, materials, or equipment not conforming to these requirements may be considered defective. The Contractor's warranty excludes remedy for damage or defect caused by abuse, alterations to the Work not executed by the Contractor, improper or insufficient maintenance, improper operation, or normal wear and tear and normal usage. If required by the Architect, the Contractor shall furnish satisfactory evidence as to the kind and quality of materials and equipment.

§ 3.5.2 All material, equipment, or other special warranties required by the Contract Documents shall be issued in the name of the Owner, or shall be transferable to the Owner, and shall commence in accordance with Section 9.8.4.

§ 3.6 Taxes

The Contractor shall pay sales, consumer, use and similar taxes for the Work provided by the Contractor that are legally enacted when bids are received or negotiations concluded, whether or not yet effective or merely scheduled to go into effect.

§ 3.7 Permits, Fees, Notices and Compliance with Laws

§ 3.7.1 Unless otherwise provided in the Contract Documents, the Contractor shall secure and pay for the building permit as well as for other permits, fees, licenses, and inspections by government agencies necessary for proper execution and completion of the Work that are customarily secured after execution of the Contract and legally required at the time bids are received or negotiations concluded.

§ 3.7.2 The Contractor shall comply with and give notices required by applicable laws, statutes, ordinances, codes, rules and regulations, and lawful orders of public authorities applicable to performance of the Work.

§ 3.7.3 If the Contractor performs Work knowing it to be contrary to applicable laws, statutes, ordinances, codes, rules and regulations, or lawful orders of public authorities, the Contractor shall assume appropriate responsibility for such Work and shall bear the costs attributable to correction.

§ 3.7.4 Concealed or Unknown Conditions

If the Contractor encounters conditions at the site that are (1) subsurface or otherwise concealed physical conditions that differ materially from those indicated in the Contract Documents or (2) unknown physical conditions of an unusual nature~~,~~ that differ materially from those ordinarily found to exist and generally recognized as inherent in construction activities of the character provided for in the Contract Documents, the Contractor shall promptly provide notice to the Owner and the Architect before conditions are disturbed and in no event later than ~~21~~14 days after first observance of the conditions. The Architect will promptly investigate such conditions and, if the Architect determines that they differ materially and cause an increase or decrease in the Contractor's cost of, or time required for, performance of any part of the Work, will recommend that an equitable adjustment be made in the Contract Sum or Contract Time, or both. If the Architect determines that the conditions at the site are not materially different from those indicated in the Contract Documents and that no change in the terms of the Contract is justified, the Architect shall promptly notify the Owner and Contractor~~ in writing~~, stating the reasons. If either party disputes the Architect's determination or recommendation, that party may ~~proceed~~submit a Claim as provided in Article 15.

§ 3.7.5 If, in the course of the Work, the Contractor encounters human remains or recognizes the existence of burial markers, archaeological sites or wetlands not indicated in the Contract Documents, the Contractor shall immediately suspend any operations that would affect them and shall notify the Owner and Architect. Upon receipt of such notice, the Owner shall promptly take any action necessary to obtain governmental authorization required to resume the operations. The Contractor shall continue to suspend such operations until otherwise instructed by the Owner but shall continue with all other operations that do not affect those remains or features. Requests for adjustments in the Contract Sum and Contract Time arising from the existence of such remains or features may be made as provided in Article 15.

§ 3.8 Allowances

§ 3.8.1 The Contractor shall include in the Contract Sum all allowances stated in the Contract Documents. Items covered by allowances shall be supplied for such amounts and by such persons or entities as the Owner may direct, but the Contractor shall not be required to employ persons or entities to whom the Contractor has reasonable objection.

§ 3.8.2 Unless otherwise provided in the Contract Documents,

- **.1** allowances shall cover the cost to the Contractor of materials and equipment delivered at the site and all required taxes, less applicable trade discounts;
- **.2** Contractor's costs for unloading and handling at the site, labor, installation costs, overhead, profit, and other expenses contemplated for stated allowance amounts shall be included in the Contract Sum but not in the allowances; and
- **.3** whenever costs are more than or less than allowances, the Contract Sum shall be adjusted accordingly by Change Order. The amount of the Change Order shall reflect (1) the difference between actual costs and the allowances under Section 3.8.2.1 and (2) changes in Contractor's costs under Section 3.8.2.2.

§ 3.8.3 Materials and equipment under an allowance shall be selected by the Owner with reasonable promptness.

§ 3.9 Superintendent

§ 3.9.1 The Contractor shall employ a competent superintendent and necessary assistants who shall be in attendance at the Project site during performance of the Work. The superintendent shall represent the Contractor, and communications given to the superintendent shall be as binding as if given to the Contractor.

§ 3.9.2 The Contractor, as soon as practicable after award of the Contract, shall ~~furnish in writing to~~notify the Owner ~~through the~~and Architect of the name and qualifications of a proposed superintendent. ~~The~~ Within 14 days of receipt of the information, the Architect may ~~reply within 14 days to~~notify the Contractor ~~in writing~~, stating ~~(1)~~ whether the Owner or the Architect (1) has reasonable objection to the proposed superintendent or (2) ~~that the Architect~~ requires additional time ~~to~~for review. Failure of the Architect to ~~reply~~ provide notice within the 14-day period shall constitute notice of no reasonable objection.

§ 3.9.3 The Contractor shall not employ a proposed superintendent to whom the Owner or Architect has made reasonable and timely objection. The Contractor shall not change the superintendent without the Owner's consent, which shall not unreasonably be withheld or delayed.

~~**§ 3.10 CONTRACTOR'S CONSTRUCTION SCHEDULES**~~

§ 3.10 Contractor's Construction and Submittal Schedules

§ 3.10.1 The Contractor, promptly after being awarded the Contract, shall ~~prepare and~~ submit for the Owner's and Architect's information a Contractor's construction schedule for the Work. The schedule shall contain detail appropriate for the Project, including (1) the date of commencement of the Work, interim schedule milestone dates, and the date of Substantial Completion; (2) an apportionment of the Work by construction activity; and (3) the time required for completion of each portion of the Work. The schedule shall provide for the orderly progression of the Work to completion and shall not exceed time limits current under the Contract Documents~~,~~. The schedule shall be revised at appropriate intervals as required by the conditions of the Work and Project~~, shall be related to the entire Project to the extent required by the Contract Documents, and shall provide for expeditious and practicable execution of the Work~~.

§ 3.10.2 The Contractor ~~shall prepare a submittal schedule~~, promptly after being awarded the Contract and thereafter as necessary to maintain a current submittal schedule, ~~and~~ shall submit ~~the~~a submittal schedule~~(s)~~ for the Architect's approval. The Architect's approval shall not be unreasonably ~~be~~ delayed or withheld. The submittal schedule shall (1) be coordinated with the Contractor's construction schedule, and (2) allow the Architect reasonable time to review submittals. If the Contractor fails to submit a submittal schedule, or fails to provide submittals in accordance with the approved submittal schedule, the Contractor shall not be entitled to any increase in Contract Sum or extension of Contract Time based on the time required for review of submittals.

§ 3.10.3 The Contractor shall perform the Work in general accordance with the most recent schedules submitted to the Owner and Architect.

§ 3.11 Documents and Samples at the Site

The Contractor shall ~~maintain~~make available, at ~~the site for~~ the ~~Owner one copy of~~Project site, the ~~Drawings, Specifications, Addenda,~~Contract Documents, including Change Orders, Construction Change Directives, and other Modifications, in good order and marked currently to indicate field changes and selections made during construction, and ~~one copy of~~the approved Shop Drawings, Product Data, Samples, and similar required submittals. These shall be in electronic form or paper copy, available to the Architect and ~~shall be~~Owner, and delivered to the Architect for submittal to the Owner upon completion of the Work as a record of the Work as constructed.

§ 3.12 Shop Drawings, Product Data and Samples

§ 3.12.1 Shop Drawings are drawings, diagrams, schedules, and other data specially prepared for the Work by the Contractor or a Subcontractor, Sub-subcontractor, manufacturer, supplier, or distributor to illustrate some portion of the Work.

§ 3.12.2 Product Data are illustrations, standard schedules, performance charts, instructions, brochures, diagrams, and other information furnished by the Contractor to illustrate materials or equipment for some portion of the Work.

§ 3.12.3 Samples are physical examples that illustrate materials, equipment, or workmanship, and establish standards by which the Work will be judged.

§ 3.12.4 Shop Drawings, Product Data, Samples, and similar submittals are not Contract Documents. Their purpose is to demonstrate ~~the way by which~~how the Contractor proposes to conform to the information given and the design concept expressed in the Contract Documents for those portions of the Work for which the Contract Documents require submittals. Review by the Architect is subject to the limitations of Section 4.2.7. Informational submittals upon which the Architect is not expected to take responsive action may be so identified in the Contract Documents. Submittals that are not required by the Contract Documents may be returned by the Architect without action.

§ 3.12.5 The Contractor shall review for compliance with the Contract Documents, approve, and submit to the Architect, Shop Drawings, Product Data, Samples, and similar submittals required by the Contract Documents, in accordance with the submittal schedule approved by the Architect or, in the absence of an approved submittal schedule, with reasonable promptness and in such sequence as to cause no delay in the Work or in the activities of the Owner or of ~~separate contractors~~Separate Contractors.

§ 3.12.6 By submitting Shop Drawings, Product Data, Samples, and similar submittals, the Contractor represents to the Owner and Architect that the Contractor has (1) reviewed and approved them, (2) determined and verified materials, field measurements and field construction criteria related thereto, or will do so, and (3) checked and coordinated the information contained within such submittals with the requirements of the Work and of the Contract Documents.

§ 3.12.7 The Contractor shall perform no portion of the Work for which the Contract Documents require submittal and review of Shop Drawings, Product Data, Samples, or similar submittals, until the respective submittal has been approved by the Architect.

§ 3.12.8 The Work shall be in accordance with approved submittals except that the Contractor shall not be relieved of responsibility for deviations from the requirements of the Contract Documents by the Architect's approval of Shop Drawings, Product Data, Samples, or similar submittals, unless the Contractor has specifically ~~informed~~notified the Architect ~~in writing~~ of such deviation at the time of submittal and (1) the Architect has given written approval to the specific deviation as a minor change in the Work, or (2) a Change Order or Construction Change Directive has been issued authorizing the deviation. The Contractor shall not be relieved of responsibility for errors or omissions in Shop Drawings, Product Data, Samples, or similar submittals, by the Architect's approval thereof.

§ 3.12.9 The Contractor shall direct specific attention, in writing or on resubmitted Shop Drawings, Product Data, Samples, or similar submittals, to revisions other than those requested by the Architect on previous submittals. In the absence of such ~~written~~ notice, the Architect's approval of a resubmission shall not apply to such revisions.

§ 3.12.10 The Contractor shall not be required to provide professional services that constitute the practice of architecture or engineering unless such services are specifically required by the Contract Documents for a portion of the Work or unless the Contractor needs to provide such services in order to carry out the Contractor's responsibilities for construction means, methods, techniques, sequences, and procedures. The Contractor shall not be required to provide professional services in violation of applicable law.

§ 3.12.10.1 If professional design services or certifications by a design professional related to systems, materials, or equipment are specifically required of the Contractor by the Contract Documents, the Owner and the Architect will specify all performance and design criteria that such services must satisfy. The Contractor shall be entitled to rely upon the adequacy and accuracy of the performance and design criteria provided in the Contract Documents. The Contractor shall cause such services or certifications to be provided by ~~a properly~~an appropriately licensed design professional, whose signature and seal shall appear on all drawings, calculations, specifications, certifications, Shop Drawings, and other submittals prepared by such professional. Shop Drawings, and other submittals related to the

Work, designed or certified by such professional, if prepared by others, shall bear such professional's written approval when submitted to the Architect. The Owner and the Architect shall be entitled to rely upon the adequacy, and accuracy and completeness of the services, certifications, and approvals performed or provided by such design professionals, provided the Owner and Architect have specified to the Contractor allthe performance and design criteria that such services must satisfy. Pursuant to this Section 3.12.10, the Architect will review, and approve or take other appropriate action on submittals only for the limited purpose of checking for conformance with information given and the design concept expressed in the Contract Documents. The Contractor shall not be responsible for the adequacy of the performance and design criteria specified in the Contract Documents.

§ 3.12.10.2 If the Contract Documents require the Contractor's design professional to certify that the Work has been performed in accordance with the design criteria, the Contractor shall furnish such certifications to the Architect at the time and in the form specified by the Architect.

§ 3.13 Use of Site
The Contractor shall confine operations at the site to areas permitted by applicable laws, statutes, ordinances, codes, rules and regulations, and lawful orders of public authorities, and the Contract Documents and shall not unreasonably encumber the site with materials or equipment.

§ 3.14 Cutting and Patching
§ 3.14.1 The Contractor shall be responsible for cutting, fitting, or patching required to complete the Work or to make its parts fit together properly. All areas requiring cutting, fitting and, or patching shall be restored to the condition existing prior to the cutting, fitting and, or patching, unless otherwise required by the Contract Documents.

§ 3.14.2 The Contractor shall not damage or endanger a portion of the Work or fully or partially completed construction of the Owner or separate contractorsSeparate Contractors by cutting, patching, or otherwise altering such construction, or by excavation. The Contractor shall not cut or otherwise alter such construction by the Owner or a separate contractor Separate Contractor except with written consent of the Owner and of such separate contractor; such consentthe Separate Contractor. Consent shall not be unreasonably withheld. The Contractor shall not unreasonably withhold, from the Owner or a separate contractor the Contractor'sSeparate Contractor, its consent to cutting or otherwise altering the Work.

§ 3.15 Cleaning Up
§ 3.15.1 The Contractor shall keep the premises and surrounding area free from accumulation of waste materials orand rubbish caused by operations under the Contract. At completion of the Work, the Contractor shall remove waste materials, rubbish, the Contractor's tools, construction equipment, machinery, and surplus materials from and about the Project.

§ 3.15.2 If the Contractor fails to clean up as provided in the Contract Documents, the Owner may do so and the Owner shall be entitled to reimbursement from the Contractor.

§ 3.16 Access to Work
The Contractor shall provide the Owner and Architect with access to the Work in preparation and progress wherever located.

§ 3.17 Royalties, Patents and Copyrights
The Contractor shall pay all royalties and license fees. The Contractor shall defend suits or claims for infringement of copyrights and patent rights and shall hold the Owner and Architect harmless from loss on account thereof, but shall not be responsible for such defense or loss when a particular design, process, or product of a particular manufacturer or manufacturers is required by the Contract Documents, or where the copyright violations are contained in Drawings, Specifications, or other documents prepared by the Owner or Architect. However, if the Contractor has reason to believe that the required design, process or product is an infringement of a copyright or a patent is discovered by, or made known to, the Contractor, the Contractor shall be responsible for suchthe loss unless suchthe information is promptly furnished to the Architect.

§ 3.18 Indemnification
§ 3.18.1 To the fullest extent permitted by law, the Contractor shall indemnify and hold harmless the Owner, Architect, Architect's consultants, and agents and employees of any of them from and against claims, damages, losses, and expenses, including but not limited to attorneys' fees, arising out of or resulting from performance of the Work, provided that such claim, damage, loss, or expense is attributable to bodily injury, sickness, disease or death, or to injury to or destruction of tangible property (other than the Work itself), but only to the extent caused by the

negligent acts or omissions of the Contractor, a Subcontractor, anyone directly or indirectly employed by them, or anyone for whose acts they may be liable, regardless of whether or not such claim, damage, loss, or expense is caused in part by a party indemnified hereunder. Such obligation shall not be construed to negate, abridge, or reduce other rights or obligations of indemnity that would otherwise exist as to a party or person described in this Section 3.18.

§ 3.18.2 In claims against any person or entity indemnified under this Section 3.18 by an employee of the Contractor, a Subcontractor, anyone directly or indirectly employed by them, or anyone for whose acts they may be liable, the indemnification obligation under Section 3.18.1 shall not be limited by a limitation on amount or type of damages, compensation, or benefits payable by or for the Contractor or a Subcontractor under workers' compensation acts, disability benefit acts, or other employee benefit acts.

Article 4 Architect

§ 4.1 General
~~§ 4.1.1 The Owner shall retain an architect lawfully licensed to practice architecture or an entity lawfully practicing architecture in the jurisdiction where the Project is located. That person or entity is identified as the Architect in the Agreement and is referred to throughout the Contract Documents as if singular in number.~~

Note: *Portions of Section 4.1.1 are relocated to Section 2.3.2.*

§ 4.1.1 The Architect is the person or entity retained by the Owner pursuant to Section 2.3.2 and identified as such in the Agreement.

§ 4.1.2 Duties, responsibilities, and limitations of authority of the Architect as set forth in the Contract Documents shall not be restricted, modified, or extended without written consent of the Owner, Contractor, and Architect. Consent shall not be unreasonably withheld.

~~§ 4.1.3 If the employment of the Architect is terminated, the Owner shall employ a successor architect as to whom the Contractor has no reasonable objection and whose status under the Contract Documents shall be that of the Architect.~~

Note: *Portions of Section 4.1.3 are relocated to Section 2.3.3.*

§ 4.2 Administration of the Contract
§ 4.2.1 The Architect will provide administration of the Contract as described in the Contract Documents and will be an Owner's representative during construction until the date the Architect issues the final Certificate for Payment. The Architect will have authority to act on behalf of the Owner only to the extent provided in the Contract Documents.

§ 4.2.2 The Architect will visit the site at intervals appropriate to the stage of construction, or as otherwise agreed with the Owner, to become generally familiar with the progress and quality of the portion of the Work completed, and to determine in general if the Work observed is being performed in a manner indicating that the Work, when fully completed, will be in accordance with the Contract Documents. However, the Architect will not be required to make exhaustive or continuous on-site inspections to check the quality or quantity of the Work. The Architect will not have control over, charge of, or responsibility for~~,~~ the construction means, methods, techniques, sequences or procedures, or for the safety precautions and programs in connection with the Work, since these are solely the Contractor's rights and responsibilities under the Contract Documents~~, except as provided in Section 3.3.1~~.

§ 4.2.3 On the basis of the site visits, the Architect will keep the Owner reasonably informed about the progress and quality of the portion of the Work completed, and promptly report to the Owner (1) known deviations from the Contract Documents ~~and~~, (2) known deviations from the most recent construction schedule submitted by the Contractor, and (~~2~~3) defects and deficiencies observed in the Work. The Architect will not be responsible for the Contractor's failure to perform the Work in accordance with the requirements of the Contract Documents. The Architect will not have control over or charge of, and will not be responsible for acts or omissions of, the Contractor, Subcontractors, or their agents or employees, or any other persons or entities performing portions of the Work.

§ 4.2.4 Communications

The Owner and Contractor shall include the Architect in all communications that relate to or affect the Architect's services or professional responsibilities. The Owner shall promptly notify the Architect of the substance of any direct communications between the Owner and the Contractor otherwise relating to the Project. Communications by and with the Architect's consultants shall be through the Architect. Communications by and with Subcontractors and suppliers shall be through the Contractor. Communications by and with Separate Contractors shall be through the Owner. The Contract Documents may specify other communication protocols.

§ 4.2.5 Based on the Architect's evaluations of the Contractor's Applications for Payment, the Architect will review and certify the amounts due the Contractor and will issue Certificates for Payment in such amounts.

§ 4.2.6 The Architect has authority to reject Work that does not conform to the Contract Documents. Whenever the Architect considers it necessary or advisable, the Architect will have authority to require inspection or testing of the Work in accordance with Sections 13.4.2 and 13.4.3, whether or not the Work is fabricated, installed or completed. However, neither this authority of the Architect nor a decision made in good faith either to exercise or not to exercise such authority shall give rise to a duty or responsibility of the Architect to the Contractor, Subcontractors, suppliers, their agents or employees, or other persons or entities performing portions of the Work.

§ 4.2.7 The Architect will review and approve, or take other appropriate action upon, the Contractor's submittals such as Shop Drawings, Product Data, and Samples, but only for the limited purpose of checking for conformance with information given and the design concept expressed in the Contract Documents. The Architect's action will be taken in accordance with the submittal schedule approved by the Architect or, in the absence of an approved submittal schedule, with reasonable promptness while allowing sufficient time in the Architect's professional judgment to permit adequate review. Review of such submittals is not conducted for the purpose of determining the accuracy and completeness of other details such as dimensions and quantities, or for substantiating instructions for installation or performance of equipment or systems, all of which remain the responsibility of the Contractor as required by the Contract Documents. The Architect's review of the Contractor's submittals shall not relieve the Contractor of the obligations under Sections 3.3, 3.5, and 3.12. The Architect's review shall not constitute approval of safety precautions or of any construction means, methods, techniques, sequences, or procedures. The Architect's approval of a specific item shall not indicate approval of an assembly of which the item is a component.

§ 4.2.8 The Architect will prepare Change Orders and Construction Change Directives, and may order minor changes in the Work as provided in Section 7.4. The Architect will investigate and make determinations and recommendations regarding concealed and unknown conditions as provided in Section 3.7.4.

§ 4.2.9 The Architect will conduct inspections to determine the date or dates of Substantial Completion and the date of final completion; issue Certificates of Substantial Completion pursuant to Section 9.8; receive and forward to the Owner, for the Owner's review and records, written warranties and related documents required by the Contract and assembled by the Contractor pursuant to Section 9.10; and issue a final Certificate for Payment pursuant to Section 9.10.

§ 4.2.10 If the Owner and Architect agree, the Architect will provide one or more Project representatives to assist in carrying out the Architect's responsibilities at the site. The Owner shall notify the Contractor of any change in the duties, responsibilities and limitations of authority of the Project representatives.

§ 4.2.11 The Architect will interpret and decide matters concerning performance under, and requirements of, the Contract Documents on written request of either the Owner or Contractor. The Architect's response to such requests will be made in writing within any time limits agreed upon or otherwise with reasonable promptness.

§ 4.2.12 Interpretations and decisions of the Architect will be consistent with the intent of, and reasonably inferable from, the Contract Documents and will be in writing or in the form of drawings. When making such interpretations and decisions, the Architect will endeavor to secure faithful performance by both Owner and Contractor, will not show partiality to either, and will not be liable for results of interpretations or decisions rendered in good faith.

§ 4.2.13 The Architect's decisions on matters relating to aesthetic effect will be final if consistent with the intent expressed in the Contract Documents.

§ 4.2.14 The Architect will review and respond to requests for information about the Contract Documents. The Architect's response to such requests will be made in writing within any time limits agreed upon or otherwise with reasonable promptness. If appropriate, the Architect will prepare and issue supplemental Drawings and Specifications in response to the requests for information.

Article 5 Subcontractors

§ 5.1 Definitions
§ 5.1.1 A Subcontractor is a person or entity who has a direct contract with the Contractor to perform a portion of the Work at the site. The term "Subcontractor" is referred to throughout the Contract Documents as if singular in number and means a Subcontractor or an authorized representative of the Subcontractor. The term "Subcontractor" does not include a ~~separate contractor~~Separate Contractor or the subcontractors of a ~~separate contractor~~Separate Contractor.

§ 5.1.2 A Sub-subcontractor is a person or entity who has a direct or indirect contract with a Subcontractor to perform a portion of the Work at the site. The term "Sub-subcontractor" is referred to throughout the Contract Documents as if singular in number and means a Sub-subcontractor or an authorized representative of the Sub-subcontractor.

§ 5.2 Award of Subcontracts and Other Contracts for Portions of the Work
§ 5.2.1 Unless otherwise stated in the Contract Documents~~ or the bidding requirements~~, the Contractor, as soon as practicable after award of the Contract, shall ~~furnish in writing to~~notify the Owner ~~through the~~and Architect of the ~~names of~~ persons or entities ~~(~~proposed for each principal portion of the Work, including those who are to furnish materials or equipment fabricated to a special design~~) proposed for each principal portion~~. Within 14 days of receipt of the ~~Work. The~~information, the Architect may ~~reply within 14 days to~~ notify the Contractor ~~in writing stating (1)~~ whether the Owner or the Architect (1) has reasonable objection to any such proposed person or entity or (2) ~~that the Architect~~ requires additional time for review. Failure of the ~~Owner or~~ Architect to ~~reply~~provide notice within the 14-day period shall constitute notice of no reasonable objection.

§ 5.2.2 The Contractor shall not contract with a proposed person or entity to whom the Owner or Architect has made reasonable and timely objection. The Contractor shall not be required to contract with anyone to whom the Contractor has made reasonable objection.

§ 5.2.3 If the Owner or Architect has reasonable objection to a person or entity proposed by the Contractor, the Contractor shall propose another to whom the Owner or Architect has no reasonable objection. If the proposed but rejected Subcontractor was reasonably capable of performing the Work, the Contract Sum and Contract Time shall be increased or decreased by the difference, if any, occasioned by such change, and an appropriate Change Order shall be issued before commencement of the substitute Subcontractor's Work. However, no increase in the Contract Sum or Contract Time shall be allowed for such change unless the Contractor has acted promptly and responsively in submitting names as required.

§ 5.2.4 The Contractor shall not substitute a Subcontractor, person, or entity for one previously selected if the Owner or Architect makes reasonable objection to such substitution.

§ 5.3 Subcontractual Relations
By appropriate written agreement~~, written where legally required for validity~~, the Contractor shall require each Subcontractor, to the extent of the Work to be performed by the Subcontractor, to be bound to the Contractor by terms of the Contract Documents, and to assume toward the Contractor all the obligations and responsibilities, including the responsibility for safety of the Subcontractor's Work~~, which~~ that the Contractor, by these Contract Documents, assumes toward the Owner and Architect. Each subcontract agreement shall preserve and protect the rights of the Owner and Architect under the Contract Documents with respect to the Work to be performed by the Subcontractor so that subcontracting thereof will not prejudice such rights, and shall allow to the Subcontractor, unless specifically provided otherwise in the subcontract agreement, the benefit of all rights, remedies, and redress against the Contractor that the Contractor, by the Contract Documents, has against the Owner. Where appropriate, the Contractor shall require each Subcontractor to enter into similar agreements with Sub-subcontractors. The Contractor shall make available to each proposed Subcontractor, prior to the execution of the subcontract agreement, copies of the Contract Documents to which the Subcontractor will be bound, and, upon written request of the

Subcontractor, identify to the Subcontractor terms and conditions of the proposed subcontract agreement that may be at variance with the Contract Documents. Subcontractors will similarly make copies of applicable portions of such documents available to their respective proposed Sub-subcontractors.

§ 5.4 Contingent Assignment of Subcontracts

§ 5.4.1 Each subcontract agreement for a portion of the Work is assigned by the Contractor to the Owner, provided that

- **.1** assignment is effective only after termination of the Contract by the Owner for cause pursuant to Section 14.2 and only for those subcontract agreements that the Owner accepts by notifying the Subcontractor and Contractor~~ in writing~~; and
- **.2** assignment is subject to the prior rights of the surety, if any, obligated under bond relating to the Contract.

When the Owner accepts the assignment of a subcontract agreement, the Owner assumes the Contractor's rights and obligations under the subcontract.

§ 5.4.2 Upon such assignment, if the Work has been suspended for more than 30 days, the Subcontractor's compensation shall be equitably adjusted for increases in cost resulting from the suspension.

§ 5.4.3 Upon ~~such~~ assignment to the Owner under this Section 5.4, the Owner may further assign the subcontract to a successor contractor or other entity. If the Owner assigns the subcontract to a successor contractor or other entity, the Owner shall nevertheless remain legally responsible for all of the successor contractor's obligations under the subcontract.

Article 6 Construction by Owner or by Separate Contractors

§ 6.1 Owner's Right to Perform Construction and to Award Separate Contracts

§ 6.1.1 The term "Separate Contractor(s)" shall mean other contractors retained by the Owner under separate agreements. The Owner reserves the right to perform construction or operations related to the Project with the Owner's own forces, and ~~to award separate contracts in connection~~ with ~~other portions of the Project or other construction or operations on the site~~Separate Contractors retained under Conditions of the Contract ~~identical or~~ substantially similar to ~~these~~those of this Contract, including those ~~portions~~ provisions of the Conditions of the Contract related to insurance and waiver of subrogation~~. If the Contractor claims that delay or additional cost is involved because of such action by the Owner, the Contractor shall make such Claim as provided in Article 15~~.

§ 6.1.2 When separate contracts are awarded for different portions of the Project or other construction or operations on the site, the term "Contractor" in the Contract Documents in each case shall mean the Contractor who executes each separate Owner-Contractor Agreement.

§ 6.1.3 The Owner shall provide for coordination of the activities of the Owner's own forces and of each Separate Contractor with the Work of the Contractor, who shall cooperate with them. The Contractor shall participate with ~~other~~any Separate Contractors and the Owner in reviewing their construction schedules. The Contractor shall make any revisions to ~~the~~its construction schedule deemed necessary after a joint review and mutual agreement. The construction schedules shall then constitute the schedules to be used by the Contractor, Separate Contractors, and the Owner until subsequently revised.

§ 6.1.4 Unless otherwise provided in the Contract Documents, when the Owner performs construction or operations related to the Project with the Owner's own forces or with Separate Contractors, the Owner or its Separate Contractors shall ~~be deemed to be subject to~~have the same obligations and ~~to have the same~~ rights that ~~apply to~~ the Contractor has under the Conditions of the Contract, including, without excluding others, those stated in Article 3, this Article 6, and Articles 10, 11, and 12.

§ 6.2 Mutual Responsibility

§ 6.2.1 The Contractor shall afford the Owner and Separate Contractors reasonable opportunity for introduction and storage of their materials and equipment and performance of their activities, and shall connect and coordinate the Contractor's construction and operations with theirs as required by the Contract Documents.

§ 6.2.2 If part of the Contractor's Work depends for proper execution or results upon construction or operations by the Owner or a Separate Contractor, the Contractor shall, prior to proceeding with that portion of the Work, promptly ~~report to~~notify the Architect of apparent discrepancies or defects in ~~such other~~the construction or operations by the Owner or Separate Contractor that would render it unsuitable for such proper execution and results. of the Contractor's Work. Failure of the Contractor soto notify the Architect of apparent discrepancies or defects prior to reportproceeding with the Work shall constitute an acknowledgment that the Owner's or Separate Contractor's completed or partially completed construction is fit and proper to receive the Contractor's Work, except as to defects not then reasonably discoverable. The Contractor shall not be responsible for discrepancies or defects in the construction or operations by the Owner or Separate Contractor that are not apparent.

§ 6.2.3 The Contractor shall reimburse the Owner for costs the Owner incurs that are payable to a Separate Contractor because of the Contractor's delays, improperly timed activities or defective construction. The Owner shall be responsible to the Contractor for costs the Contractor incurs because of a Separate Contractor's delays, improperly timed activities, damage to the Work or defective construction.

§ 6.2.4 The Contractor shall promptly remedy damage that the Contractor wrongfully causes to completed or partially completed construction or to property of the Owner or Separate Contractor as provided in Section 10.2.5.

§ 6.2.5 The Owner and each Separate Contractor shall have the same responsibilities for cutting and patching as are described for the Contractor in Section 3.14.

§ 6.3 Owner's Right to Clean Up
If a dispute arises among the Contractor, Separate Contractors, and the Owner as to the responsibility under their respective contracts for maintaining the premises and surrounding area free from waste materials and rubbish, the Owner may clean up and the Architect will allocate the cost among those responsible.

Article 7 Changes in the Work

§ 7.1 General
§ 7.1.1 Changes in the Work may be accomplished after execution of the Contract, and without invalidating the Contract, by Change Order, Construction Change Directive or order for a minor change in the Work, subject to the limitations stated in this Article 7 and elsewhere in the Contract Documents.

§ 7.1.2 A Change Order shall be based upon agreement among the Owner, Contractor, and Architect; a. A Construction Change Directive requires agreement by the Owner and Architect and may or may not be agreed to by the Contractor; an. An order for a minor change in the Work may be issued by the Architect alone.

§ 7.1.3 Changes in the Work shall be performed under applicable provisions of the Contract Documents, and the. The Contractor shall proceed promptly with changes in the Work, unless otherwise provided in the Change Order, Construction Change Directive, or order for a minor change in the Work.

§ 7.2 Change Orders
§ 7.2.1 A Change Order is a written instrument prepared by the Architect and signed by the Owner, Contractor, and Architect stating their agreement upon all of the following:

- **.1** The change in the Work;
- **.2** The amount of the adjustment, if any, in the Contract Sum; and
- **.3** The extent of the adjustment, if any, in the Contract Time.

§ 7.3 Construction Change Directives
§ 7.3.1 A Construction Change Directive is a written order prepared by the Architect and signed by the Owner and Architect, directing a change in the Work prior to agreement on adjustment, if any, in the Contract Sum or Contract Time, or both. The Owner may by Construction Change Directive, without invalidating the Contract, order changes in the Work within the general scope of the Contract consisting of additions, deletions, or other revisions, the Contract Sum and Contract Time being adjusted accordingly.

§ 7.3.2 A Construction Change Directive shall be used in the absence of total agreement on the terms of a Change Order.

§ 7.3.3 If the Construction Change Directive provides for an adjustment to the Contract Sum, the adjustment shall be based on one of the following methods:

- **.1** Mutual acceptance of a lump sum properly itemized and supported by sufficient substantiating data to permit evaluation;
- **.2** Unit prices stated in the Contract Documents or subsequently agreed upon;

.3 Cost to be determined in a manner agreed upon by the parties and a mutually acceptable fixed or percentage fee; or

.4 As provided in Section 7.3.~~7~~4.

~~§ 7.3.4 If unit prices are stated in the Contract Documents or subsequently agreed upon, and if quantities originally contemplated are materially changed in a proposed Change Order or Construction Change Directive so that application of such unit prices to quantities of Work proposed will cause substantial inequity to the Owner or Contractor, the applicable unit prices shall be equitably adjusted.~~

~~§ 7.3.5 Upon receipt of a Construction Change Directive, the Contractor shall promptly proceed with the change in the Work involved and advise the Architect of the Contractor's agreement or disagreement with the method, if any, provided in the Construction Change Directive for determining the proposed adjustment in the Contract Sum or Contract Time.~~

~~§ 7.3.6 A Construction Change Directive signed by the Contractor indicates the Contractor's agreement therewith, including adjustment in Contract Sum and Contract Time or the method for determining them. Such agreement shall be effective immediately and shall be recorded as a Change Order.~~

> ***Note:*** *Section 7.3.4 formerly appeared at Section 7.3.7. Former Section 7.3.4 has been relocated to Section 9.1.2.*

§ 7.3.~~7~~4 If the Contractor does not respond promptly or disagrees with the method for adjustment in the Contract Sum, the Architect shall determine the ~~method and the~~ adjustment on the basis of reasonable expenditures and savings of those performing the Work attributable to the change, including, in case of an increase in the Contract Sum, an amount for overhead and profit as set forth in the Agreement, or if no such amount is set forth in the Agreement, a reasonable amount. In such case, and also under Section 7.3.3.3, the Contractor shall keep and present, in such form as the Architect may prescribe, an itemized accounting together with appropriate supporting data. Unless otherwise provided in the Contract Documents, costs for the purposes of this Section 7.3.~~7~~4 shall be limited to the following:

.1 Costs of labor, including ~~social security, old age and unemployment insurance~~applicable payroll taxes, fringe benefits required by agreement or custom, ~~and~~ workers' compensation insurance, and other employee costs approved by the Architect;

.2 Costs of materials, supplies, and equipment, including cost of transportation, whether incorporated or consumed;

.3 Rental costs of machinery and equipment, exclusive of hand tools, whether rented from the Contractor or others;

.4 Costs of premiums for all bonds and insurance, permit fees, and sales, use, or similar taxes, directly related to the ~~Work~~change; and

.5 ~~Additional costs~~Costs of supervision and field office personnel directly attributable to the change.

§ 7.3.5 If the Contractor disagrees with the adjustment in the Contract Time, the Contractor may make a Claim in accordance with applicable provisions of Article 15.

§ 7.3.6 Upon receipt of a Construction Change Directive, the Contractor shall promptly proceed with the change in the Work involved and advise the Architect of the Contractor's agreement or disagreement with the method, if any, provided in the Construction Change Directive for determining the proposed adjustment in the Contract Sum or Contract Time.

§ 7.3.7 A Construction Change Directive signed by the Contractor indicates the Contractor's agreement therewith, including adjustment in Contract Sum and Contract Time or the method for determining them. Such agreement shall be effective immediately and shall be recorded as a Change Order.

§ 7.3.8 The amount of credit to be allowed by the Contractor to the Owner for a deletion or change that results in a net decrease in the Contract Sum shall be actual net cost as confirmed by the Architect. When both additions and credits covering related Work or substitutions are involved in a change, the allowance for overhead and profit shall be figured on the basis of net increase, if any, with respect to that change.

§ 7.3.9 Pending final determination of the total cost of a Construction Change Directive to the Owner, the Contractor may request payment for Work completed under the Construction Change Directive in Applications for Payment.

The Architect will make an interim determination for purposes of monthly certification for payment for those costs and certify for payment the amount that the Architect determines, in the Architect's professional judgment, to be reasonably justified. The Architect's interim determination of cost shall adjust the Contract Sum on the same basis as a Change Order, subject to the right of either party to disagree and assert a Claim in accordance with Article 15.

§ 7.3.10 When the Owner and Contractor agree with a determination made by the Architect concerning the adjustments in the Contract Sum and Contract Time, or otherwise reach agreement upon the adjustments, such agreement shall be effective immediately and the Architect will prepare a Change Order. Change Orders may be issued for all or any part of a Construction Change Directive.

§ 7.4 Minor Changes in the Work
The Architect ~~has authority to~~ may order minor changes in the Work that are consistent with the intent of the Contract Documents and do not ~~involving~~involve an adjustment in the Contract Sum or an extension of the Contract Time ~~and not inconsistent with the intent of the Contract Documents. Such~~. The Architect's order for minor changes shall be in writing. If the Contractor believes that the proposed minor change in the Work will ~~be effected by written order signed by~~affect the Contract Sum or Contract Time, the Contractor shall notify the Architect and shall ~~be binding on the Owner and Contractor.~~ not proceed to implement the change in the Work. If the Contractor performs the Work set forth in the Architect's order for a minor change without prior notice to the Architect that such change will affect the Contract Sum or Contract Time, the Contractor waives any adjustment to the Contract Sum or extension of the Contract Time.

Article 8 Time

§ 8.1 Definitions
§ 8.1.1 Unless otherwise provided, Contract Time is the period of time, including authorized adjustments, allotted in the Contract Documents for Substantial Completion of the Work.

§ 8.1.2 The date of commencement of the Work is the date established in the Agreement.

§ 8.1.3 The date of Substantial Completion is the date certified by the Architect in accordance with Section 9.8.

§ 8.1.4 The term "day" as used in the Contract Documents shall mean calendar day unless otherwise specifically defined.

§ 8.2 Progress and Completion
§ 8.2.1 Time limits stated in the Contract Documents are of the essence of the Contract. By executing the Agreement, the Contractor confirms that the Contract Time is a reasonable period for performing the Work.

§ 8.2.2 The Contractor shall not knowingly, except by agreement or instruction of the Owner in writing, ~~prematurely~~ commence ~~operations on~~ the ~~site or elsewhere~~Work prior to the effective date of insurance required ~~by Article 11~~ to be furnished by the Contractor and Owner. ~~The date of commencement of the Work shall not be changed by the effective date of such insurance.~~

§ 8.2.3 The Contractor shall proceed expeditiously with adequate forces and shall achieve Substantial Completion within the Contract Time.

§ 8.3 Delays and Extensions of Time
§ 8.3.1 If the Contractor is delayed at any time in the commencement or progress of the Work by (1) an act or neglect of the Owner or Architect, ~~or~~ of an employee of either, or of a ~~separate contractor employed by the Owner; or~~ Separate Contractor; (2) by changes ordered in the Work; ~~or~~(3) by labor disputes, fire, unusual delay in deliveries, unavoidable casualties ~~or~~, adverse weather conditions documented in accordance with Section 15.1.6.2, or other causes beyond the Contractor's control; ~~or~~(4) by delay authorized by the Owner pending mediation and ~~arbitration; or~~binding dispute resolution; or (5) by other causes that the Contractor asserts, and the Architect determines ~~may~~, justify delay, then the Contract Time shall be extended ~~by Change Order~~ for such reasonable time as the Architect may determine.

§ 8.3.2 Claims relating to time shall be made in accordance with applicable provisions of Article 15.

§ 8.3.3 This Section 8.3 does not preclude recovery of damages for delay by either party under other provisions of the Contract Documents.

Article 9 Payments and Completion

§ 9.1 Contract Sum

§ 9.1.1 The Contract Sum is stated in the Agreement and, including authorized adjustments, is the total amount payable by the Owner to the Contractor for performance of the Work under the Contract Documents.

§ 9.1.2 If unit prices are stated in the Contract Documents or subsequently agreed upon, and if quantities originally contemplated are materially changed so that application of such unit prices to the actual quantities causes substantial inequity to the Owner or Contractor, the applicable unit prices shall be equitably adjusted.

> ***Note:*** *Section 9.1.2 formerly appeared at Section 7.3.4.*

§ 9.2 Schedule of Values

Where the Contract is based on a stipulated sum or Guaranteed Maximum Price, the Contractor shall submit a schedule of values to the Architect~~,~~ before the first Application for Payment, ~~a schedule of values~~ allocating the entire Contract Sum to the various portions of the Work~~ and~~. The schedule of values shall be prepared in ~~such~~the form, and supported by ~~such~~the data to substantiate its accuracy~~ as~~, required by the Architect~~ may require~~. This schedule, unless objected to by the Architect, shall be used as a basis for reviewing the Contractor's Applications for Payment. Any changes to the schedule of values shall be submitted to the Architect and supported by such data to substantiate its accuracy as the Architect may require, and unless objected to by the Architect, shall be used as a basis for reviewing the Contractor's subsequent Applications for Payment.

§ 9.3 Applications for Payment

§ 9.3.1 At least ten days before the date established for each progress payment, the Contractor shall submit to the Architect an itemized Application for Payment prepared in accordance with the schedule of values, if required under Section 9.2, for completed portions of the Work. ~~Such~~The application shall be notarized, if required, and supported by ~~such~~all data substantiating the Contractor's right to payment ~~as~~that the Owner or Architect ~~may~~ require, such as copies of requisitions, and releases and waivers of liens from Subcontractors and~~ material~~ suppliers, and shall reflect retainage if provided for in the Contract Documents.

§ 9.3.1.1 As provided in Section 7.3.9, such applications may include requests for payment on account of changes in the Work that have been properly authorized by Construction Change Directives, or by interim determinations of the Architect, but not yet included in Change Orders.

§ 9.3.1.2 Applications for Payment shall not include requests for payment for portions of the Work for which the Contractor does not intend to pay a Subcontractor or ~~material~~ supplier, unless such Work has been performed by others whom the Contractor intends to pay.

§ 9.3.2 Unless otherwise provided in the Contract Documents, payments shall be made on account of materials and equipment delivered and suitably stored at the site for subsequent incorporation in the Work. If approved in advance by the Owner, payment may similarly be made for materials and equipment suitably stored off the site at a location agreed upon in writing. Payment for materials and equipment stored on or off the site shall be conditioned upon compliance by the Contractor with procedures satisfactory to the Owner to establish the Owner's title to such materials and equipment or otherwise protect the Owner's interest, and shall include the costs of applicable insurance, storage, and transportation to the site, for such materials and equipment stored off the site.

§ 9.3.3 The Contractor warrants that title to all Work covered by an Application for Payment will pass to the Owner no later than the time of payment. The Contractor further warrants that upon submittal of an Application for Payment all Work for which Certificates for Payment have been previously issued and payments received from the Owner shall, to the best of the Contractor's knowledge, information, and belief, be free and clear of liens, claims, security interests, or encumbrances, in favor of the Contractor, Subcontractors, ~~material~~ suppliers, or other persons or entities ~~making a claim by reason of having~~that provided labor, materials, and equipment relating to the Work.

§ 9.4 Certificates for Payment

§ 9.4.1 The Architect will, within seven days after receipt of the Contractor's Application for Payment, either (1) issue to the Owner a Certificate for Payment in the full amount of the Application for Payment, with a copy to the Contractor~~,~~; or (2) issue to the Owner a Certificate for Payment for such amount as the Architect determines is properly due, ~~or~~and notify the Contractor and Owner ~~in writing~~ of the Architect's reasons for withholding certification in part as provided in Section 9.5.1; or (3) withhold certification of the entire Application for Payment, and notify the Contractor and Owner of the Architect's reason for withholding certification in whole~~ or in part~~ as provided in Section 9.5.1.

§ 9.4.2 The issuance of a Certificate for Payment will constitute a representation by the Architect to the Owner, based on the Architect's evaluation of the Work and the data ~~comprising~~in the Application for Payment, that, to the best of the Architect's knowledge, information, and belief, the Work has progressed to the point indicated~~ and that~~, the quality of the Work is in accordance with the Contract Documents, and that the Contractor is entitled to payment in the amount certified. The foregoing representations are subject to an evaluation of the Work for conformance with the Contract Documents upon Substantial Completion, to results of subsequent tests and inspections, to correction of minor deviations from the Contract Documents prior to completion, and to specific qualifications expressed by the Architect.~~ The issuance of a Certificate for Payment will further constitute a representation that the Contractor is entitled to payment in the amount certified.~~ However, the issuance of a Certificate for Payment will not be a representation that the Architect has (1) made exhaustive or continuous on-site inspections to check the quality or quantity of the Work~~,~~; (2) reviewed construction means, methods, techniques, sequences, or procedures~~,~~; (3) reviewed copies of requisitions received from Subcontractors and ~~material~~ suppliers and other data requested by the Owner to substantiate the Contractor's right to payment~~,~~; or (4) made examination to ascertain how or for what purpose the Contractor has used money previously paid on account of the Contract Sum.

§ 9.5 Decisions to Withhold Certification

§ 9.5.1 The Architect may withhold a Certificate for Payment in whole or in part, to the extent reasonably necessary to protect the Owner, if in the Architect's opinion the representations to the Owner required by Section 9.4.2 cannot be made. If the Architect is unable to certify payment in the amount of the Application, the Architect will notify the Contractor and Owner as provided in Section 9.4.1. If the Contractor and Architect cannot agree on a revised amount, the Architect will promptly issue a Certificate for Payment for the amount for which the Architect is able to make such representations to the Owner. The Architect may also withhold a Certificate for Payment or, because of subsequently discovered evidence, may nullify the whole or a part of a Certificate for Payment previously issued, to such extent as may be necessary in the Architect's opinion to protect the Owner from loss for which the Contractor is responsible, including loss resulting from acts and omissions described in Section 3.3.2, because of

- **.1** defective Work not remedied;
- **.2** third party claims filed or reasonable evidence indicating probable filing of such claims, unless security acceptable to the Owner is provided by the Contractor;
- **.3** failure of the Contractor to make payments properly to Subcontractors or suppliers for labor, materials or equipment;
- **.4** reasonable evidence that the Work cannot be completed for the unpaid balance of the Contract Sum;
- **.5** damage to the Owner or a ~~separate contractor~~Separate Contractor;
- **.6** reasonable evidence that the Work will not be completed within the Contract Time, and that the unpaid balance would not be adequate to cover actual or liquidated damages for the anticipated delay; or
- **.7** repeated failure to carry out the Work in accordance with the Contract Documents.

§ 9.5.2 When ~~the above~~either party disputes the Architect's decision regarding a Certificate for Payment under Section 9.5.1, in whole or in part, that party may submit a Claim in accordance with Article 15.

§ 9.5.3 When the reasons for withholding certification are removed, certification will be made for amounts previously withheld.

§ 9.5.~~3~~4 If the Architect withholds certification for payment under Section 9.5.1.3, the Owner may, at its sole option, issue joint checks to the Contractor and to any Subcontractor or ~~material or equipment suppliers~~supplier to whom the Contractor failed to make payment for Work properly performed or material or equipment suitably delivered. If the Owner makes payments by joint check, the Owner shall notify the Architect and the ~~Architect will~~Contractor shall reflect such payment on ~~the~~its next ~~Certificate~~Application for Payment.

§ 9.6 Progress Payments

§ 9.6.1 After the Architect has issued a Certificate for Payment, the Owner shall make payment in the manner and within the time provided in the Contract Documents, and shall so notify the Architect.

§ 9.6.2 The Contractor shall pay each Subcontractor, no later than seven days after receipt of payment from the Owner, the amount to which the Subcontractor is entitled, reflecting percentages actually retained from payments to the Contractor on account of the Subcontractor's portion of the Work. The Contractor shall, by appropriate agreement with each Subcontractor, require each Subcontractor to make payments to Sub-subcontractors in a similar manner.

§ 9.6.3 The Architect will, on request, furnish to a Subcontractor, if practicable, information regarding percentages of completion or amounts applied for by the Contractor and action taken thereon by the Architect and Owner on account of portions of the Work done by such Subcontractor.

§ 9.6.4 The Owner has the right to request written evidence from the Contractor that the Contractor has properly paid Subcontractors and ~~material and equipment~~ suppliers amounts paid by the Owner to the Contractor for subcontracted Work. If the Contractor fails to furnish such evidence within seven days, the Owner shall have the right to contact Subcontractors and suppliers to ascertain whether they have been properly paid. Neither the Owner nor Architect shall have an obligation to pay, or to see to the payment of money to, a Subcontractor or supplier, except as may otherwise be required by law.

§ 9.6.5 ~~Contractor~~The Contractor's payments to ~~material and equipment~~ suppliers shall be treated in a manner similar to that provided in Sections 9.6.2, 9.6.3 and 9.6.4.

§ 9.6.6 A Certificate for Payment, a progress payment, or partial or entire use or occupancy of the Project by the Owner shall not constitute acceptance of Work not in accordance with the Contract Documents.

§ 9.6.7 Unless the Contractor provides the Owner with a payment bond in the full penal sum of the Contract Sum, payments received by the Contractor for Work properly performed by Subcontractors ~~and~~or provided by suppliers shall be held by the Contractor for those Subcontractors or suppliers who performed Work or furnished materials, or both, under contract with the Contractor for which payment was made by the Owner. Nothing contained herein shall require money to be placed in a separate account and not commingled with money of the Contractor, ~~shall~~ create any fiduciary liability or tort liability on the part of the Contractor for breach of trust, or ~~shall~~ entitle any person or entity to an award of punitive damages against the Contractor for breach of the requirements of this provision.

§ 9.6.8 Provided the Owner has fulfilled its payment obligations under the Contract Documents, the Contractor shall defend and indemnify the Owner from all loss, liability, damage or expense, including reasonable attorney's fees and litigation expenses, arising out of any lien claim or other claim for payment by any Subcontractor or supplier of any tier. Upon receipt of notice of a lien claim or other claim for payment, the Owner shall notify the Contractor. If approved by the applicable court, when required, the Contractor may substitute a surety bond for the property against which the lien or other claim for payment has been asserted.

§ 9.7 Failure of Payment

If the Architect does not issue a Certificate for Payment, through no fault of the Contractor, within seven days after receipt of the Contractor's Application for Payment, or if the Owner does not pay the Contractor within seven days after the date established in the Contract Documents, the amount certified by the Architect or awarded by binding dispute resolution, then the Contractor may, upon seven additional days' ~~written~~ notice to the Owner and Architect, stop the Work until payment of the amount owing has been received. The Contract Time shall be extended appropriately and the Contract Sum shall be increased by the amount of the Contractor's reasonable costs of ~~shut-down~~shutdown, delay and start-up, plus interest as provided for in the Contract Documents.

§ 9.8 Substantial Completion

§ 9.8.1 Substantial Completion is the stage in the progress of the Work when the Work or designated portion thereof is sufficiently complete in accordance with the Contract Documents so that the Owner can occupy or utilize the Work for its intended use.

§ 9.8.2 When the Contractor considers that the Work, or a portion thereof which the Owner agrees to accept separately, is substantially complete, the Contractor shall prepare and submit to the Architect a comprehensive list of items to be completed or corrected prior to final payment. Failure to include an item on such list does not alter the responsibility of the Contractor to complete all Work in accordance with the Contract Documents.

§ 9.8.3 Upon receipt of the Contractor's list, the Architect will make an inspection to determine whether the Work or designated portion thereof is substantially complete. If the Architect's inspection discloses any item, whether or not included on the Contractor's list, which is not sufficiently complete in accordance with the Contract Documents so that the Owner can occupy or utilize the Work or designated portion thereof for its intended use, the Contractor shall, before issuance of the Certificate of Substantial Completion, complete or correct such item upon notification by the Architect. In such case, the Contractor shall then submit a request for another inspection by the Architect to determine Substantial Completion.

§ 9.8.4 When the Work or designated portion thereof is substantially complete, the Architect will prepare a Certificate of Substantial Completion that shall establish the date of Substantial Completion~~, shall~~; establish responsibilities of the Owner and Contractor for security, maintenance, heat, utilities, damage to the Work and insurance~~,~~; and~~ shall~~ fix the time within which the Contractor shall finish all items on the list accompanying the Certificate. Warranties required by the Contract Documents shall commence on the date of Substantial Completion of the Work or designated portion thereof unless otherwise provided in the Certificate of Substantial Completion.

§ 9.8.5 The Certificate of Substantial Completion shall be submitted to the Owner and Contractor for their written acceptance of responsibilities assigned to them in ~~such~~the Certificate. Upon such acceptance, and consent of surety~~,~~ if any, the Owner shall make payment of retainage applying to ~~such~~the Work or designated portion thereof. Such payment shall be adjusted for Work that is incomplete or not in accordance with the requirements of the Contract Documents.

§ 9.9 Partial Occupancy or Use

§ 9.9.1 The Owner may occupy or use any completed or partially completed portion of the Work at any stage when such portion is designated by separate agreement with the Contractor, provided such occupancy or use is consented to by the insurer ~~as required under Section 11.3.1.5~~ and authorized by public authorities having jurisdiction over the Project. Such partial occupancy or use may commence whether or not the portion is substantially complete, provided the Owner and Contractor have accepted in writing the responsibilities assigned to each of them for payments, retainage, if any, security, maintenance, heat, utilities, damage to the Work and insurance, and have agreed in writing concerning the period for correction of the Work and commencement of warranties required by the Contract Documents. When the Contractor considers a portion substantially complete, the Contractor shall prepare and submit a list to the Architect as provided under Section 9.8.2. Consent of the Contractor to partial occupancy or use shall not be unreasonably withheld. The stage of the progress of the Work shall be determined by written agreement between the Owner and Contractor or, if no agreement is reached, by decision of the Architect.

§ 9.9.2 Immediately prior to such partial occupancy or use, the Owner, Contractor, and Architect shall jointly inspect the area to be occupied or portion of the Work to be used in order to determine and record the condition of the Work.

§ 9.9.3 Unless otherwise agreed upon, partial occupancy or use of a portion or portions of the Work shall not constitute acceptance of Work not complying with the requirements of the Contract Documents.

§ 9.10 Final Completion and Final Payment

§ 9.10.1 Upon receipt of the Contractor's~~ written~~ notice that the Work is ready for final inspection and acceptance and upon receipt of a final Application for Payment, the Architect will promptly make such inspection~~ and, when~~. When the Architect finds the Work acceptable under the Contract Documents and the Contract fully performed, the Architect will promptly issue a final Certificate for Payment stating that to the best of the Architect's knowledge, information and belief, and on the basis of the Architect's on-site visits and inspections, the Work has been completed in accordance with ~~terms and conditions of~~ the Contract Documents and that the entire balance found to be due the Contractor and noted in the final Certificate is due and payable. The Architect's final Certificate for Payment will constitute a further representation that conditions listed in Section 9.10.2 as precedent to the Contractor's being entitled to final payment have been fulfilled.

§ 9.10.2 Neither final payment nor any remaining retained percentage shall become due until the Contractor submits to the Architect (1) an affidavit that payrolls, bills for materials and equipment, and other indebtedness connected with the Work for which the Owner or the Owner's property might be responsible or encumbered (less amounts withheld by Owner) have been paid or otherwise satisfied, (2) a certificate evidencing that insurance required by the Contract Documents to remain in force after final payment is currently in effect~~ and will not be canceled or allowed to expire until at least 30 days' prior written notice has been given to the Owner~~, (3) a written statement that the Contractor knows of no~~ substantial~~ reason that the insurance will not be renewable to cover the period required by the Contract Documents, (4) consent of surety, if any, to final payment~~ and (5)~~, (5) documentation of any special warranties, such as manufacturers' warranties or specific Subcontractor warranties, and (6) if required by the Owner, other data establishing payment or satisfaction of obligations, such as receipts~~,~~ and releases and waivers of liens, claims, security interests, or encumbrances arising out of the Contract, to the extent and in such form as may be designated by the Owner. If a Subcontractor refuses to furnish a release or waiver required by the Owner, the Contractor may furnish a bond satisfactory to the Owner to indemnify the Owner against such lien~~. If such lien~~, claim, security interest, or encumbrance. If a lien, claim, security interest, or encumbrance remains unsatisfied after payments are made, the Contractor shall refund to the Owner all money that the Owner may be compelled to pay in discharging ~~such lien~~the lien, claim, security interest, or encumbrance, including all costs and reasonable attorneys' fees.

§ 9.10.3 If, after Substantial Completion of the Work, final completion thereof is materially delayed through no fault of the Contractor or by issuance of Change Orders affecting final completion, and the Architect so confirms, the Owner shall, upon application by the Contractor and certification by the Architect, and without terminating the Contract, make payment of the balance due for that portion of the Work fully completed, corrected, and accepted. If the remaining balance for Work not fully completed or corrected is less than retainage stipulated in the Contract Documents, and if bonds have been furnished, the written consent of the surety to payment of the balance due for that portion of the Work fully completed and accepted shall be submitted by the Contractor to the Architect prior to certification of such payment. Such payment shall be made under terms and conditions governing final payment, except that it shall not constitute a waiver of ~~claims~~Claims.

§ 9.10.4 The making of final payment shall constitute a waiver of Claims by the Owner except those arising from

- **.1** liens, Claims, security interests, or encumbrances arising out of the Contract and unsettled;
- **.2** failure of the Work to comply with the requirements of the Contract Documents; ~~or~~
- **.3** terms of special warranties required by the Contract Documents~~.~~; or
- **.4** audits performed by the Owner, if permitted by the Contract Documents, after final payment.

§ 9.10.5 Acceptance of final payment by the Contractor, a Subcontractor, or ~~material~~a supplier, shall constitute a waiver of claims by that payee except those previously made in writing and identified by that payee as unsettled at the time of final Application for Payment.

Article 10 Protection of Persons and Property

§ 10.1 Safety Precautions and Programs

The Contractor shall be responsible for initiating, maintaining, and supervising all safety precautions and programs in connection with the performance of the Contract.

§ 10.2 Safety of Persons and Property

§ 10.2.1 The Contractor shall take reasonable precautions for safety of, and shall provide reasonable protection to prevent damage, injury, or loss to

- **.1** employees on the Work and other persons who may be affected thereby;
- **.2** the Work and materials and equipment to be incorporated therein, whether in storage on or off the site, under care, custody, or control of the Contractor, a Subcontractor, or ~~the Contractor's Subcontractors or~~a Sub-~~subcontractors~~subcontractor; and
- **.3** other property at the site or adjacent thereto, such as trees, shrubs, lawns, walks, pavements, roadways, structures, and utilities not designated for removal, relocation, or replacement in the course of construction.

§ 10.2.2 The Contractor shall comply with, and give notices required by applicable laws, statutes, ordinances, codes, rules and regulations, and lawful orders of public authorities, bearing on safety of persons or property or their protection from damage, injury, or loss.

§ 10.2.3 The Contractor shall implement, erect, and maintain, as required by existing conditions and performance of the Contract, reasonable safeguards for safety and protection, including posting danger signs and other warnings against hazards~~,~~; promulgating safety regulations; and notifying the owners and users of adjacent sites and utilities of the safeguards.

§ 10.2.4 When use or storage of explosives or other hazardous materials or equipment, or unusual methods are necessary for execution of the Work, the Contractor shall exercise utmost care and carry on such activities under supervision of properly qualified personnel.

§ 10.2.5 The Contractor shall promptly remedy damage and loss (other than damage or loss insured under property insurance required by the Contract Documents) to property referred to in Sections 10.2.1.2 and 10.2.1.3 caused in whole or in part by the Contractor, a Subcontractor, a Sub-subcontractor, or anyone directly or indirectly employed by any of them, or by anyone for whose acts they may be liable and for which the Contractor is responsible under Sections 10.2.1.2 and 10.2.1.3~~, except~~. The Contractor may make a Claim for the cost to remedy the damage or loss to the extent such damage or loss is attributable to acts or omissions of the Owner or Architect or anyone directly or indirectly employed by either of them, or by anyone for whose acts either of them may be liable, and not attributable to the fault or negligence of the Contractor. The foregoing obligations of the Contractor are in addition to the Contractor's obligations under Section 3.18.

§ 10.2.6 The Contractor shall designate a responsible member of the Contractor's organization at the site whose duty shall be the prevention of accidents. This person shall be the Contractor's superintendent unless otherwise designated by the Contractor in writing to the Owner and Architect.

§ 10.2.7 The Contractor shall not permit any part of the construction or site to be loaded so as to cause damage or create an unsafe condition.

§ 10.2.8 Injury or Damage to Person or Property
If either party suffers injury or damage to person or property because of an act or omission of the other party, or of others for whose acts such party is legally responsible, ~~written~~ notice of ~~such~~the injury or damage, whether or not insured, shall be given to the other party within a reasonable time not exceeding 21 days after discovery. The notice shall provide sufficient detail to enable the other party to investigate the matter.

§ 10.3 Hazardous Materials and Substances
§ 10.3.1 The Contractor is responsible for compliance with any requirements included in the Contract Documents regarding hazardous materials~~.~~ or substances. If the Contractor encounters a hazardous material or substance not addressed in the Contract Documents and if reasonable precautions will be inadequate to prevent foreseeable bodily injury or death to persons resulting from a material or substance, including but not limited to asbestos or polychlorinated biphenyl (PCB), encountered on the site by the Contractor, the Contractor shall, upon recognizing the condition, immediately stop Work in the affected area and ~~report the condition to~~notify the Owner and Architect ~~in writing~~of the condition.

§ 10.3.2 Upon receipt of the Contractor's ~~written~~ notice, the Owner shall obtain the services of a licensed laboratory to verify the presence or absence of the material or substance reported by the Contractor and, in the event such material or substance is found to be present, to cause it to be rendered harmless. Unless otherwise required by the Contract Documents, the Owner shall furnish in writing to the Contractor and Architect the names and qualifications of persons or entities who are to perform tests verifying the presence or absence of ~~such~~the material or substance or who are to perform the task of removal or safe containment of ~~such~~the material or substance. The Contractor and the Architect will promptly reply to the Owner in writing stating whether or not either has reasonable objection to the persons or entities proposed by the Owner. If either the Contractor or Architect has an objection to a person or entity proposed by the Owner, the Owner shall propose another to whom the Contractor and the Architect have no reasonable objection. When the material or substance has been rendered harmless, Work in the affected area shall resume upon written agreement of the Owner and Contractor. By Change Order, the Contract Time shall be extended appropriately and the Contract Sum shall be increased ~~in~~by the amount of the Contractor's reasonable additional costs of ~~shut-down~~shutdown, delay, and start-up.

§ 10.3.3 To the fullest extent permitted by law, the Owner shall indemnify and hold harmless the Contractor, Subcontractors, Architect, Architect's consultants, and agents and employees of any of them from and against claims, damages, losses, and expenses, including but not limited to attorneys' fees, arising out of or resulting from performance of the Work in the affected area if in fact the material or substance presents the risk of bodily injury or death as described in Section 10.3.1 and has not been rendered harmless, provided that such claim, damage, loss, or expense is attributable to bodily injury, sickness, disease or death, or to injury to or destruction of tangible property (other than the Work itself), except to the extent that such damage, loss, or expense is due to the fault or negligence of the party seeking indemnity.

§ 10.3.4 The Owner shall not be responsible under this Section 10.3 for hazardous materials or substances the Contractor brings to the site unless such materials or substances are required by the Contract Documents. The Owner shall be responsible for hazardous materials or substances required by the Contract Documents, except to the extent of the Contractor's fault or negligence in the use and handling of such materials or substances.

§ 10.3.5 The Contractor shall ~~indemnify~~reimburse the Owner for the cost and expense the Owner incurs (1) for remediation of ~~a material~~hazardous materials or ~~substance~~substances the Contractor brings to the site and negligently handles, or (2) where the Contractor fails to perform its obligations under Section 10.3.1, except to the extent that the cost and expense are due to the Owner's fault or negligence.

§ 10.3.6 If, without negligence on the part of the Contractor, the Contractor is held liable by a government agency for the cost of remediation of a hazardous material or substance solely by reason of performing Work as required by the

Contract Documents, the Owner shall ~~indemnify~~reimburse the Contractor for all cost and expense thereby incurred.

§ 10.4 Emergencies
In an emergency affecting safety of persons or property, the Contractor shall act, at the Contractor's discretion, to prevent threatened damage, injury, or loss. Additional compensation or extension of time claimed by the Contractor on account of an emergency shall be determined as provided in Article 15 and Article 7.

Article 11 Insurance and Bonds

> ***Note:*** *Many of the terms and provisions previously located in Article 11 are relocated to a new Insurance and Bonds Exhibit, which is Exhibit A to AIA Documents A101™–2017, A102™–2017, and A103™–2017. The text of the Insurance and Bonds Exhibit immediately follows this comparative.*

~~**§ 11.1 CONTRACTOR'S LIABILITY INSURANCE**~~
~~**§ 11.1.1** The Contractor shall purchase from and maintain in a company or companies lawfully authorized to do business in the jurisdiction in which the Project is located such insurance as will protect the Contractor from claims set forth below which may arise out of or result from the Contractor's operations and completed operations under the Contract and for which the Contractor may be legally liable, whether such operations be by the Contractor or by a Subcontractor or by anyone directly or indirectly employed by any of them, or by anyone for whose acts any of them may be liable:~~

~~**.1** Claims under workers' compensation, disability benefit and other similar employee benefit acts that are applicable to the Work to be performed;~~
~~**.2** Claims for damages because of bodily injury, occupational sickness or disease, or death of the Contractor's employees;~~
~~**.3** Claims for damages because of bodily injury, sickness or disease, or death of any person other than the Contractor's employees;~~
~~**.4** Claims for damages insured by usual personal injury liability coverage;~~
~~**.5** Claims for damages, other than to the Work itself, because of injury to or destruction of tangible property, including loss of use resulting therefrom;~~
~~**.6** Claims for damages because of bodily injury, death of a person or property damage arising out of ownership, maintenance or use of a motor vehicle;~~
~~**.7** Claims for bodily injury or property damage arising out of completed operations; and~~
~~**.8** Claims involving contractual liability insurance applicable to the Contractor's obligations under Section 3.18.~~

~~**§ 11.1.2** The insurance required by Section 11.1.1 shall be written for not less than limits of liability specified in the Contract Documents or required by law, whichever coverage is greater. Coverages, whether written on an occurrence or claims-made basis, shall be maintained without interruption from the date of commencement of the Work until the date of final payment and termination of any coverage required to be maintained after final payment, and, with respect to the Contractor's completed operations coverage, until the expiration of the period for correction of Work or for such other period for maintenance of completed operations coverage as specified in the Contract Documents.~~

~~**§ 11.1.3** Certificates of insurance acceptable to the Owner shall be filed with the Owner prior to commencement of the Work and thereafter upon renewal or replacement of each required policy of insurance. These certificates and the insurance policies required by this Section 11.1 shall contain a provision that coverages afforded under the policies will not be canceled or allowed to expire until at least 30 days' prior written notice has been given to the Owner. An additional certificate evidencing continuation of liability coverage, including coverage for completed operations, shall be submitted with the final Application for Payment as required by Section 9.10.2 and thereafter upon renewal or replacement of such coverage until the expiration of the time required by Section 11.1.2. Information concerning reduction of coverage on account of revised limits or claims paid under the General Aggregate, or both, shall be furnished by the Contractor with reasonable promptness.~~

§ 11.1.4 ~~The Contractor shall cause the~~ commercial ~~liability coverage required by the Contract Documents to include (1) the Owner, the Architect and the Architect's consultants as additional insureds for claims caused in whole or in part by the Contractor~~'s neglig~~ent acts or omissions during the Contractor's operations; and (2)~~ the Owner ~~as an additional~~ insur~~ed for claims caused in whole or in part by the Contractor's negligent acts or omissions during the Contractor's completed operations.~~

§ 11.1 Contractor's Insurance and Bonds
§ 11.1.1 The Contractor shall purchase and maintain insurance of the types and limits of liability, containing the endorsements, and subject to the terms and conditions, as described in the Agreement or elsewhere in the Contract Documents. The Contractor shall purchase and maintain the required insurance from an insurance company or insurance companies lawfully authorized to issue insurance in the jurisdiction where the Project is located. The Owner, Architect, and Architect's consultants shall be named as additional insureds under the Contractor's commercial general liability policy or as otherwise described in the Contract Documents.

§ 11.1.2 The Contractor shall provide surety bonds of the types, for such penal sums, and subject to such terms and conditions as required by the Contract Documents. The Contractor shall purchase and maintain the required bonds from a company or companies lawfully authorized to issue surety bonds in the jurisdiction where the Project is located.

§ 11.1.3 Upon the request of any person or entity appearing to be a potential beneficiary of bonds covering payment of obligations arising under the Contract, the Contractor shall promptly furnish a copy of the bonds or shall authorize a copy to be furnished.

§ 11.1.4 Notice of Cancellation or Expiration of Contractor's Required Insurance. Within three (3) business days of the date the Contractor becomes aware of an impending or actual cancellation or expiration of any insurance required by the Contract Documents, the Contractor shall provide notice to the Owner of such impending or actual cancellation or expiration. Upon receipt of notice from the Contractor, the Owner shall, unless the lapse in coverage arises from an act or omission of the Owner, have the right to stop the Work until the lapse in coverage has been cured by the procurement of replacement coverage by the Contractor. The furnishing of notice by the Contractor shall not relieve the Contractor of any contractual obligation to provide any required coverage.

***Note:** Portions of Sections 11.1.2 and 11.1.3 formerly appeared at Section 11.4.*

~~**§ 11.2 OWNER'S LIABILITY INSURANCE**~~
~~The Owner shall be responsible for purchasing and maintaining the Owner's usual liability insurance.~~

§ 11.2 Owner's Insurance
§ 11.2.1 The Owner shall purchase and maintain insurance of the types and limits of liability, containing the endorsements, and subject to the terms and conditions, as described in the Agreement or elsewhere in the Contract Documents. The Owner shall purchase and maintain the required insurance from an insurance company or insurance companies lawfully authorized to issue insurance in the jurisdiction where the Project is located.

§ 11.2.2 Failure to Purchase Required Property Insurance. If the Owner fails to purchase and maintain the required property insurance, with all of the coverages and in the amounts described in the Agreement or elsewhere in the Contract Documents, the Owner shall inform the Contractor in writing prior to commencement of the Work. Upon receipt of notice from the Owner, the Contractor may delay commencement of the Work and may obtain insurance that will protect the interests of the Contractor, Subcontractors, and Sub-Subcontractors in the Work. When the failure to provide coverage has been cured or resolved, the Contract Sum and Contract Time shall be equitably adjusted. In the event the Owner fails to procure coverage, the Owner waives all rights against the Contractor, Subcontractors, and Sub-subcontractors to the extent the loss to the Owner would have been covered by the insurance to have been procured by the Owner. The cost of the insurance shall be charged to the Owner by a Change Order. If the Owner does not provide written notice, and the Contractor is damaged by the failure or neglect of the Owner to purchase or maintain the required insurance, the Owner shall reimburse the Contractor for all reasonable costs and damages attributable thereto.

***Note:** Section 11.2.2 formerly appeared at Section 11.3.1.2.*

§ 11.2.3 Notice of Cancellation or Expiration of Owner's Required Property Insurance. Within three (3) business days of the date the Owner becomes aware of an impending or actual cancellation or expiration of any property insurance required by the Contract Documents, the Owner shall provide notice to the Contractor of such impending or actual cancellation or expiration. Unless the lapse in coverage arises from an act or omission of the Contractor: (1) the Contractor, upon receipt of notice from the Owner, shall have the right to stop the Work until the lapse in coverage has been cured by the procurement of replacement coverage by either the Owner or the Contractor; (2) the Contract Time and Contract Sum shall be equitably adjusted; and (3) the Owner waives all rights against the Contractor, Subcontractors, and Sub-subcontractors to the extent any loss to the Owner would have been covered by the insurance had it not expired or been cancelled. If the Contractor purchases replacement coverage, the cost of the insurance shall be charged to the Owner by an appropriate Change Order. The furnishing of notice by the Owner shall not relieve the Owner of any contractual obligation to provide required insurance.

§ 11.3 PROPERTY INSURANCE

§ 11.3.1 Unless otherwise provided, the Owner shall purchase and maintain, in a company or companies lawfully authorized to do business in the jurisdiction in which the Project is located, property insurance written on a builder's risk "all-risk" or equivalent policy form in the amount of the initial Contract Sum, plus value of subsequent Contract Modifications and cost of materials supplied or installed by others, comprising total value for the entire Project at the site on a replacement cost basis without optional deductibles. Such property insurance shall be maintained, unless otherwise provided in the Contract Documents or otherwise agreed in writing by all persons and entities who are beneficiaries of such insurance, until final payment has been made as provided in Section 9.10 or until no person or entity other than the Owner has an insurable interest in the property required by this Section 11.3 to be covered, whichever is later. This insurance shall include interests of the Owner, the Contractor, Subcontractors and Sub-subcontractors in the Project.

§ 11.3.1.1 Property insurance shall be on an "all-risk" or equivalent policy form and shall include, without limitation, insurance against the perils of fire (with extended coverage) and physical loss or damage including, without duplication of coverage, theft, vandalism, malicious mischief, collapse, earthquake, flood, windstorm, falsework, testing and startup, temporary buildings and debris removal including demolition occasioned by enforcement of any applicable legal requirements, and shall cover reasonable compensation for Architect's and Contractor's services and expenses required as a result of such insured loss.

§ 11.3.1.2 If the Owner does not intend to purchase such property insurance required by the Contract and with all of the coverages in the amount described above, the Owner shall so inform the Contractor in writing prior to commencement of the Work. The Contractor may then effect insurance that will protect the interests of the Contractor, Subcontractors and Sub-subcontractors in the Work, and by appropriate Change Order the cost thereof shall be charged to the Owner. If the Contractor is damaged by the failure or neglect of the Owner to purchase or maintain insurance as described above, without so notifying the Contractor in writing, then the Owner shall bear all reasonable costs properly attributable thereto.

§ 11.3.1.3 If the property insurance requires deductibles, the Owner shall pay costs not covered because of such deductibles.

§ 11.3.1.4 This property insurance shall cover portions of the Work stored off the site, and also portions of the Work in transit.

§ 11.3.1.5 Partial occupancy or use in accordance with Section 9.9 shall not commence until the insurance company or companies providing property insurance have consented to such partial occupancy or use by endorsement or otherwise. The Owner and the Contractor shall take reasonable steps to obtain consent of the insurance company or companies and shall, without mutual written consent, take no action with respect to partial occupancy or use that would cause cancellation, lapse or reduction of insurance.

§ 11.3.2 BOILER AND MACHINERY INSURANCE

The Owner shall purchase and maintain boiler and machinery insurance required by the Contract Documents or by law, which shall specifically cover such insured objects during installation and until final acceptance by the Owner; this insurance shall include interests of the Owner, Contractor, Subcontractors and Sub-subcontractors in the Work, and the Owner and Contractor shall be named insureds.

§ 11.3.3 LOSS OF USE INSURANCE

The Owner, at the Owner's option, may purchase and maintain such insurance as will insure the Owner against loss of use of the Owner's property due to fire or other hazards, however caused. The Owner waives all rights of action against the Contractor for loss of use of the Owner's property, including consequential losses due to fire or other hazards however caused.

§ 11.3.4 If the Contractor requests in writing that insurance for risks other than those described herein or other special causes of loss be included in the property insurance policy, the Owner shall, if possible, include such insurance, and the cost thereof shall be charged to the Contractor by appropriate Change Order.

§ 11.3.5 If during the Project construction period the Owner insures properties, real or personal or both, at or adjacent to the site by property insurance under policies separate from those insuring the Project, or if after final payment

~~property insurance is to be provided on the completed Project through a policy or policies other than those insuring the Project during the construction period, the Owner shall waive all rights in accordance with the terms of Section 11.3.7 for damages caused by fire or other causes of loss covered by this separate property insurance. All separate policies shall provide this waiver of subrogation by endorsement or otherwise.~~

~~**§ 11.3.6** Before an exposure to loss may occur, the Owner shall file with the Contractor a copy of each policy that includes insurance coverages required by this Section 11.3. Each policy shall contain all generally applicable conditions, definitions, exclusions and endorsements related to this Project. Each policy shall contain a provision that the policy will not be canceled or allowed to expire, and that its limits will not be reduced, until at least 30 days' prior written notice has been given to the Contractor.~~

~~**§ 11.3.7 WAIVERS OF SUBROGATION**~~
~~The Owner and Contractor waive all rights against (1) each other and any of their subcontractors, sub-subcontractors, agents and employees, each of the other, and (2) the Architect, Architect's consultants, separate contractors described in Article 6, if any, and any of their subcontractors, sub-subcontractors, agents and employees, for damages caused by fire or other causes of loss to the extent covered by property insurance obtained pursuant to this Section 11.3 or other property insurance applicable to the Work, except such rights as they have to proceeds of such insurance held by the Owner as fiduciary. The Owner or Contractor, as appropriate, shall require of the Architect, Architect's consultants, separate contractors described in Article 6, if any, and the subcontractors, sub-subcontractors, agents and employees of any of them, by appropriate agreements, written where legally required for validity, similar waivers each in favor of other parties enumerated herein. The policies shall provide such waivers of subrogation by endorsement or otherwise. A waiver of subrogation shall be effective as to a person or entity even though that person or entity would otherwise have a duty of indemnification, contractual or otherwise, did not pay the insurance premium directly or indirectly, and whether or not the person or entity had an insurable interest in the property damaged.~~

~~**§ 11.3.8** A loss insured under the Owner's property insurance shall be adjusted by the Owner as fiduciary and made payable to the Owner as fiduciary for the insureds, as their interests may appear, subject to requirements of any applicable mortgagee clause and of Section 11.3.10. The Contractor shall pay Subcontractors their just shares of insurance proceeds received by the Contractor, and by appropriate agreements, written where legally required for validity, shall require Subcontractors to make payments to their Sub-subcontractors in similar manner.~~

~~**§ 11.3.9** If required in writing by a party in interest, the Owner as fiduciary shall, upon occurrence of an insured loss, give bond for proper performance of the Owner's duties. The cost of required bonds shall be charged against proceeds received as fiduciary. The Owner shall deposit in a separate account proceeds so received, which the Owner shall distribute in accordance with such agreement as the parties in interest may reach, or as determined in accordance with the method of binding dispute resolution selected in the Agreement between the Owner and Contractor. If after such loss no other special agreement is made and unless the Owner terminates the Contract for convenience, replacement of damaged property shall be performed by the Contractor after notification of a Change in the Work in accordance with Article 7.~~

~~**§ 11.3.10** The Owner as fiduciary shall have power to adjust and settle a loss with insurers unless one of the parties in interest shall object in writing within five days after occurrence of loss to the Owner's exercise of this power; if such objection is made, the dispute shall be resolved in the manner selected by the Owner and Contractor as the method of binding dispute resolution in the Agreement. If the Owner and Contractor have selected arbitration as the method of binding dispute resolution, the Owner as fiduciary shall make settlement with insurers or, in the case of a dispute over distribution of insurance proceeds, in accordance with the directions of the arbitrators.~~

§ 11.3 Waivers of Subrogation
§ 11.3.1 The Owner and Contractor waive all rights against (1) each other and any of their subcontractors, sub-subcontractors, agents, and employees, each of the other; (2) the Architect and Architect's consultants; and (3) Separate Contractors, if any, and any of their subcontractors, sub-subcontractors, agents, and employees, for damages caused by fire, or other causes of loss, to the extent those losses are covered by property insurance required by the Agreement or other property insurance applicable to the Project, except such rights as they have to proceeds of such insurance. The Owner or Contractor, as appropriate, shall require similar written waivers in favor of the individuals and entities identified above from the Architect, Architect's consultants, Separate Contractors, subcontractors, and sub-subcontractors. The policies of insurance purchased and maintained by each person or entity agreeing to waive claims pursuant to this section 11.3.1 shall not prohibit this waiver of subrogation. This waiver of subrogation shall be effective as to a person or entity (1) even though that person or entity would otherwise have a duty of indemnification, contractual or otherwise, (2) even though that person or entity did not pay the insurance

premium directly or indirectly, or (3) whether or not the person or entity had an insurable interest in the damaged property.

§ 11.3.2 If during the Project construction period the Owner insures properties, real or personal or both, at or adjacent to the site by property insurance under policies separate from those insuring the Project, or if after final payment property insurance is to be provided on the completed Project through a policy or policies other than those insuring the Project during the construction period, to the extent permissible by such policies, the Owner waives all rights in accordance with the terms of Section 11.3.1 for damages caused by fire or other causes of loss covered by this separate property insurance.

***Note:** Section 11.3 formerly appeared at Section 11.3.7.*

~~**§ 11.4 PERFORMANCE BOND AND PAYMENT BOND**~~

~~**§ 11.4.1** The Owner shall have the right to require the Contractor to furnish bonds covering faithful performance of the Contract and payment of obligations arising thereunder as stipulated in bidding requirements or specifically required in the Contract Documents on the date of execution of the Contract.~~

~~**§ 11.4.2** Upon the request of any person or entity appearing to be a potential beneficiary of bonds covering payment of obligations arising under the Contract, the Contractor shall promptly furnish a copy of the bonds or shall authorize a copy to be furnished.~~

§ 11.4 Loss of Use, Business Interruption, and Delay in Completion Insurance

The Owner, at the Owner's option, may purchase and maintain insurance that will protect the Owner against loss of use of the Owner's property, or the inability to conduct normal operations, due to fire or other causes of loss. The Owner waives all rights of action against the Contractor and Architect for loss of use of the Owner's property, due to fire or other hazards however caused.

***Note:** Section 11.4 formerly appeared at Section 11.3.3.*

§11.5 Adjustment and Settlement of Insured Loss

§ 11.5.1 A loss insured under the property insurance required by the Agreement shall be adjusted by the Owner as fiduciary and made payable to the Owner as fiduciary for the insureds, as their interests may appear, subject to requirements of any applicable mortgagee clause and of Section 11.5.2. The Owner shall pay the Architect and Contractor their just shares of insurance proceeds received by the Owner, and by appropriate agreements the Architect and Contractor shall make payments to their consultants and Subcontractors in similar manner.

***Note:** Section 11.5.1 formerly appeared at Section 11.3.8.*

§ 11.5.2 Prior to settlement of an insured loss, the Owner shall notify the Contractor of the terms of the proposed settlement as well as the proposed allocation of the insurance proceeds. The Contractor shall have 14 days from receipt of notice to object to the proposed settlement or allocation of the proceeds. If the Contractor does not object, the Owner shall settle the loss and the Contractor shall be bound by the settlement and allocation. Upon receipt, the Owner shall deposit the insurance proceeds in a separate account and make the appropriate distributions. Thereafter, if no other agreement is made or the Owner does not terminate the Contract for convenience, the Owner and Contractor shall execute a Change Order for reconstruction of the damaged or destroyed Work in the amount allocated for that purpose. If the Contractor timely objects to either the terms of the proposed settlement or the allocation of the proceeds, the Owner may proceed to settle the insured loss, and any dispute between the Owner and Contractor arising out of the settlement or allocation of the proceeds shall be resolved pursuant to Article 15. Pending resolution of any dispute, the Owner may issue a Construction Change Directive for the reconstruction of the damaged or destroyed Work.

***Note:** Section 11.5.2 combines language formerly appearing at Section 11.3.9 and Section 11.3.10.*

Article 12 Uncovering and Correction of Work

§ 12.1 Uncovering of Work
§ 12.1.1 If a portion of the Work is covered contrary to the Architect's request or to requirements specifically expressed in the Contract Documents, it must, if requested in writing by the Architect, be uncovered for the Architect's examination and be replaced at the Contractor's expense without change in the Contract Time.

§ 12.1.2 If a portion of the Work has been covered that the Architect has not specifically requested to examine prior to its being covered, the Architect may request to see such Work and it shall be uncovered by the Contractor. If such Work is in accordance with the Contract Documents, ~~costs of uncovering and replacement~~the Contractor shall~~, by~~ be entitled to an equitable adjustment to the Contract Sum and Contract Time as may be appropriate~~ Change Order, be at the Owner's expense.~~. If such Work is not in accordance with the Contract Documents, ~~such~~the costs of uncovering the Work, and the cost of correction, shall be at the Contractor's expense~~ unless the condition was caused by the Owner or a separate contractor in which event the Owner shall be responsible for payment of such costs~~.

§ 12.2 Correction of Work
§ 12.2.1 ~~BEFORE OR AFTER SUBSTANTIAL COMPLETION~~Before Substantial Completion
The Contractor shall promptly correct Work rejected by the Architect or failing to conform to the requirements of the Contract Documents, ~~whether~~ discovered before~~ or after~~ Substantial Completion and whether or not fabricated, installed or completed. Costs of correcting such rejected Work, including additional testing and inspections, the cost of uncovering and replacement, and compensation for the Architect's services and expenses made necessary thereby, shall be at the Contractor's expense.

§ 12.2.2 After Substantial Completion
§ 12.2.2.1 In addition to the Contractor's obligations under Section 3.5, if, within one year after the date of Substantial Completion of the Work or designated portion thereof or after the date for commencement of warranties established under Section 9.9.1, or by terms of ~~an~~any applicable special warranty required by the Contract Documents, any of the Work is found to be not in accordance with the requirements of the Contract Documents, the Contractor shall correct it promptly after receipt of ~~written~~ notice from the Owner to do so, unless the Owner has previously given the Contractor a written acceptance of such condition. The Owner shall give such notice promptly after discovery of the condition. During the one-year period for correction of Work, if the Owner fails to notify the Contractor and give the Contractor an opportunity to make the correction, the Owner waives the rights to require correction by the Contractor and to make a claim for breach of warranty. If the Contractor fails to correct nonconforming Work within a reasonable time during that period after receipt of notice from the Owner or Architect, the Owner may correct it in accordance with Section 2.4~~5~~.

§ 12.2.2.2 The one-year period for correction of Work shall be extended with respect to portions of Work first performed after Substantial Completion by the period of time between Substantial Completion and the actual completion of that portion of the Work.

§ 12.2.2.3 The one-year period for correction of Work shall not be extended by corrective Work performed by the Contractor pursuant to this Section 12.2.

§ 12.2.3 The Contractor shall remove from the site portions of the Work that are not in accordance with the requirements of the Contract Documents and are neither corrected by the Contractor nor accepted by the Owner.

§ 12.2.4 The Contractor shall bear the cost of correcting destroyed or damaged construction of the Owner or Separate Contractors, whether completed or partially completed,~~ of the Owner or separate contractors~~ caused by the Contractor's correction or removal of Work that is not in accordance with the requirements of the Contract Documents.

§ 12.2.5 Nothing contained in this Section 12.2 shall be construed to establish a period of limitation with respect to other obligations the Contractor has under the Contract Documents. Establishment of the one-year period for correction of Work as described in Section 12.2.2 relates only to the specific obligation of the Contractor to correct the Work, and has no relationship to the time within which the obligation to comply with the Contract Documents may be sought to be enforced, nor to the time within which proceedings may be commenced to establish the Contractor's liability with respect to the Contractor's obligations other than specifically to correct the Work.

§ 12.3 Acceptance of Nonconforming Work
If the Owner prefers to accept Work that is not in accordance with the requirements of the Contract Documents, the Owner may do so instead of requiring its removal and correction, in which case the Contract Sum will be reduced as appropriate and equitable. Such adjustment shall be effected whether or not final payment has been made.

Article 13 Miscellaneous Provisions

§ 13.1 Governing Law
The Contract shall be governed by the law of the place where the Project is located ~~except~~, excluding that, ~~if~~ jurisdiction's choice of law rules. If the parties have selected arbitration as the method of binding dispute resolution, the Federal Arbitration Act shall govern Section 15.4.

§ 13.2 Successors and Assigns
§ 13.2.1 The Owner and Contractor respectively bind themselves, their partners, successors, assigns, and legal representatives to covenants, agreements, and obligations contained in the Contract Documents. Except as provided in Section 13.2.2, neither party to the Contract shall assign the Contract as a whole without written consent of the other. If either party attempts to make ~~such~~ an assignment without such consent, that party shall nevertheless remain legally responsible for all obligations under the Contract.

§ 13.2.2 The Owner may, without consent of the Contractor, assign the Contract to a lender providing construction financing for the Project, if the lender assumes the Owner's rights and obligations under the Contract Documents. The Contractor shall execute all consents reasonably required to facilitate ~~such~~the assignment.

§ 13.3 ~~WRITTEN NOTICE~~
~~Written notice shall be deemed to have been duly served if delivered in person to the individual, to a member of the firm or entity, or to an officer of the corporation for which it was intended; or if delivered at, or sent by registered or certified mail or by courier service providing proof of delivery to, the last business address known to the party giving notice.~~

Note: Former Section 13.3 is moved to Section 1.6.

§ 13.~~4~~ Rights and Remedies
§ ~~13.4~~3.1 Duties and obligations imposed by the Contract Documents and rights and remedies available thereunder shall be in addition to and not a limitation of duties, obligations, rights, and remedies otherwise imposed or available by law.

§ 13.~~4~~3.2 No action or failure to act by the Owner, Architect, or Contractor shall constitute a waiver of a right or duty afforded them under the Contract, nor shall such action or failure to act constitute approval of or acquiescence in a breach ~~there under~~thereunder, except as may be specifically agreed upon in writing.

§ 13.~~5~~4 Tests and Inspections
§ 13.~~5~~4.1 Tests, inspections, and approvals of portions of the Work shall be made as required by the Contract Documents and by applicable laws, statutes, ordinances, codes, rules, and regulations or lawful orders of public authorities. Unless otherwise provided, the Contractor shall make arrangements for such tests, inspections, and approvals with an independent testing laboratory or entity acceptable to the Owner, or with the appropriate public authority, and shall bear all related costs of tests, inspections, and approvals. The Contractor shall give the Architect timely notice of when and where tests and inspections are to be made so that the Architect may be present for such procedures. The Owner shall bear costs of ~~(1)~~ tests, inspections, or approvals that do not become requirements until after bids are received or negotiations concluded~~,~~. The Owner shall directly arrange and ~~(2)~~pay for tests, inspections, or approvals where building codes or applicable laws or regulations ~~prohibit the Owner from delegating their cost to the Contractor~~so require.

§ 13.~~5~~4.2 If the Architect, Owner, or public authorities having jurisdiction determine that portions of the Work require additional testing, inspection, or approval not included under Section 13.~~5~~4.1, the Architect will, upon written authorization from the Owner, instruct the Contractor to make arrangements for such additional testing, inspection, or approval, by an entity acceptable to the Owner, and the Contractor shall give timely notice to the Architect of when and where tests and inspections are to be made so that the Architect may be present for such procedures. Such costs, except as provided in Section 13.~~5~~4.3, shall be at the Owner's expense.

§ 13.~~5~~4.3 If ~~such~~ procedures for testing, inspection, or approval under Sections 13.~~5~~4.1 and 13.~~5~~4.2 reveal failure of the portions of the Work to comply with requirements established by the Contract Documents, all costs made necessary by such failure, including those of repeated procedures and compensation for the Architect's services and expenses, shall be at the Contractor's expense.

§ 13.~~5~~4.4 Required certificates of testing, inspection, or approval shall, unless otherwise required by the Contract Documents, be secured by the Contractor and promptly delivered to the Architect.

§ 13.~~5~~4.5 If the Architect is to observe tests, inspections, or approvals required by the Contract Documents, the Architect will do so promptly and, where practicable, at the normal place of testing.

§ 13.~~5~~4.6 Tests or inspections conducted pursuant to the Contract Documents shall be made promptly to avoid unreasonable delay in the Work.

§ 13.~~6~~5 Interest
Payments due and unpaid under the Contract Documents shall bear interest from the date payment is due at ~~such~~the rate ~~as~~ the parties ~~may~~ agree upon in writing or, in the absence thereof, at the legal rate prevailing from time to time at the place where the Project is located.

~~**§ 13.7 TIME LIMITS ON CLAIMS**~~
~~The Owner and Contractor shall commence all claims and causes of action, whether in contract, tort, breach of warranty or otherwise, against the other arising out of or related to the Contract in accordance with the requirements of the final dispute resolution method selected in the Agreement within the time period specified by applicable law, but in any case not more than 10 years after the date of Substantial Completion of the Work. The Owner and Contractor waive all claims and causes of action not commenced in accordance with this Section 13.7.~~

Note: Former Section 13.7 moved to Section 15.1.2.

Article 14 Termination or Suspension of the Contract

§ 14.1 Termination by the Contractor
§ 14.1.1 The Contractor may terminate the Contract if the Work is stopped for a period of 30 consecutive days through no act or fault of the Contractor ~~or~~, a Subcontractor, a Sub-subcontractor ~~or~~, their agents or employees, or any other persons or entities performing portions of the Work ~~under direct or indirect contract with the Contractor~~, for any of the following reasons:

- **.1** Issuance of an order of a court or other public authority having jurisdiction that requires all Work to be stopped;
- **.2** An act of government, such as a declaration of national emergency, that requires all Work to be stopped;
- **.3** Because the Architect has not issued a Certificate for Payment and has not notified the Contractor of the reason for withholding certification as provided in Section 9.4.1, or because the Owner has not made payment on a Certificate for Payment within the time stated in the Contract Documents; or
- **.4** The Owner has failed to furnish to the Contractor ~~promptly, upon the Contractor's request,~~ reasonable evidence as required by Section 2.2~~.1~~.

§ 14.1.2 The Contractor may terminate the Contract if, through no act or fault of the Contractor ~~or~~, a Subcontractor, a Sub-subcontractor ~~or~~, their agents or employees, or any other persons or entities performing portions of the Work ~~under direct or indirect contract with the Contractor~~, repeated suspensions, delays, or interruptions of the entire Work by the Owner as described in Section 14.3, constitute in the aggregate more than 100 percent of the total number of days scheduled for completion, or 120 days in any 365-day period, whichever is less.

§ 14.1.3 If one of the reasons described in Section 14.1.1 or 14.1.2 exists, the Contractor may, upon seven days' ~~written~~ notice to the Owner and Architect, terminate the Contract and recover from the Owner payment for Work executed, ~~including~~as well as reasonable overhead and profit~~,~~ on Work not executed, and costs incurred by reason of such termination~~, and damages~~.

§ 14.1.4 If the Work is stopped for a period of 60 consecutive days through no act or fault of the Contractor ~~or~~, a Subcontractor, a Sub-subcontractor, or their agents or employees or any other persons or entities performing portions of the Work~~ under contract with the Contractor~~ because the Owner has repeatedly failed to fulfill the Owner's obligations under the Contract Documents with respect to matters important to the progress of the Work, the Contractor may, upon seven additional days' ~~written~~ notice to the Owner and the Architect, terminate the Contract and recover from the Owner as provided in Section 14.1.3.

§ 14.2 Termination by the Owner for Cause

§ 14.2.1 The Owner may terminate the Contract if the Contractor

.1 repeatedly refuses or fails to supply enough properly skilled workers or proper materials;

.2 fails to make payment to Subcontractors ~~for materials~~ or ~~labor~~suppliers in accordance with the respective agreements between the Contractor and the Subcontractors or Suppliers;

.3 repeatedly disregards applicable laws, statutes, ordinances, codes, rules and regulations, or lawful orders of a public authority; or

.4 otherwise is guilty of substantial breach of a provision of the Contract Documents.

§ 14.2.2 When any of the ~~above~~ reasons described in Section 14.2.1 exist, ~~the Owner,~~and upon certification by the ~~Initial Decision Maker~~Architect that sufficient cause exists to justify such action, the Owner may, without prejudice to any other rights or remedies of the Owner and after giving the Contractor and the Contractor's surety, if any, seven days' ~~written~~ notice, terminate employment of the Contractor and may, subject to any prior rights of the surety:

.1 Exclude the Contractor from the site and take possession of all materials, equipment, tools, and construction equipment and machinery thereon owned by the Contractor;

.2 Accept assignment of subcontracts pursuant to Section 5.4; and

.3 Finish the Work by whatever reasonable method the Owner may deem expedient. Upon written request of the Contractor, the Owner shall furnish to the Contractor a detailed accounting of the costs incurred by the Owner in finishing the Work.

§ 14.2.3 When the Owner terminates the Contract for one of the reasons stated in Section 14.2.1, the Contractor shall not be entitled to receive further payment until the Work is finished.

§ 14.2.4 If the unpaid balance of the Contract Sum exceeds costs of finishing the Work, including compensation for the Architect's services and expenses made necessary thereby, and other damages incurred by the Owner and not expressly waived, such excess shall be paid to the Contractor. If such costs and damages exceed the unpaid balance, the Contractor shall pay the difference to the Owner. The amount to be paid to the Contractor or Owner, as the case may be, shall be certified by the Initial Decision Maker, upon application, and this obligation for payment shall survive termination of the Contract.

§ 14.3 Suspension by the Owner for Convenience

§ 14.3.1 The Owner may, without cause, order the Contractor in writing to suspend, delay or interrupt the Work, in whole or in part for such period of time as the Owner may determine.

§ 14.3.2 The Contract Sum and Contract Time shall be adjusted for increases in the cost and time caused by suspension, delay, or interruption ~~as described in~~under Section 14.3.1. Adjustment of the Contract Sum shall include profit. No adjustment shall be made to the extent

.1 that performance is, was, or would have been, so suspended, delayed, or interrupted, by another cause for which the Contractor is responsible; or

.2 that an equitable adjustment is made or denied under another provision of the Contract.

§ 14.4 Termination by the Owner for Convenience

§ 14.4.1 The Owner may, at any time, terminate the Contract for the Owner's convenience and without cause.

§ 14.4.2 Upon receipt of~~ written~~ notice from the Owner of such termination for the Owner's convenience, the Contractor shall

.1 cease operations as directed by the Owner in the notice;

.2 take actions necessary, or that the Owner may direct, for the protection and preservation of the Work; and

.3 except for Work directed to be performed prior to the effective date of termination stated in the notice, terminate all existing subcontracts and purchase orders and enter into no further subcontracts and purchase orders.

§ 14.4.3 In case of such termination for the Owner's convenience, the Owner shall pay the Contractor ~~shall be entitled to receive payment~~ for Work properly executed~~, and~~; costs incurred by reason of ~~such~~the termination, ~~along with reasonable overhead and profit on the Work not executed~~including costs attributable to termination of Subcontracts; and the termination fee, if any, set forth in the Agreement.

Article 15 Claims and Disputes

§ 15.1 Claims

§ 15.1.1 Definition

A Claim is a demand or assertion by one of the parties seeking, as a matter of right, payment of money, a change in the Contract Time, or other relief with respect to the terms of the Contract. The term "Claim" also includes other disputes and matters in question between the Owner and Contractor arising out of or relating to the Contract. The responsibility to substantiate Claims shall rest with the party making the Claim. This Section 15.1.1 does not require the Owner to file a Claim in order to impose liquidated damages in accordance with the Contract Documents.

§ 15.1.2 ~~NOTICE OF CLAIMS~~ Time Limits on Claims

The Owner and Contractor shall commence all Claims and causes of action against the other and arising out of or related to the Contract, whether in contract, tort, breach of warranty or otherwise, in accordance with the requirements of the binding dispute resolution method selected in the Agreement and within the period specified by applicable law, but in any case not more than 10 years after the date of Substantial Completion of the Work. The Owner and Contractor waive all Claims and causes of action not commenced in accordance with this Section 15.1.2.

> ***Note:*** *Section 15.1.2 relocated from Section 13.7.*

§ 15.1.3 Notice of Claims

§ 15.1.3.1 Claims by either the Owner or Contractor ~~must~~, where the condition giving rise to the Claim is first discovered prior to expiration of the period for correction of the Work set forth in Section 12.2.2, shall be initiated by ~~written~~ notice to the other party and to the Initial Decision Maker with a copy sent to the Architect, if the Architect is not serving as the Initial Decision Maker. Claims by either party ~~must~~under this Section 15.1.3.1 shall be initiated within 21 days after occurrence of the event giving rise to such Claim or within 21 days after the claimant first recognizes the condition giving rise to the Claim, whichever is later.

~~**§ 15.1.3 CONTINUING CONTRACT PERFORMANCE**~~

§ 15.1.3.2 Claims by either the Owner or Contractor, where the condition giving rise to the Claim is first discovered after expiration of the period for correction of the Work set forth in Section 12.2.2, shall be initiated by notice to the other party. In such event, no decision by the Initial Decision Maker is required.

§ 15.1.4 Continuing Contract Performance

§ 15.1.4.1 Pending final resolution of a Claim, except as otherwise agreed in writing or as provided in Section 9.7 and Article 14, the Contractor shall proceed diligently with performance of the Contract and the Owner shall continue to make payments in accordance with the Contract Documents. ~~The Architect will prepare Change Orders and issue Certificates for Payment in accordance with the decisions of the Initial Decision Maker.~~

~~**§ 15.1.4 CLAIMS FOR ADDITIONAL COST**~~

§ 15.1.4.2 The Contract Sum and Contract Time shall be adjusted in accordance with the Initial Decision Maker's decision, subject to the right of either party to proceed in accordance with this Article 15. The Architect will issue Certificates for Payment in accordance with the decision of the Initial Decision Maker.

§ 15.1.5 Claims for Additional Cost

If the Contractor wishes to make a Claim for an increase in the Contract Sum, ~~written~~ notice as provided ~~herein~~in Section 15.1.3 shall be given before proceeding to execute the portion of the Work that is the subject of the Claim. Prior notice is not required for Claims relating to an emergency endangering life or property arising under Section 10.4.

§ 15.1.~~5~~6 ~~CLAIMS FOR ADDITIONAL TIME~~ Claims for Additional Time

§ 15.1.~~5~~6.1 If the Contractor wishes to make a Claim for an increase in the Contract Time, ~~written~~ notice as provided ~~herein~~in Section 15.1.3 shall be given. The Contractor's Claim shall include an estimate of cost and of probable effect of delay on progress of the Work. In the case of a continuing delay, only one Claim is necessary.

§ 15.1.56.2 If adverse weather conditions are the basis for a Claim for additional time, such Claim shall be documented by data substantiating that weather conditions were abnormal for the period of time, could not have been reasonably anticipated, and had an adverse effect on the scheduled construction.

§ 15.1.67 CLAIMS FOR CONSEQUENTIAL DAMAGES Waiver of Claims for Consequential Damages
The Contractor and Owner waive Claims against each other for consequential damages arising out of or relating to this Contract. This mutual waiver includes

.1 damages incurred by the Owner for rental expenses, for losses of use, income, profit, financing, business and reputation, and for loss of management or employee productivity or of the services of such persons; and

.2 damages incurred by the Contractor for principal office expenses including the compensation of personnel stationed there, for losses of financing, business and reputation, and for loss of profit, except anticipated profit arising directly from the Work.

This mutual waiver is applicable, without limitation, to all consequential damages due to either party's termination in accordance with Article 14. Nothing contained in this Section 15.1.67 shall be deemed to preclude an awardassessment of liquidated damages, when applicable, in accordance with the requirements of the Contract Documents.

§ 15.2 Initial Decision
§ 15.2.1 Claims, excluding those where the condition giving rise to the Claim is first discovered after expiration of the period for correction of the Work set forth in Section 12.2.2 or arising under Sections 10.3, 10.4, 11.3.9, and 11.3.105, shall be referred to the Initial Decision Maker for initial decision. The Architect will serve as the Initial Decision Maker, unless otherwise indicated in the Agreement. Except for those Claims excluded by this Section 15.2.1, an initial decision shall be required as a condition precedent to mediation of any Claim arising prior to the date final payment is due, unless. If an initial decision has not been rendered within 30 days have passed after the Claim has been referred to the Initial Decision Maker with no, the party asserting the Claim may demand mediation and binding dispute resolution without a decision having been rendered. Unless the Initial Decision Maker and all affected parties agree, the Initial Decision Maker will not decide disputes between the Contractor and persons or entities other than the Owner.

§ 15.2.2 The Initial Decision Maker will review Claims and within ten days of the receipt of a Claim take one or more of the following actions: (1) request additional supporting data from the claimant or a response with supporting data from the other party, (2) reject the Claim in whole or in part, (3) approve the Claim, (4) suggest a compromise, or (5) advise the parties that the Initial Decision Maker is unable to resolve the Claim if the Initial Decision Maker lacks sufficient information to evaluate the merits of the Claim or if the Initial Decision Maker concludes that, in the Initial Decision Maker's sole discretion, it would be inappropriate for the Initial Decision Maker to resolve the Claim.

§ 15.2.3 In evaluating Claims, the Initial Decision Maker may, but shall not be obligated to, consult with or seek information from either party or from persons with special knowledge or expertise who may assist the Initial Decision Maker in rendering a decision. The Initial Decision Maker may request the Owner to authorize retention of such persons at the Owner's expense.

§ 15.2.4 If the Initial Decision Maker requests a party to provide a response to a Claim or to furnish additional supporting data, such party shall respond, within ten days after receipt of suchthe request, and shall either (1) provide a response on the requested supporting data, (2) advise the Initial Decision Maker when the response or supporting data will be furnished, or (3) advise the Initial Decision Maker that no supporting data will be furnished. Upon receipt of the response or supporting data, if any, the Initial Decision Maker will either reject or approve the Claim in whole or in part.

§ 15.2.5 The Initial Decision Maker will render an initial decision approving or rejecting the Claim, or indicating that the Initial Decision Maker is unable to resolve the Claim. This initial decision shall (1) be in writing; (2) state the reasons therefor; and (3) notify the parties and the Architect, if the Architect is not serving as the Initial Decision Maker, of any change in the Contract Sum or Contract Time or both. The initial decision shall be final and binding on the parties but subject to mediation and, if the parties fail to resolve their dispute through mediation, to binding dispute resolution.

§ 15.2.6 Either party may file for mediation of an initial decision at any time, subject to the terms of Section 15.2.6.1.

§ 15.2.6.1 Either party may, within 30 days from the date of receipt of an initial decision, demand in writing that the other party file for mediation ~~within 60 days of the initial decision.~~. If such a demand is made and the party receiving the demand fails to file for mediation within ~~the time required~~30 days after receipt thereof, then both parties waive their rights to mediate or pursue binding dispute resolution proceedings with respect to the initial decision.

§ 15.2.7 In the event of a Claim against the Contractor, the Owner may, but is not obligated to, notify the surety, if any, of the nature and amount of the Claim. If the Claim relates to a possibility of a Contractor's default, the Owner may, but is not obligated to, notify the surety and request the surety's assistance in resolving the controversy.

§ 15.2.8 If a Claim relates to or is the subject of a mechanic's lien, the party asserting such Claim may proceed in accordance with applicable law to comply with the lien notice or filing deadlines.

§ 15.3 Mediation

§ 15.3.1 Claims, disputes, or other matters in controversy arising out of or related to the Contract, except those waived as provided for in Sections 9.10.4, 9.10.5, and 15.1.~~6~~7, shall be subject to mediation as a condition precedent to binding dispute resolution.

§ 15.3.2 The parties shall endeavor to resolve their Claims by mediation which, unless the parties mutually agree otherwise, shall be administered by the American Arbitration Association in accordance with its Construction Industry Mediation Procedures in effect on the date of the Agreement. A request for mediation shall be made in writing, delivered to the other party to the Contract, and filed with the person or entity administering the mediation. The request may be made concurrently with the filing of binding dispute resolution proceedings but, in such event, mediation shall proceed in advance of binding dispute resolution proceedings, which shall be stayed pending mediation for a period of 60 days from the date of filing, unless stayed for a longer period by agreement of the parties or court order. If an arbitration is stayed pursuant to this Section 15.3.2, the parties may nonetheless proceed to the selection of the arbitrator(s) and agree upon a schedule for later proceedings.

~~§ 15.3.3~~**§ 15.3.3** Either party may, within 30 days from the date that mediation has been concluded without resolution of the dispute or 60 days after mediation has been demanded without resolution of the dispute, demand in writing that the other party file for binding dispute resolution. If such a demand is made and the party receiving the demand fails to file for binding dispute resolution within 60 days after receipt thereof, then both parties waive their rights to binding dispute resolution proceedings with respect to the initial decision.

§ 15.3.4 The parties shall share the mediator's fee and any filing fees equally. The mediation shall be held in the place where the Project is located, unless another location is mutually agreed upon. Agreements reached in mediation shall be enforceable as settlement agreements in any court having jurisdiction thereof.

§ 15.4 Arbitration

§ 15.4.1 If the parties have selected arbitration as the method for binding dispute resolution in the Agreement, any Claim subject to, but not resolved by, mediation shall be subject to arbitration which, unless the parties mutually agree otherwise, shall be administered by the American Arbitration Association in accordance with its Construction Industry Arbitration Rules in effect on the date of the Agreement. The Arbitration shall be conducted in the place where the Project is located, unless another location is mutually agreed upon. A demand for arbitration shall be made in writing, delivered to the other party to the Contract, and filed with the person or entity administering the arbitration. The party filing a notice of demand for arbitration must assert in the demand all Claims then known to that party on which arbitration is permitted to be demanded.

§ 15.4.1.1 A demand for arbitration shall be made no earlier than concurrently with the filing of a request for mediation, but in no event shall it be made after the date when the institution of legal or equitable proceedings based on the Claim would be barred by the applicable statute of limitations. For statute of limitations purposes, receipt of a written demand for arbitration by the person or entity administering the arbitration shall constitute the institution of legal or equitable proceedings based on the Claim.

§ 15.4.2 The award rendered by the arbitrator or arbitrators shall be final, and judgment may be entered upon it in accordance with applicable law in any court having jurisdiction thereof.

§ 15.4.3 The foregoing agreement to arbitrate and other agreements to arbitrate with an additional person or entity duly consented to by parties to the Agreement, shall be specifically enforceable under applicable law in any court having jurisdiction thereof.

§ 15.4.4 Consolidation or Joinder

~~**§ 15.4.4.1** Either party, at its sole discretion,~~**§ 15.4.4.1** Subject to the rules of the American Arbitration Association or other applicable arbitration rules, either party may consolidate an arbitration conducted under this Agreement with any other arbitration to which it is a party provided that (1) the arbitration agreement governing the other arbitration permits consolidation, (2) the arbitrations to be consolidated substantially involve common questions of law or fact, and (3) the arbitrations employ materially similar procedural rules and methods for selecting arbitrator(s).

~~**§ 15.4.4.2** Either party, at its sole discretion,~~**§ 15.4.4.2** Subject to the rules of the American Arbitration Association or other applicable arbitration rules, either party may include by joinder persons or entities substantially involved in a common question of law or fact whose presence is required if complete relief is to be accorded in arbitration, provided that the party sought to be joined consents in writing to such joinder. Consent to arbitration involving an additional person or entity shall not constitute consent to arbitration of any claim, dispute or other matter in question not described in the written consent.

§ 15.4.4.3 The Owner and Contractor grant to any person or entity made a party to an arbitration conducted under this Section 15.4, whether by joinder or consolidation, the same rights of joinder and consolidation as those of the Owner and Contractor under this Agreement.

AIA® Document A101™ – 2017 Exhibit A

Insurance and Bonds

Note: This Insurance and Bonds Exhibit, which is Exhibit A to the 2017 versions of A101, A102 and A103, is attached to the A201-2017 Comparative for reference because it is intended to be used in conjunction with A201, Article 11.

This document has important legal consequences. Consultation with an attorney is encouraged with respect to its completion or modification.

This document is intended to be used in conjunction with AIA Document A201™–2017, General Conditions of the Contract for Construction. Article 11 of A201™–2017 contains additional insurance provisions.

This Insurance and Bonds Exhibit is part of the Agreement, between the Owner and the Contractor, dated the ________ day of ________ in the year ________
(In words, indicate day, month and year.)

for the following **PROJECT**:
(Name and location or address)

THE OWNER:
(Name, legal status and address)

THE CONTRACTOR:
(Name, legal status and address)

TABLE OF ARTICLES

ARTICLE A.1 GENERAL

The Owner and Contractor shall purchase and maintain insurance, and provide bonds, as set forth in this Exhibit. As used in this Exhibit, the term General Conditions refers to AIA Document A201™–2017, General Conditions of the Contract for Construction.

ARTICLE A.2 OWNER'S INSURANCE

§ A.2.1 General

Prior to commencement of the Work, the Owner shall secure the insurance, and provide evidence of the coverage, required under this Article A.2 and, upon the Contractor's request, provide a copy of the property insurance policy or policies required by Section A.2.3. The copy of the policy or policies provided shall contain all applicable conditions, definitions, exclusions, and endorsements.

§ A.2.2 Liability Insurance

The Owner shall be responsible for purchasing and maintaining the Owner's usual general liability insurance.

§ A.2.3 Required Property Insurance

§ A.2.3.1 Unless this obligation is placed on the Contractor pursuant to Section A.3.3.2.1, the Owner shall purchase and maintain, from an insurance company or insurance companies lawfully authorized to issue insurance in the jurisdiction where the Project is located, property insurance written on a builder's risk "all-risks" completed value or equivalent policy form and sufficient to cover the total value of the entire Project on a replacement cost basis. The Owner's property insurance coverage shall be no less than the amount of the initial Contract Sum, plus the value of subsequent Modifications and labor performed and materials or equipment supplied by others. The property insurance shall be maintained until Substantial Completion and thereafter as provided in Section A.2.3.1.3, unless otherwise provided in the Contract Documents or otherwise agreed in writing by the parties to this Agreement. This insurance shall include the interests of the Owner, Contractor, Subcontractors, and Sub-subcontractors in the Project as insureds. This insurance shall include the interests of mortgagees as loss payees.

§ A.2.3.1.1 Causes of Loss. The insurance required by this Section A.2.3.1 shall provide coverage for direct physical loss or damage, and shall not exclude the risks of fire, explosion, theft, vandalism, malicious mischief, collapse, earthquake, flood, or windstorm. The insurance shall also provide coverage for ensuing loss or resulting damage from error, omission, or deficiency in construction methods, design, specifications, workmanship, or materials. Sub-limits, if any, are as follows:
(Indicate below the cause of loss and any applicable sub-limit.)

Cause of Loss	Sub-Limit

§ A.2.3.1.2 Specific Required Coverages. The insurance required by this Section A.2.3.1 shall provide coverage for loss or damage to falsework and other temporary structures, and to building systems from testing and startup. The insurance shall also cover debris removal, including demolition occasioned by enforcement of any applicable legal requirements, and reasonable compensation for the Architect's and Contractor's services and expenses required as a result of such insured loss, including claim preparation expenses. Sub-limits, if any, are as follows:
(Indicate below type of coverage and any applicable sub-limit for specific required coverages.)

Coverage	Sub-Limit

§ A.2.3.1.3 Unless the parties agree otherwise, upon Substantial Completion, the Owner shall continue the insurance required by Section A.2.3.1 or, if necessary, replace the insurance policy required under Section A.2.3.1 with property insurance written for the total value of the Project that shall remain in effect until expiration of the period for correction of the Work set forth in Section 12.2.2 of the General Conditions.

§ A.2.3.1.4 Deductibles and Self-Insured Retentions. If the insurance required by this Section A.2.3 is subject to deductibles or self-insured retentions, the Owner shall be responsible for all loss not covered because of such deductibles or retentions.

§ A.2.3.2 Occupancy or Use Prior to Substantial Completion. The Owner's occupancy or use of any completed or partially completed portion of the Work prior to Substantial Completion shall not commence until the insurance company or companies providing the insurance under Section A.2.3.1 have consented in writing to the continuance of coverage. The Owner and the Contractor shall take no action with respect to partial occupancy or use that would cause cancellation, lapse, or reduction of insurance, unless they agree otherwise in writing.

§ A.2.3.3 Insurance for Existing Structures
If the Work involves remodeling an existing structure or constructing an addition to an existing structure, the Owner shall purchase and maintain, until the expiration of the period for correction of Work as set forth in Section 12.2.2 of the General Conditions, "all-risks" property insurance, on a replacement cost basis, protecting the existing structure against direct physical loss or damage from the causes of loss identified in Section A.2.3.1, notwithstanding the undertaking of the Work. The Owner shall be responsible for all co-insurance penalties.

§ A.2.4 Optional Extended Property Insurance.
The Owner shall purchase and maintain the insurance selected and described below.
(Select the types of insurance the Owner is required to purchase and maintain by placing an X in the box(es) next to the description(s) of selected insurance. For each type of insurance selected, indicate applicable limits of coverage or other conditions in the fill point below the selected item.)

☐ **§ A.2.4.1 Loss of Use, Business Interruption, and Delay in Completion Insurance,** to reimburse the Owner for loss of use of the Owner's property, or the inability to conduct normal operations due to a covered cause of loss.

☐ **§ A.2.4.2 Ordinance or Law Insurance,** for the reasonable and necessary costs to satisfy the minimum requirements of the enforcement of any law or ordinance regulating the demolition, construction, repair, replacement or use of the Project.

☐ **§ A.2.4.3 Expediting Cost Insurance**, for the reasonable and necessary costs for the temporary repair of damage to insured property, and to expedite the permanent repair or replacement of the damaged property.

☐ **§ A.2.4.4 Extra Expense Insurance,** to provide reimbursement of the reasonable and necessary excess costs incurred during the period of restoration or repair of the damaged property that are over and above the total costs that would normally have been incurred during the same period of time had no loss or damage occurred.

☐ **§ A.2.4.5 Civil Authority Insurance,** for losses or costs arising from an order of a civil authority prohibiting access to the Project, provided such order is the direct result of physical damage covered under the required property insurance.

☐ **§ A.2.4.6 Ingress/Egress Insurance,** for loss due to the necessary interruption of the insured's business due to physical prevention of ingress to, or egress from, the Project as a direct result of physical damage.

☐ **§ A.2.4.7 Soft Costs Insurance,** to reimburse the Owner for costs due to the delay of completion of the Work, arising out of physical loss or damage covered by the required property insurance: including construction loan fees; leasing and marketing expenses; additional fees, including those of architects, engineers, consultants, attorneys and accountants, needed for the completion of the construction, repairs, or reconstruction; and carrying costs such as property taxes, building permits, additional interest on loans, realty taxes, and insurance premiums over and above normal expenses.

§ A.2.5 Other Optional Insurance.
The Owner shall purchase and maintain the insurance selected below.
(Select the types of insurance the Owner is required to purchase and maintain by placing an X in the box(es) next to the description(s) of selected insurance.)

☐ **§ A.2.5.1 Cyber Security Insurance** for loss to the Owner due to data security and privacy breach, including costs of investigating a potential or actual breach of confidential or private information. *(Indicate applicable limits of coverage or other conditions in the fill point below.)*

☐ **§ A.2.5.2 Other Insurance**
(List below any other insurance coverage to be provided by the Owner and any applicable limits.)

Coverage	**Limits**

ARTICLE A.3 CONTRACTOR'S INSURANCE AND BONDS
§ A.3.1 General
§ A.3.1.1 Certificates of Insurance. The Contractor shall provide certificates of insurance acceptable to the Owner evidencing compliance with the requirements in this Article A.3 at the following times: (1) prior to commencement of the Work; (2) upon renewal or replacement of each required policy of insurance; and (3) upon the Owner's written request. An additional certificate evidencing continuation of commercial liability coverage, including coverage for completed operations, shall be submitted with the final Application for Payment and thereafter upon renewal or replacement of such coverage until the expiration of the periods required by Section A.3.2.1 and Section A.3.3.1. The certificates will show the Owner as an additional insured on the Contractor's Commercial General Liability and excess or umbrella liability policy or policies.

§ A.3.1.2 Deductibles and Self-Insured Retentions. The Contractor shall disclose to the Owner any deductible or self-insured retentions applicable to any insurance required to be provided by the Contractor.

§ A.3.1.3 Additional Insured Obligations. To the fullest extent permitted by law, the Contractor shall cause the commercial general liability coverage to include (1) the Owner, the Architect, and the Architect's consultants as additional insureds for claims caused in whole or in part by the Contractor's negligent acts or omissions during the Contractor's operations; and (2) the Owner as an additional insured for claims caused in whole or in part by the Contractor's negligent acts or omissions for which loss occurs during completed operations. The additional insured coverage shall be primary and non-contributory to any of the Owner's general liability insurance policies and shall apply to both ongoing and completed operations. To the extent commercially available, the additional insured coverage shall be no less than that provided by Insurance Services Office, Inc. (ISO) forms CG 20 10 07 04, CG 20 37 07 04, and, with respect to the Architect and the Architect's consultants, CG 20 32 07 04.

§ A.3.2 Contractor's Required Insurance Coverage
§ A.3.2.1 The Contractor shall purchase and maintain the following types and limits of insurance from an insurance company or insurance companies lawfully authorized to issue insurance in the jurisdiction where the Project is located. The Contractor shall maintain the required insurance until the expiration of the period for correction of Work as set forth in Section 12.2.2 of the General Conditions, unless a different duration is stated below:
(If the Contractor is required to maintain insurance for a duration other than the expiration of the period for correction of Work, state the duration.)

§ A.3.2.2 Commercial General Liability
§ A.3.2.2.1 Commercial General Liability insurance for the Project written on an occurrence form with policy limits of not less than ________ ($__) each occurrence, ________ ($__) general aggregate, and ________ ($__) aggregate for products-completed operations hazard, providing coverage for claims including

- **.1** damages because of bodily injury, sickness or disease, including occupational sickness or disease, and death of any person;
- **.2** personal injury and advertising injury;
- **.3** damages because of physical damage to, or destruction of, tangible property, including the loss of use of such property;
- **.4** bodily injury or property damage arising out of completed operations; and
- **.5** the Contractor's indemnity obligations under Section 3.18 of the General Conditions.

§ A.3.2.2.2 The Contractor's Commercial General Liability policy under this Section A.3.2.2 shall not contain an exclusion or restriction of coverage for the following:

.1 Claims by one insured against another insured, if the exclusion or restriction is based solely on the fact that the claimant is an insured, and there would otherwise be coverage for the claim.
.2 Claims for property damage to the Contractor's Work arising out of the products-completed operations hazard where the damaged Work or the Work out of which the damage arises was performed by a Subcontractor.
.3 Claims for bodily injury other than to employees of the insured.
.4 Claims for indemnity under Section 3.18 of the General Conditions arising out of injury to employees of the insured
.5 Claims or loss excluded under a prior work endorsement or other similar exclusionary language.
.6 Claims or loss due to physical damage under a prior injury endorsement or similar exclusionary language.
.7 Claims related to residential, multi-family, or other habitational projects, if the Work is to be performed on such a project.
.8 Claims related to roofing, if the Work involves roofing.
.9 Claims related to exterior insulation finish systems (EIFS), synthetic stucco or similar exterior coatings or surfaces, if the Work involves such coatings or surfaces.
.10 Claims related to earth subsidence or movement, where the work involves such hazards.
.11 Claims related to explosion, collapse, and underground hazards, where the Work involves such hazards.

§ A.3.2.3 Automobile Liability covering vehicles owned, and non-owned vehicles used, by the Contractor, with policy limits of not less than ________ ($__) per accident, for bodily injury, death of any person, and property damage arising out of the ownership, maintenance and use of those motor vehicles along with any other statutorily required automobile coverage.

§ A.3.2.4 The Contractor may achieve the required limits and coverage for Commercial General Liability and Automobile Liability through a combination of primary and excess or umbrella liability insurance, provided such primary and excess or umbrella insurance policies result in the same or greater coverage as the coverages required under Section A.3.2.2 and A.3.2.3, and in no event shall any excess or umbrella liability insurance provide narrower coverage than the primary policy. The excess policy shall not require the exhaustion of the underlying limits only through the actual payment by the underlying insurers.

§ A.3.2.5 Workers' Compensation at statutory limits.

§ A.3.2.6 Employers' Liability with policy limits not less than ________ ($__) each accident, ________ ($__) each employee, and ________ ($__) policy limit.

§ A.3.2.7 Jones Act, and the Longshore & Harbor Workers' Compensation Act, as required, if the Work involves hazards arising from work on or near navigable waterways, including vessels and docks

§ A.3.2.8 If the Contractor is required to furnish professional services as part of the Work, the Contractor shall procure Professional Liability insurance covering performance of the professional services, with policy limits of not less than ________ ($__) per claim and ________ ($__) in the aggregate.

§ A.3.2.9 If the Work involves the transport, dissemination, use, or release of pollutants, the Contractor shall procure Pollution Liability insurance, with policy limits of not less than ________ ($__) per claim and ________ ($__) in the aggregate.

§ A.3.2.10 Coverage under Sections A.3.2.8 and A.3.2.9 may be procured through a Combined Professional Liability and Pollution Liability insurance policy, with combined policy limits of not less than ________ ($__) per claim and ________ ($__) in the aggregate.

§ A.3.2.11 Insurance for maritime liability risks associated with the operation of a vessel, if the Work requires such activities, with policy limits of not less than________ ($__) per claim and________ ($__) in the aggregate.

§ A.3.2.12 Insurance for the use or operation of manned or unmanned aircraft, if the Work requires such activities, with policy limits of not less than________ ($__) per claim and________ ($__) in the aggregate.

§ A.3.3 Contractor's Other Insurance Coverage

§ A.3.3.1 Insurance selected and described in this Section A.3.3 shall be purchased from an insurance company or insurance companies lawfully authorized to issue insurance in the jurisdiction where the Project is located. The Contractor shall maintain the required insurance until the expiration of the period for correction of Work as set forth in Section 12.2.2 of the General Conditions, unless a different duration is stated below:
(If the Contractor is required to maintain any of the types of insurance selected below for a duration other than the expiration of the period for correction of Work, state the duration.)

§ A.3.3.2 The Contractor shall purchase and maintain the following types and limits of insurance in accordance with Section A.3.3.1.
(Select the types of insurance the Contractor is required to purchase and maintain by placing an X in the box(es) next to the description(s) of selected insurance. Where policy limits are provided, include the policy limit in the appropriate fill point.)

☐ **§ A.3.3.2.1** Property insurance of the same type and scope satisfying the requirements identified in Section A.2.3, which, if selected in this section A.3.3.2.1, relieves the Owner of the responsibility to purchase and maintain such insurance except insurance required by Section A.2.3.1.3 and Section A.2.3.3. The Contractor shall comply with all obligations of the Owner under Section A.2.3 except to the extent provided below. The Contractor shall disclose to the Owner the amount of any deductible, and the Owner shall be responsible for losses within the deductible. Upon request, the Contractor shall provide the Owner with a copy of the property insurance policy or policies required. The Owner shall adjust and settle the loss with the insurer and be the trustee of the proceeds of the property insurance in accordance with Article 11 of the General Conditions unless otherwise set forth below.

(Where the Contractor's obligation to provide property insurance differs from the Owner's obligations as described under Section A.2.3, indicate such differences in the space below. Additionally, if a party other than the Owner will be responsible for adjusting and settling a loss with the insurer and acting as the trustee of the proceeds of property insurance in accordance with Article 11 of the General Conditions, indicate the responsible party below.)

☐ **§ A.3.3.2.2 Railroad Protective Liability Insurance,** with policy limits of not less than ________ ($__) per claim and ________ ($__) in the aggregate, for Work within fifty (50) feet of railroad property.

☐ **§ A.3.3.2.3 Asbestos Abatement Liability Insurance,** with policy limits of not less than ________ ($__) per claim and ________ ($__) in the aggregate, for liability arising from the encapsulation, removal, handling, storage, transportation, and disposal of asbestos-containing materials.

☐ **§ A.3.3.2.4** Insurance for physical damage to property while it is in storage and in transit to the construction site on an "all-risks" completed value form.

☐ **§ A.3.3.2.5** Property insurance on an "all-risks" completed value form, covering property owned by the Contractor and used on the Project, including scaffolding and other equipment.

☐ **§ A.3.3.2.6 Other Insurance**
(List below any other insurance coverage to be provided by the Contractor and any applicable limits.)

Coverage	**Limits**

Init.

/

§ A.3.4 Performance Bond and Payment Bond
The Contractor shall provide surety bonds, from a company or companies lawfully authorized to issue surety bonds in the jurisdiction where the Project is located, as follows:
(Specify type and penal sum of bonds.)

Type	**Penal Sum ($0.00)**
Payment Bond	
Performance Bond	

Payment and Performance Bonds shall be AIA Document A312™, Payment Bond and Performance Bond, or contain provisions identical to AIA Document A312™, current as of the date of this Agreement.

ARTICLE A.4 SPECIAL TERMS AND CONDITIONS
Special terms and conditions that modify this Insurance and Bonds Exhibit, if any, are as follows:

Index

www.ingramcontent.com/pod-product-compliance
Lightning Source LLC
LaVergne TN
LVHW080636011026
849160LV00014B/1233